Springer Proceedings in Mathematics & Statistics

Volume 378

This book series features volumes composed of selected contributions from workshops and conferences in all areas of current research in mathematics and statistics, including data science, operations research and optimization. In addition to an overall evaluation of the interest, scientific quality, and timeliness of each proposal at the hands of the publisher, individual contributions are all refereed to the high quality standards of leading journals in the field. Thus, this series provides the research community with well-edited, authoritative reports on developments in the most exciting areas of mathematical and statistical research today.

More information about this series at https://link.springer.com/bookseries/10533

Stefania Ugolini · Marco Fuhrman ·
Elisa Mastrogiacomo · Paola Morando ·
Barbara Rüdiger
Editors

Geometry and Invariance in Stochastic Dynamics

Verona, Italy, March 25–29, 2019

Springer

Editors
Stefania Ugolini
Department of Mathematics
Università degli Studi di Milano
Milan, Italy

Elisa Mastrogiacomo
Department of Economics
Università degli Studi dell'Insubria
Varese, Italy

Barbara Rüdiger
Mathematic and Informatic Department
Bergische Universität Wuppertal
Wuppertal, Germany

Marco Fuhrman
Department of Mathematics
Università degli Studi di Milano
Milan, Italy

Paola Morando
Department of Agricultural
and Environmental Sciences
Università degli Studi di Milano
Milan, Italy

ISSN 2194-1009 ISSN 2194-1017 (electronic)
Springer Proceedings in Mathematics & Statistics
ISBN 978-3-030-87434-6 ISBN 978-3-030-87432-2 (eBook)
https://doi.org/10.1007/978-3-030-87432-2

Mathematics Subject Classification: 60HXX, 60H15, 34C15, 35B06, 37HXX

© Springer Nature Switzerland AG 2021

This work is subject to copyright. All rights are reserved by the Publisher, whether the whole or part of the material is concerned, specifically the rights of translation, reprinting, reuse of illustrations, recitation, broadcasting, reproduction on microfilms or in any other physical way, and transmission or information storage and retrieval, electronic adaptation, computer software, or by similar or dissimilar methodology now known or hereafter developed.

The use of general descriptive names, registered names, trademarks, service marks, etc. in this publication does not imply, even in the absence of a specific statement, that such names are exempt from the relevant protective laws and regulations and therefore free for general use.

The publisher, the authors and the editors are safe to assume that the advice and information in this book are believed to be true and accurate at the date of publication. Neither the publisher nor the authors or the editors give a warranty, expressed or implied, with respect to the material contained herein or for any errors or omissions that may have been made. The publisher remains neutral with regard to jurisdictional claims in published maps and institutional affiliations.

This Springer imprint is published by the registered company Springer Nature Switzerland AG
The registered company address is: Gewerbestrasse 11, 6330 Cham, Switzerland

Proceedings CONFERENCE VERONA 2019 dedicated to Albeverio's 80th birthday

Geometry and Invariance in Stochastic Dynamics

Sala Capitolare, Chiostro di San Giorgio in Braida, 25–29 March 2019

Preface

Geometry and Invariance in Stochastic Dynamics

The study of symmetries and invariance properties of ordinary and partial differential equations (ODE and PDE resp.) is a classical and well-developed area of research and provides a powerful tool for computing some explicit solutions to the equations and analyzing their qualitative behavior.

In the last decades, the fruitful notions of symmetry and invariance have been extended beyond classical mechanical systems. In fact, the development of geometric mechanics allowed the generalization of these notions to more complex (finite and infinite dimensional) systems arising in many different areas of physics.

It is well known that variational principles, and their associated Hamiltonian formulations, provide a natural framework for both classical and modern physics, such as general relativity and quantum mechanics, and constitute one of the most useful tools in mathematical physics. In particular, the physics of the XXth century was deeply influenced by the fundamental theorem of Noether, which associates symmetries of a classical dynamical system with its invariants, i.e., with quantities that remain constant during the time evolution of the system.

The modern theory of symmetry for ODEs and PDEs, due to S. Lie, is based on the extension of the original concept of discrete group introduced by E. Galois to continuous groups of transformations. Nowadays, Lie's symmetry theory is widely applied both to ODEs and PDEs in order to reduce the original system to a simpler form, exploiting symmetry-adapted coordinates.

From a numerical point of view, the theory of geometric numerical integration for ODEs focuses on the preservation of continuous geometric structures under time discretization. It provides a powerful alternative to standard discretization methods with a significant impact on the theoretical and practical aspects of modern computational mathematics.

Another research area involving symmetry and invariance arises from the interaction of quantum theory and probability. Indeed, since the first quarter of last century, the development of the theory of general relativity enhanced the traditional links between classical deterministic mechanics and analysis and set off emerging research fields between algebra and geometry. At the same time, quantum mechanics connected analysis and probability and in the second half of last century, these interactions spread to quantum field theory and to the study of singular partial differential equations.

Despite the big achievements obtained in the deterministic setting, the importance of the study of invariance properties and geometric structure of finite- or infinite-dimensional *stochastic differential equations* (SDEs and SPDEs resp.) has been overlooked for a long time and a systematic generalization of the deterministic results to the stochastic framework is much needed from both practical and theoretical point of view.

The purpose of this book is to collect contributions in this direction and to provide an overview that can inspire further researches aiming at generalizing to the stochastic framework results involving geometric structures and invariance properties of deterministic ODEs and PDEs. For this reasons, the papers included in the volume range from theoretical probability to the study of geometric and algebraic structures, offering an extraordinary opportunity to approach this promising research field from different perspectives.

In the following, without claiming to be complete, we try to outline the main research topics that are tackled in the book.

From a geometric point of view, the generalization of geometric mechanics, which is essentially based on group-invariant variational principles, to (Stratonovich) stochastic setting gave rise to the new research area called *stochastic geometric mechanics* (a beautiful introduction to the subject can be found in the Springer Volume titled Stochastic Geometric Mechanics, 2017). Important achievements of this stochastic extension are the variational formulation of SDEs and the Euler-Poincarè reduction of stochastic infinite-dimensional variational systems in stochastic fluid-dynamics. In particular, recent theoretical results in this framework turned out to be extremely useful in order to obtain advanced numerical analysis techniques.

On the other hand, a direct (but non-trivial) extension of Lie symmetries approach to the stochastic setting can be successfully exploited in order to determine explicit solutions to SDEs, to reduce and reconstruct symmetric SDEs as well as to find finite-dimensional solutions to SPDEs.

Moreover, recent advances unveil the central role played by algebraic structures such as pre-Lie and post-Lie algebras, and their enveloping algebras which permit to join B-series and Lie-series into Munthe-Kaas' Lie-Butcher series on manifolds. These structures turn out to be associated with Euclidean geometry and with homogeneous manifolds and Lie groups, and recent results in this framework provide an interesting extension of the theory of Lie group integration to nonlinear SDEs. Moreover, the Lie and Hopf algebraic setting underlying Lie group integration was recently adapted to Lyons' theory of rough paths, extending the notion of rough differential equations to homogeneous spaces. In this area advanced combinatorial methods have been successfully applied.

Furthermore, the interest in the study of singular PDEs, renormalization theory for quantum fields and critical phenomena in statistical mechanics gave a big boost to new researches on SPDEs that experienced striking new developments in recent years.

On the other side, the investigation of *invariance properties* for stochastic processes provides an interesting and well-established research topic in theoretical probability.

When stochastic processes take their values in manifolds, the standard probabilistic tools have to interact with differential geometry techniques, originating a research field called *stochastic differential geometry*. Since Lie groups combine the algebraic structure of a group with the geometric notion of differential manifold, an interesting challenge is the study of diffusions, Markov processes, and Lévy processes on Lie groups. In particular, the analysis of the invariance properties of this kind of processes under the action of the group gives promising results in the study of SDEs driven by Lévy processes (also in the jump case) as well as in their characterization in terms of stochastic variational principles. Moreover, the investigation of a stochastic process by means of its invariance under random transformations provides useful characterizations of the process itself.

Since many equations are not perfectly symmetric, another interesting theoretical emerging area is the perturbations of symmetric or Hamiltonian systems. In this setting, the symmetries of a diffusion equation can also be exploited in order to obtain precise analytical properties of the related semigroup. These results have important applications in the case of symmetric spaces with invariant Riemannian structures.

Finally, the knowledge of some closed formula is also crucial in many applications of stochastic processes, since it permits to develop faster and cheaper numerical algorithms for the simulation of the process or to evaluate interesting quantities related to it, such as martingales which can be individuated through stochastic Noether theorem as counterparts of the deterministic conserved quantities.

The Conference, titled *Random Transformations and Invariance in Stochastic Dynamics* (25–29 March 2019), held in the cloister of San Giorgio in Braida in Verona (Italy) and was dedicated to Sergio Albeverio for his 80th birthday.

In connection with the conference two volumes are being published. A first volume consists of contributions directly related to the extraordinarily rich and exciting human and scientific adventure of Sergio Albeverio; the present second volume, as recalled above, contains the main research lines on geometry and invariance in stochastic dynamics.

The editors of the present volume thank the authors who, besides actively participating to the conference, accepted the invitation to write up their contributions. We also thank other lecturers at the conference who greatly collaborated to the success of the event and all the participants, for their presence and their active contribute to create a really agreeable and inspiring atmosphere.

Many thanks to Marina Reizakis of Springer-Verlag, for accepting our invitation to attend the Conference and to put on display books related to the conference. Her competent and stimulating advice during the preparation of these proceedings was greatly appreciated. We also thank Banu Dhayalan of Springer-Verlag for their technical support in the preparation of the printing process.

Special thanks to the Rector of San Giorgio in Braida, Don Giorgio Marchesi, for the permission to enjoy the Renaissance chapter house and the beautiful cloister on the banks of Adige river who constituted the venue of the Conference.

The editors are also grateful to the following institutions for the financial support: University of Milano (Transition grant to S. Ugolini), GNAMPA (INDAM), University of Trento, Linneaus University, University of Wuppertal, and University of Insubria.

Milan, Italy	Stefania Ugolini
Milan, Italy	Marco Fuhrman
Varese, Italy	Elisa Mastrogiacomo
Milan, Italy	Paola Morando
Wuppertal, Germany	Barbara Rüdiger

Random Transformations and Invariance in Stochastic Dynamics

Sala Capitolare, Chiostro di San Giorgio in Braida, Verona (Italy), 25–29 March 2019

Conference in Verona: List of Participants

Sergio Albeverio, University of Bonn, HCM, Germany, albeverio(at)iam-bonn.de
Ana Bela Cruzeiro, University of Lisboa, Portugal, ana.cruzeiro(at)tecnico.ulisboa.pt
Francesco De Vecchi, University of Bonn, HCM, Germany, fdevecchi(at)uni-bonn.de
David Elworthy, University of Warwick, England, K. D. Elworthy(at)warwick.ac.uk
Darryl Holm, Imperial College, England, d.holm(at)imperial.ac.uk
Paul Lescot, Université de Rouen, France, paul.lescot(at)univ-rouen.fr
Xue-Mei Li, Imperial College, England, xue-mei.li(at)imperial.ac.uk
Juan Pablo Ortega, Universitat Sankt Gallen, Switzerland, juan-pablo.ortega(at)unisg.ch
Frédric Patras, Université Côte d'Azur, France, patras(at)unice.fr
Nicolas Privault, Nanyang Technological University, Singapore, nprivaul(at)ntu.edu.sg
Tudor Ratiu, University of California, USA, tratiu(at)scsc.edu
Francesco Russo, Institut Polytecnique de Paris, ENSTA, France, francesco.russo(at)ensta-paris.fr
Laurene Valade, Université de Rouen, France, laurene.valade(at)etu.univ-rouen.fr
Jean Claude Zambrini, University of Lisboa, Portugal, jczambrini(at)fc.ul.pt
Michael Rochner, University of Bielefeld, Germany, roechner(at)math.uni-bielefeld.de
Philippe Blanchard, University of Bielefeld, Germany, blanchard(at)physik.uni-bielefeld.de
Minoru Yoshida, Kanagawa University, Japan, washizuminoru(at)Hotmail.com

Claudio Cacciapuoti, University of Insubria, Italy,
Claudio.cacciapuoti(at)uninsubria.it
Fabio Cipriani, Politecnico di Milano, Italy, fabio.cipriani(at)polimi.it
Alexei Daletskii, University of York, England, alex.daletskii(at)york.ac.uk
Gianfausto Dell'Antonio, SISSA (Scuola Internazionale Superiore di Studi Avanzati), Italy, gianfa(at)sissa.it
Luca Di Persio, University of Verona, Italy, luca.dipersio(at)univr.it
Benedetta Ferrario, University of Pavia, Italy, benedetta.ferrario(at)unipv.it
Rodolfo Figari, University of Napoli, Italy, figari(at)na.infn.it
Francesco Guerra, University of Roma "La Sapienza", Italy,
francesco.guerra(at)roma1.infn.it
Nadia Robotti, University of Genova, Italy, robotti(at)fisica.unige.it
Hanno Gottschalk, University of Wuppertal, Germany, hanno.gottschalk(at)uni-wuppertal.de
Yuri Kondratiev, University of Bielefeld, Germany, kondrat(at)math.uni-bielefeld.de
Pavel Kurasov, University of Stockholm, Norvey, kurasov(at)math.su.se
Vidadhar Mandrekar, University of Michigan, USA, mandrekar(at)stt.msu.edu
Tom Lindstrom, University of Oslo, lindstro(at)math.uio.no
Carlo Marinelli, University College London, England, c.marinelli(at)ucl.ac.uk
Danilo Merlini, CERFIM, Switzerland, merlini(at)cerfim.ch
Andrea Posilicano, University of Insubria, Italy, andrea.posilicano(at)uninsubria.it
Alessandro Teta, University of Roma "La Sapienza", Italy, teta(at)mat.uniroma1.it
Luciano Tubaro, University of Trento, Italy, luciano.tubaro(at)unitn.it
Andrea Romano, MatemUpper Association, andrea.23.romano(at)gmail.com
Marco Tarsia, University of Uninsubria, Italy, masia1(at)uninsubria.it
Francesco Cordoni, University of Verona, Italy,
francescogiuseppe.cordoni(at)univr.it
Walter Moretti, University of Trento, Italy, valter.moretti(at)unitn.it
Giulia Basti, GSSI L'Aquila, Italy, giulia.basti(at)gssi.it
Luigi Galgani, University of Milano, Italy, luigi.galgani(at)unimi.it
Diego Noja, University of Milano Bicocca, Italy, diego.noja(at)unimib.it

Conference organizers:

Stefania Ugolini, University of Milano, Italy, stefania.ugolini(at)unimi.it
Marco Fuhrman, University of Milano, Italy, marco.fuhrman(at)unimi.it
Astrid Hilbert, Linnaeus University, astrid.hilbert(at)lnu.se
Elisa Mastrogiacomo, University of Insubria, Italy,
elisa.mastrogiacomo(at)uninsubria.it
Sonia Mazzucchi, University of Trento, Italy, sonia.mazzucchi(at)unitn.it
Paola Morando, University of Milano, Italy, paola.morando(at)unimi.it
Barbara Rüdiger, University of Wuppertal, Germany, ruediger(at)uni-wuppertal.de

Contents

Some Recent Developments on Lie Symmetry Analysis of Stochastic Differential Equations 1
Sergio Albeverio and Francesco C. De Vecchi

Markov Processes with Jumps on Manifolds and Lie Groups 25
David Applebaum and Ming Liao

Asymptotic Expansion for a Black–Scholes Model with Small Noise Stochastic Jump-Diffusion Interest Rate 47
Francesco Cordoni and Luca Di Persio

Stochastic Geodesics 59
Ana Bela Cruzeiro and Jean-Claude Zambrini

A Note on Supersymmetry and Stochastic Differential Equations 71
Francesco C. De Vecchi and Massimiliano Gubinelli

Quasi-shuffle Algebras in Non-commutative Stochastic Calculus 89
Kurusch Ebrahimi-Fard and Frédéric Patras

Higher Order Derivatives of Heat Semigroups on Spheres and Riemannian Symmetric Spaces 113
K. David Elworthy

Rough Homogenisation with Fractional Dynamics 137
Johann Gehringer and Xue-Mei Li

Stochastic Geometric Mechanics with Diffeomorphisms 169
Darryl D. Holm and Erwin Luesink

McKean Feynman-Kac Probabilistic Representations of Non-linear Partial Differential Equations 187
Lucas Izydorczyk, Nadia Oudjane, and Francesco Russo

Bernstein Processes, Isovectors and Mechanics 213
Paul Lescot and Laurène Valade

On the Positivity of Local Mild Solutions to Stochastic Evolution Equations 231
Carlo Marinelli and Luca Scarpa

Invariance of Poisson Point Processes by Moment Identities with Statistical Applications 247
Nicolas Privault

Some Recent Developments on Lie Symmetry Analysis of Stochastic Differential Equations

Sergio Albeverio and Francesco C. De Vecchi

Abstract We present a brief review of recent progresses on Lie symmetry analysis of stochastic differential equations (SDEs). In particular, we consider some general definitions of symmetries for Brownian motion driven SDEs, as well as of weak and gauge symmetries of SDEs driven by discrete-time semimartingales. Some applications of Lie symmetry analysis to reduction and reconstruction of SDEs, Kolmogorov equation and numerical schemes for SDEs are discussed. Studies on random symmetries of SDEs, as well as extension of Noether theorem on invariants to stochastic systems and the finding of finite-dimensional solutions to SPDEs are also briefly reviewed.

Keywords Lie symmetry analysis · Stochastic differential equations · Kolmogorov equation · Noether stochastic theorem

AMS: 60H10 · 58D19 · 60J76 · 60H35 · 60H15

1 Introduction

The study of infinitesimal (respectively global) symmetries of ordinary differential equations (ODEs) and partial differential equations (PDEs) has a long tradition in classical analysis, especially since the groundbreaking work of S. Lie. In fact he introduced the very concepts that, since the turn of the XIX century, one associates with the denominations of Lie algebra and Lie groups, acting as transformations algebras (respectively groups) on solutions of ODEs and PDEs.

S. Albeverio · F. C. De Vecchi (✉)
Institute for Applied Mathematics & Hausdorff Center for Mathematics, University of Bonn, Bonn, Germany
e-mail: francesco.devecchi@uni-bonn.de

S. Albeverio
e-mail: albeverio@iam.uni-bonn.de

Lie's work has its roots in the study of discrete symmetries of algebraic and geometric structures especially since N. H. Abel and E. Galois discovery of the notion of group and its developments, see, e.g. [36]. Another important concept in the study of symmetries of ODEs and PDEs comes from the connection of these equations with the calculus of variations, especially since E. Noether's fundamental work on the relation of invariants associated with ODEs and PDEs and the transformation groups, see, e.g. [40, 53–55].

A possible application of these works is the reduction of interesting ODEs and PDEs to simpler forms that can be more easily classified and studied. Furthermore, exploiting the invariants associated with transformation laws, it is often possible to provide concrete solutions to the associated differential equations. Due to these and many other interesting applications, there has been an important development in the theory of symmetries of deterministic differential equations. We refer the interested reader to, e.g. [4, 5, 39, 51, 56, 57, 61].

Only in more relatively recent years this field of investigation has been extended to its stochastic counterpart, especially in the form of studies concerning finite and infinite dimensional stochastic differential equations (SDEs). The most studied case in the one of finite dimensional SDEs driven by Brownian motion. In this case we can find in the literature two approaches. The first one exploits the fact that the solutions to a Brownian-motion-driven SDEs are Markov processes associated with a second order differential operator L, depending on the SDE, which is an analytic and deterministic object. In this way one can apply the usual notion of symmetry, coming from the deterministic Lie symmetry analysis. This kind of research has been developed in [6, 7, 34, 35, 48–50]. The second approach is instead to generalize directly the original Lie idea of symmetry as a transformation leaving invariant the set of solutions. This idea was developed in [26, 29, 32, 41, 42, 52, 60, 63]; see also [30] for a partial review on this subject and [18–21, 27, 28, 31, 33, 43, 44] for some more recent results.

Contrary to what happens for the Brownian case, the literature in the setting of SDEs driven by general càdlàg semimartingales (thus where the driving stochastic term can contain a jump component) is very scarce. The only references are the works of [34, 35, 35, 48–50] (already quoted) which, dealing with the general case of Markov process, cover the setting where the driving process is a Lévy process. Furthermore there is the work [46] covering the case of SDE driven by general continuous semimartingales, and [2, 3, 14] discussing the case of SDEs driven by general semimartingales with jumps.

There are also some interesting works on SDEs driven by Brownian motion and admitting a variational structure, and a generalization of Noether theorem has been obtained in this case (see [12, 45, 47, 58, 62, 64] and also [1, 11, 38]).

For the case of infinite dimension SDEs (or stochastic partial differential equations SPDEs) there is no work addressing directly the problem of the formulation of the concept of symmetry, with the notable exception of [10] in the particular case of Zakai equation (arising in filter theory). Nevertheless there are some articles about the reduction of an SPDE to a finite dimensional SDE looking for explicit solutions, see [8, 9, 13, 25].

A detailed overview of the literature on symmetries of SDEs would require too much space, for this reason we shall limit ourselves to make a particular choice of some of the main directions of research and refer to given references for complements. In particular we shall review our own contributions on the subject [2, 3, 18–22] and we limit ourselves to present, in Sect. 5, some other recent interesting ideas, in connection with other authors, that are not covered by the cited articles.

In Sect. 2 we briefly recall definitions of symmetries of SDEs driven by Brownian motions contained in [18–21], and summarize some main results on them. In Sect. 3 we present a corresponding discussion for the case of SDEs driven by discrete-time semimartingale discussed in [2, 3]. In Sect. 4 we present some applications beginning, in Sect. 4.1, with the reduction and reconstruction theorem for symmetric SDEs as presented in [18]. The relation between the symmetries of SDEs and the corresponding Lie-point symmetries of Kolmogorov equations are discussed in Sect. 4.2. In Sect. 4.3 the concept of weak symmetries is discussed in relation to numerical schemes for SDEs. In Sect. 5 we present other approaches to Lie symmetry analysis of stochastic systems, from random symmetries (Sect. 5.1) to variational SDEs and their relation with Noether's theory (Sect. 5.2). We close with a short presentation of works on finite dimensional reductions for SPDEs (Sect. 5.3).

2 Symmetries of SDEs Driven by Brownian Motion

In this section we consider the following stochastic differential equation (SDE)

$$dX_t = \mu(X_t)dt + \sigma(X_t) \cdot dW_t \tag{1}$$

where $t \geq 0$, $X_t \in M$, M being an open subset of $\mathbb{R}^m$, $\mu : M \to \mathbb{R}^m$ and $\sigma : M \to \mathbb{R}^{m \times n}$ are smooth functions, and W is an n-dimensional (standard) Brownian motion (i.e. an $\mathbb{R}^n$-Brownian motion). In the following we say that Eq. (1) is an SDE (μ, σ) (said to be smooth if (μ, σ) are smooth) and the process (X, W) (where X is a semimartingale on M and W is an n dimensional Brownian motion) is a solution to the SDE (μ, σ) if it satisfies Eq. (1).

Hereafter the couple (X, W), where where X is a semimartingale on M and W is an n dimensional Brownian motion adapted to the same filtration, shall be called simply a (standard) process (for the concept of semimartingale see, e.g., [37, 59]; roughly speaking semimartingales form the most general class of processes for which Itô integral can be defined).

Let L be the second order operator defined on $C^2(M)$ functions given by

$$L(f)(x) = \mu^i(x)\partial_i(f)(x) + \frac{1}{2}\sum_{\alpha=1}^{n} \sigma_\alpha^i(x)\sigma_\alpha^j(x)\partial_{ij}(f)(x) \tag{2}$$

where $x \in M$ and $f \in C^2(M)$. Here and elsewhere in the paper we are using Einstein summation convention of summing over repeated indices. Under suitable assumptions on the coefficients (μ, σ), the closure of L in the Banach space $C^0(M)$ is the generator of the (Markovian) diffusion process associated with the SDE (μ, σ) in the sense of solving (1).

2.1 Strong Symmetries

The most simple class of symmetries of Eq. (1) is given by diffeomorphisms $\Phi : M \to M$. Indeed if X is a semimartingale on M also $\Phi(X)$ is a semimartingale on M, see, e.g. [37, 59].

Definition 1 We say that Φ is a *strong (finite) symmetry* of the SDE (μ, σ) if for any solution (X, W) to (μ, σ) we have that $(\Phi(X), W)$ is another solution to the SDE (μ, σ). If Y is a vector field on M generating a one-parameter group of diffeomorphism Φ_a, where $a \in \mathbb{R}$, we say that Y is a strong (infinitesimal) symmetry of (μ, σ) if Φ_a is a strong finite symmetry of (μ, σ) for any $a \in \mathbb{R}$.

Remark 2 The name strong symmetries, for the previous kind of invariant transformations, derives from the fact that a strong symmetry Φ sends any *strong solution to the SDE* (μ, σ) (in the probabilistic sense) into a strong solution to the same SDE. Indeed (X, W) and $(\Phi(X), W)$ are solutions to (μ, σ) having the same Brownian motion W.

By Itô's Lemma we have that if (X, W) solves the SDE (μ, σ) then $(\Phi(X), W)$ solves the SDE (μ', σ') where

$$\mu'(x) = (L(\Phi)) \circ \Phi^{-1}(x) \quad \sigma'(x) = (\nabla \Phi \cdot \sigma) \circ \Phi^{-1}(x).$$

This implies the following theorem.

Theorem 3 *The diffeomorphism Φ is a strong symmetry of the SDE (μ, σ) if and only if*

$$\mu(x) = (L(\Phi)) \circ \Phi^{-1}(x) \quad \sigma(x) = (\nabla \Phi \cdot \sigma) \circ \Phi^{-1}(x),$$

where $x \in M$, with L as in (2). The vector field Y is a strong symmetry of the SDE (μ, σ) if and only if

$$Y(\mu) - L(Y) = 0 \quad \text{and} \quad [Y, \sigma_\alpha] = 0,$$

where $[\cdot, \cdot]$ stands for the Lie brackets between a vector fields, σ_α is the vector field given by the α-th colomn of the matrix σ, and $\alpha = 1, \ldots, m$.

Proof The proof can be found in [19] Proposition 3, Theorem 17 and Theorem 19. □

2.2 Weak Symmetries

The concept of strong symmetries of an SDE is quite restrictive. Indeed let us consider the following trivial SDE with $m = n = 2$, $\mu = 0$ and $\sigma = I_2$ namely

$$dX_t^1 = dW_t^1 \quad dX_t^2 = dW_t^2, \tag{3}$$

where $t \geq 0$. The solutions to Eq. (3) are the x_0-translated $\mathbb{R}^2$-Brownian motions $X_t = x_0 + W_t$ (for $x_0 \in \mathbb{R}^2$). The rotations

$$\Phi_a(x) = \begin{pmatrix} \cos(a) & -\sin(a) \\ \sin(a) & \cos(a) \end{pmatrix} \cdot x,$$

where $x \in \mathbb{R}^2$ and $a \in \mathbb{R}$, transform solutions $X_t = (X_t^1, X_t^2)$ to Eq. (3) into processes with similar laws, indeed $\Phi_a(X_t) = \Phi_a(x_0) + \Phi_a(W_t) = x_0' + W_t'$ where $x_0' \in \mathbb{R}^2$ and W_t' is a new (standard) Brownian motion $t \geq 0$. On the other hand $\Phi_a(x)$ is not a strong symmetry of (3) since $\Phi_a(X_t) \neq \Phi_a(x_0) + W_t$, i.e. $\Phi_a(X_t)$ is not a solution to the SDE (3) with respect to the same $\mathbb{R}^2$-Brownian motion W.

The reason for this phenomenon is that strong symmetries do not take into account the invariance properties of Brownian motion. Indeed the Brownian motion is invariant in law with respect to spatial rotations and space-time rescaling. These two invariances of Brownian motion allow us to introduce the following set of transformations.

2.2.1 Random Rotations

Let $B : M \to O(n)$ be a smooth function (where $O(n)$ denotes the group of orthogonal matrices on $\mathbb{R}^n$) and let (X, W) be a (standard) process. We define

$$W_t' = \int_0^t B(X_s) \cdot dW_s \tag{4}$$

where the integral is in the Itô sense. The following Proposition characterizes W'.

Proposition 4 *The process W', defined by Eq. (4), is an n dimensional Brownian motion.*

Proof The proof can be found in [19] Proposition 7. □

2.2.2 Random Time Change

Let $\eta : M \to \mathbb{R}_+ = (0, +\infty)$ be a smooth positive function and let (X, W) be a standard process then we consider, for $t \in \mathbb{R}_+$:

$$\beta_t = \int_0^t \eta(X_s)\,ds,$$

and set $\alpha_t = \inf\{s | \beta_s > t\}$. The process α_t is called a time change associated with η. If Z is a stochastic process we define $H_\eta(Z)_t = Z_{\alpha_t}$. If (X, W) is a standard process, then we use the notation:

$$W'' = H_\eta\left(\int \sqrt{\eta(X_s)}\,dW_s\right), \tag{5}$$

where the integral is in the Itô sense.

Proposition 5 *The process W'', defined by Eq. (5), is an n dimensional Brownian motion.*

Proof The proof can be found in [19] Proposition 5. □

2.2.3 Weak Stochastic Transformations

Definition 6 Let Φ be a diffeomorphism, let $B: M \to O(n)$ and $\eta: M \to \mathbb{R}_+$ be two smooth functions. We call the triad $T = (\Phi, B, \eta)$ a stochastic transformation.

A stochastic transformation admits an action on the set of smooth SDEs (μ, σ) and on the set of standard processes (X, W). The action E_T on (μ, σ) is defined as follows

$$E_T(\mu) := \left(\frac{L(\Phi)}{\eta}\right) \circ \Phi^{-1}$$

$$E_T(\sigma) := \left(\frac{1}{\sqrt{\eta}}\nabla\Phi \cdot \sigma \cdot B\right) \circ \Phi^{-1},$$

where L is defined as in (2), and the action P_T on the set of standard processes (X, W) is defined by

$$P_T(X) = \Phi(H_\eta(X))$$

$$P_T(W) = H_\eta\left(\int \sqrt{\eta(X_s)}B(X_s) \cdot dW_s\right).$$

Definition 7 A stochastic transformation T is a (finite weak) symmetry of the SDE (μ, σ) if for any solution (X, W) to (μ, σ) also $P_T(X, W)$ is solution to (μ, σ) (where $P_T(X, W) = (P_T(X), P_T(W))$).

A consequence of the previous definitions is the following theorem.

Theorem 8 *A stochastic transformation T is a symmetry of the SDE (μ, σ) if and only if $E_T(\mu, \sigma) = (\mu, \sigma)$.*

Proof The proof can be found in [19] Theorem 17. □

2.2.4 Infinitesimal Stochastic Transformations

The set $\mathcal{T}$ of stochastic transformations $T = (\Phi, B, \eta) \in \mathcal{T}$ has the natural structure of an infinite dimensional Lie group (more precisely of a Lie grupoid). Indeed we can define the following composition law

$$T_1 \circ T_2 = (\Phi_1 \circ \Phi_2, (B_1 \circ \Phi_2) \cdot B_2, (\eta_1 \circ \Phi_2)\eta_2), \tag{6}$$

for any $T_i = (\Phi_i, B_i, \eta_i) \in \mathcal{T}, i = 1, 2$.

This composition satisfies the following important property.

Theorem 9 *Consider the SDE (μ, σ) and the process (X, W), then we have*

$$E_{T_1 \circ T_2}(\mu, \sigma) = E_{T_1}(E_{T_2}(\mu, \sigma)) \qquad P_{T_1 \circ T_2}(X, W) = P_{T_1}(P_{T_2}(X, W)).$$

Proof The proof can be found in [19] Theorem 15. □

Using the composition (6) we can define the concept of one-parameter group of stochastic transformations $T_a = (\Phi_a, B_a, \eta_a)$, $a \in \mathbb{R}$. This one-parameter group T_a is generated by *an infinitesimal stochastic transformation* $V = (Y, C, \tau)$, where Y is a vector field on M, $C : M \to \mathfrak{o}(n)$ and $\tau : M \to \mathbb{R}$ are smooth functions (here $\mathfrak{o}(n)$ is the Lie algebra associated with $O(n)$, i.e. the Lie algebra of antisymmetric $n \times n$ matrices) through the following relations

$$\partial_a \Phi_a = Y \circ \Phi_a$$
$$\partial_a B_a = C \circ \Phi_a \cdot B_a$$
$$\partial_a \eta_a = \tau \circ \Phi_a \cdot \eta_a.$$

Using the composition law (6) we can also define a natural action of a finite stochastic transformation $T = (\Phi, B, \eta)$ on an infinitesimal one $V = (Y, C, \tau)$, and a Lie bracket $[\cdot, \cdot]$ between infinitesimal stochastic transformations $T_i = (Y_i, C_i, \tau_i)$, $i = 1, 2$:

$$T_*(V) = (\Phi_*(Y), (B \cdot C \cdot B^{-1} + Y(B) \cdot B^{-1}) \circ \Phi^{-1}, (\tau + Y(\eta) \cdot \eta^{-1}) \circ \Phi^{-1}) \tag{7}$$

$$[V_1, V_2] = ([Y_1, Y_2], Y_1(C_2) - Y_2(C_1) - \{C_1, C_2\}, Y_1(\tau_2) - Y_2(\tau_1)) \tag{8}$$

where $\{\cdot, \cdot\}$ is the usual commutator between matrices.

Definition 10 An infinitesimal stochastic transformation V is an infinitesimal (weak) symmetry of the SDE (μ, σ) if the one-parameter group T_a generated by V is a set of (finite weak) symmetries, in the sense of Definition 7, of (μ, σ).

We can determine a set of necessary and sufficient conditions (called *determining equations*) such that $V = (Y, C, \tau)$ is an infinitesimal symmetry of the SDE (μ, σ).

Theorem 11 *The infinitesimal stochastic transformation $V = (Y, C, \tau)$ is a symmetry of the SDE (μ, σ) if and only if the following determining equations hold*

$$Y(\mu) - L(Y) = -\tau\mu$$
$$[Y, \sigma] + \sigma \cdot C = -\frac{1}{2}\tau\sigma$$

Proof The proof can be found in [19] Theorem 19. □

2.3 Extended Symmetries

The set of weak symmetries exploit all the internal invariance properties of Brownian motion and in this way generalizes in a non-trivial way the concept of strong symmetry of a SDE. Unfortunately this is not enough for having a correspondence with transformations of the Kolmogorov equation associated with a given SDE. More precisely there is no one-to-one correspondence between weak symmetries of a given SDE and (deterministic) Lie point symmetries of the associated Kolmogorov equation (see Sect. 4.2 below). For this reason in [20] a generalization of weak symmetry is proposed considering not only the invariance of Brownian motion with respect rotations and rescaling but also its symmetries with respect to the change of measure (the well known Cameron-Martin-Girsanov theorem).

An extended stochastic transformation $T_e = (\Phi, B, \eta, h)$ is a set of four elements where $h: M \to \mathbb{R}^n$ is a smooth function. The action of T_e on the processes (X, W) is the same as the one of $T = (\Phi, B, \eta)$ on X, and on W it is the following

$$P_{T_e}(W) = H_\eta \left(\int \sqrt{\eta(X_s)} B(X_s)(dW_s - h(X_s)\,ds) \right).$$

Furthermore if $\mathbb{P}$ is the probability law of (X, W), we consider the new process $P_{T_e}(X, W)$ with respect the new probability measure $\mathbb{Q}$ given by

$$\left.\frac{d\mathbb{Q}}{d\mathbb{P}}\right|_T = \exp\left(\sum_{\alpha=1}^n \int_0^T h^\alpha(X_s) dW_s^\alpha - \frac{1}{2}\int_0^T (h^\alpha(X_s))^2 ds \right), \qquad (9)$$

for any $T \geq 0$. By Girsanov theorem, and under some additional technical conditions on h (see [20]), $P_{T_e}(W)$ is a new Brownian motion with respect to the probability measure $\mathbb{Q}$ (for Girsanov's theorem and its extension see [24]).

We can define an action of T_e on the set of SDEs (μ, σ). In particular $E_{T_e}(\sigma) = E_T(\sigma)$ and

$$E_{T_e}(\mu) = \left(\frac{L(\Phi) + \nabla \Phi \cdot \sigma \cdot h}{\eta}\right) \circ \Phi^{-1},$$

with L as in (2). It is possible to define a composition between extended stochastic transformations and, thus also a concept of infinitesimal extended stochastic transformations in the following sense. An infinitesimal stochastic transformation $V_e = (Y, C, \tau, H)$ is composed by a vector field Y, and three smooth functions $C : M \to \mathfrak{o}(n)$, $\tau : M \to \mathbb{R}$ and $H : M \to \mathbb{R}^n$. The relation, between the infinitesimal stochastic transformation V_e and the one-parameter group $T_{e,a} = (\Phi_a, B_a, \eta_a, h_a)$, $a \in \mathbb{R}$, generated by it, is the same of the one of weak stochastic transformations for Φ_a, B_a and η_a, and for h_a it is given by

$$\partial_a h_a = \sqrt{\eta_a} B_a^{-1} \cdot H \circ \Phi_a.$$

Definition 12 An extended stochastic transformation $T_e = (\Phi, B, \eta, h)$ is an (extended weak) symmetry of the SDE (μ, σ) if for any solution (X, W) to (μ, σ) under the probability $\mathbb{P}$ also $P_{T_e}(X, W)$ is solution to (μ, σ) under the probability $\mathbb{Q}$ given by (9). An infinitesimal (extended) stochastic transformation V_e is a(n extended weak) symmetry of the SDE (μ, σ) if the one-parameter group $T_{e,a}$ generated by V_e is a set of (finite extended weak) symmetries of (μ, σ).

It is possible to give an extension of Theorems 8 and 11 for extended stochastic transformations.

Theorem 13 *An extended stochastic transformation T_e is a symmetry of the SDE (μ, σ) if and only if $E_{T_e}(\mu, \sigma) = (\mu, \sigma)$. The infinitesimal extended stochastic transformation $V_e = (Y, C, \tau, H)$ is a symmetry of the SDE (μ, σ) if and only if the following determining equations hold*

$$Y(\mu) - L(Y) - \sigma \cdot H = -\tau \mu$$
$$[Y, \sigma] + \sigma \cdot C = -\frac{1}{2}\tau\sigma,$$

with L as in (2).

Proof The proof can be found in [20] Theorem 3.13, using also results of [19]. □

3 Symmetries of SDEs Driven by Discrete Time processes

In this section we consider the symmetries of discrete time processes and in particular of iterated random maps (see [23]). The theory presented here is a special case of the idea of symmetry of an SDE driven by a general semimartingale studied in [2].

We consider an open set $M \subset \mathbb{R}^m$, and a Lie group of matrices N. If Z is a discrete time stochastic process on N we define

$$\Delta Z_n = Z_n \cdot (Z_{n-1})^{-1}.$$

Let $\Psi : M \times N \to M$ be a smooth map. If X is a discrete time process defined on M we say that (X, Z) is the solution of the SDE Ψ driven by the process Z_n if

$$X_n = \Psi(X_{n-1}, \Delta Z_n). \tag{10}$$

3.1 Gauge Symmetries

In this brief section we want to discuss the concept of gauge symmetries of a discrete time process taking values in a Lie group. The concept of gauge symmetry of a general semimartingale was introduced in [3]. The concept of gauge symmetry is a generalization of the invariance of Brownian motion with respect to random rotations (in the sense of Proposition 4).

Let $\mathcal{G}$ be a Lie group of matrices, we consider $\Xi. : \mathcal{G} \times N \to N$ as a $\mathcal{G}$ action of N, namely we require that

$$\Xi_{g_1}(\Xi_{g_2}(z)) = \Xi_{g_1 \cdot g_2}(z),$$

for any $g_1, g_2 \in \mathcal{G}$, $z \in N$. Let Z be a discrete time stochastic process on N and let G be a predictable process with respect to the natural filtration of Z. We define the process Z' on N in the following way $Z'_0 = 1_N$ and

$$\Delta Z'_n = \Xi_{G_n}(\Delta Z_n).$$

Hereafter if Z' is defined as just described we write $dZ' = \Xi_G(dZ)$.

Definition 14 Using the previous notations we say that $\mathcal{G}$ is a gauge symmetry group with respect to the action Ξ_g for the process Z if for any predictable process G we have that Z' (defined as $dZ' = \Xi_G(dZ)$) has the same law as Z.

Let $\nu_n(dz, \Delta Z_1, \ldots, \Delta Z_{n-1})$ be the probability law of ΔZ_n conditioned with respect to $Z_1, \ldots, Z_{n-1}$ (where $z \in N$).

Theorem 15 *The group of matrices $\mathcal{G}$ is a gauge symmetry group for Z if and only if, for any $g_1, \ldots, g_n \in \mathcal{G}$, we have*

$$\nu_n(d\Xi_{g_1}(z), \Xi_{g_2}(\Delta Z_1), \ldots, \Xi_{g_n}(\Delta Z_{n-1})) = \nu_n(dz, \Delta Z_1, \ldots, \Delta Z_{n-1}).$$

Proof The proof can be found in [3] Theorem 4.5 (that provides a general result on transformations of semimartingales) and Sect. 4.3 (where gauge symmetries of discrete processes are discussed). □

3.2 Weak Symmetries of Discrete Time SDEs

Definition 16 Let $\Phi : M \to M$ be a diffeomorphism and let $B : M \to \mathcal{G}$ be a smooth function. We call the pair (Φ, B) a (weak finite) stochastic transformation. Let Y be a vector field on M and let $C : M \to \mathfrak{g}$ (where $\mathfrak{g}$ is the Lie algebra of $\mathcal{G}$) be a smooth function, then we call the pair (Y, C) a (weak) infinitesimal stochastic transformation.

As for the Brownian motion case, the set of stochastic transformations admits a composition $\circ$, a push-forward $\cdot_*$ and the Lie brackets $[\cdot, \cdot]$ defined as follows

$$T_1 \circ T_2 = (\Phi_1 \circ \Phi_2, (B_1 \circ \Phi_2) \cdot B_2)$$
$$T_*(V) = (\Phi_*(Y), (B \cdot C \cdot B^{-1} + Y(B) \cdot B^{-1}) \circ \Phi^{-1})$$
$$[V_1, V_2] = ([Y_1, Y_2], Y_1(C_2) - Y_2(C_1) - \{C_1, C_2\}),$$

where $T = (\Phi, B)$, $V = (Y, C)$ and $V_i = (Y_i, C_i)$, $i = 1, 2$. We can also define an action of a stochastic transformation $T = (\Phi, B)$ on the equation Ψ and on the process (X, Z)

$$E_T(\Psi)(x, z) = \Phi(\Psi(\Phi^{-1}(x), \Xi_{B(x)^{-1}}(z)))$$
$$P_T(X)_n = \Phi(X_n)$$
$$\Delta P_T(Z)_n = \Xi_{B(X_{n-1})}(\Delta Z_n),$$

where $x \in M$ and $z \in N$. Hereafter we suppose that Z has a fixed probability law.

Definition 17 A stochastic transformation T is a symmetry of the SDE Ψ if for any solution (X, Z) to Ψ we have that $P_T(X, Z)$ is solution to the same equation Ψ.

By definition we obviously have that if (X, Z) is a solution to the SDE Ψ then $P_T(X, Z)$ is a solution of the equation $E_T(\Psi)$. Using this fact it is simple to prove the following theorem.

Theorem 18 *If $E_T(\Psi) = \Psi$ then T is a symmetry of the SDE Ψ.*

Proof The proof can be found in [2] Theorem 5.8 (see also the corresponding discussion in [3]). □

Remark 19 The converse of Theorem 18 is in general not true. In particular if the law of ΔZ_n does not have support on all of N, but has support in a proper closed subset $\tilde{N} \subset N$, $\tilde{N} \neq N$, we can only prove that $E_T(\Psi)(x, z) = \Psi(x, z)$ for any $(x, z) \in M \times \tilde{N}$. When $\tilde{N} = N$ then also the converse of Theorem 18 holds (see [2] Lemma 5.9 and Theorem 5.10).

4 Applications

4.1 Reduction and Reconstruction of Symmetric SDEs

One of the standard results of Lie symmetry analysis for ODEs is the possibility of reducing an n dimensional ODE admitting a (non degenerate) *solvable r dimensional Lie algebra* of infinitesimal symmetries to a reduced $n - r$ dimensional ODE and, in addition, the possibility of reconstructing the solution to the original ODE from the solutions to the reduced one using only the integration and the inversion operation of some functions (this is called reconstruction by quadrature). It is possible to obtain a similar result for SDEs driven by general semimartingales admitting a solvable Lie algebra of weak symmetries (see [2]). For simplicity we describe here only the case of Brownian motion driven SDEs as treated in [18].

Definition 20 We say that the SDE (μ, σ) is r-triangular with respect to the first r coordinates $x^1, \ldots, x^r$ of $x \in M$, if $\mu^i(x)$ and $\sigma^i_\alpha(x)$ depend only on $x^{i+1}, \ldots, x^m$ for $i = 1, \ldots, r(\leq m)$ and $\alpha = 1, \ldots, n$.

Obviously if (X, W) is solution of a r-triangular SDE (μ, σ) we can write $X^1_t, \ldots, X^r_t, t \in \mathbb{R}_+$, in the following way

$$X^i_t = X^i_0 + \int_0^t \mu^i(X^{i+1}_s, \ldots, X^m_s)ds + \int_0^t \sigma^i_\alpha(X^{i+1}_s, \ldots, X^m_s)dW^\alpha_s,$$

for $\alpha = 1, \ldots, n$ and $i = 1, \ldots, r$, where the last integral is in the Itô sense, in other words we can reconstruct the solution to the m dimensional SDE (μ, σ) from the solution to the $m - r$ dimensional SDE $((\mu^{r+1}, \ldots \mu^m), (\sigma^{r+1}, \ldots, \sigma^m))$ using only the Riemann and Itô integral respectively.

So, if (μ, σ) is an n dimensional SDE such that there exists a stochastic transformation T for which $E_T(\mu, \sigma)$ is in an r-triangular form, we can reconstruct the solution (X, W) from $E_T(X^{r+1}), \ldots, E_T(X^n)$ computing the inverse T and using only the Riemann and respectively Itô integration.

In order to describe the reduction and reconstruction theorem for SDEs we introduce the following definition.

Definition 21 Let $Y_1, \ldots, Y_r$ be r vector fields on M we say that $Y_1, \ldots, Y_r$ are in canonical form if

$$Y^i_j(x) = \delta^i_j,$$

where $x \in M$, for $i \geq j$, and $i, j \in \{1, \ldots, r\}$.

We give now the theorem which guarantees the existence of a transformation T putting a symmetric SDE (μ, σ) in an r-triangular form.

Theorem 22 *Suppose that the SDE (μ, σ) admits an r dimensional solvable Lie algebra $V_1 = (Y_1, B_1, \tau_1), \ldots, V_r = (Y_r, B_r, \tau_r)$ of (weak) symmetries (in the sense* of Definition 7), such that for any $x \in M$ the matrix $(Y_1(x)| \ldots |Y_r(x))$ has maximal rank. Then there exists a (locally defined) transformation $T = (\Phi, B, \eta)$ (in the sense of Definition 6) satisfying

$$Y_i(B) = C_i \cdot B \quad Y_i(\eta) = \tau_i \eta,$$

$i = 1, \ldots, r$, and such that $\Phi_*(Y_1), \ldots, \Phi_*(Y_r)$ are in canonical form (in the sense of Definition 21). Furthermore for any T satisfying the previous conditions $E_T(\mu, \sigma)$ is in an r-triangular form.

Proof The proof can be found in [18] Theorem 4.4 and Theorem 4.11 (where Theorem 24 of [19] is used). See also Theorem 5.19 of [2] where a more general result for SDEs driven by semimartingales with jumps is proven. □

4.2 Symmetries of SDEs and Symmetries of Corresponding Kolmogorov equations

The solutions to a Brownian motion driven SDE (or more generally the solutions to a Lévy process driven SDE) are Markov processes. This implies that SDEs of the form (1) are closely related to the corresponding Kolmogorov equations: these are the (deterministic) PDEs defined by

$$\partial_t u = L(u) = \mu^i \partial_i(u) + \frac{1}{2} \sum_{\alpha=1}^n \sigma_\alpha^i \sigma_\alpha^j \partial_{ij}(u), \tag{11}$$

where u belongs to $C^2(\mathbb{R}_+ \times M)$. The relation between Eqs. (1) and (11) is the following: let $f : M \to \mathbb{R}$ be a C^2 function with compact support then the unique solution $u(x, t)$ to Eq. (11) such that $u(x, 0) = f(x)$ is given by

$$u(x, t) = \mathbb{E}[f(X_t^x)],$$

where $t \in \mathbb{R}_+$, X_t^x is the unique solution to Eq. (1) such that $X_0^x = x$ and $x \in M$.

Since Eq. (11) is a deterministic PDE we can study its Lie point symmetries (for this concept see [56, 57]). Here for simplicity we consider only the symmetries that do not depend on the time t (for the general case see [20]). Since (11) is a PDE we can see it as a submanifold of the second order jet bundle $J^2(\mathbb{R}^{m+1}, \mathbb{R})$ described by the equation $\partial_t u + L(u) = 0$ (see, e.g., [14, 20]). In this setting a Lie point infinitesimal symmetry of (11) is a vector field

$$\Theta = -m(x)\partial_t + \phi^i(x)\partial_{x^i} + \Psi(x, u)\partial_u \tag{12}$$

(where (x, u) are the standard coordinates on the jet space $J^0(\mathbb{R}^{m+1}, \mathbb{R})$) such that the second order prolongation $\Theta^{(2)}$ satisfies $\Theta^{(2)}(\partial_t u - L(u))|_{\partial_t u - L(u)=0} = 0$ (see [20]).

Using the previous condition we obtain

$$\begin{aligned}
\Psi(x, u) &= -k(x)u + k_0(x) \\
L(k_0) &= 0 \\
L(k) &= 0 \\
L(\phi) - \Theta(\mu) + 2 A \cdot \nabla k - L(m)\mu &= 0 \\
L(m)A + \Theta(A) - \nabla \phi \cdot A - A \cdot (\nabla \phi)^T &= 0 \\
A \cdot \nabla m &= 0,
\end{aligned} \qquad (13)$$

with L and μ as in (11), for some smooth functions k and k_0, where $A = \frac{1}{2}\sigma \cdot \sigma^T$ (being σ as in (11) and $\cdot^T$ standing for transpose) and with ϕ as in (12).

We have the following important theorem which establishes a relation between Lie point symmetries of the Kolmogorov equation (11) and extended weak symmetries of the SDE (μ, σ) (in the sense of Definition 7).

Theorem 23 *Let Θ be a Lie point symmetry of Eq. (11) and suppose that σ has maximal rank, then the extended infinitesimal stochastic transformation $V_e = (Y, C, \tau, H)$ given by*

$$\begin{aligned}
Y^i &= \phi^i \\
\tau &= L(m) \\
C &= (\sigma^T \cdot \sigma)^{-1} \cdot \sigma^T \cdot [\phi, \sigma] - \frac{1}{2}L(m) \\
H &= \sigma^T \cdot \nabla k,
\end{aligned}$$

where $i = 1, \ldots, m$, is an extended weak symmetry of the SDE (μ, σ).

Proof The proof can be found in [20] Theorem 4.6. □

Remark 24 The converse of Theorem 23 is false, because, in general, there are more extended weak symmetries for a SDE (μ, σ) than Lie point symmetries for the associated Kolmogorov equation. This is due to the fact that the condition $H = \sigma^T \cdot \nabla k$, with $L(k) = 0$, as in (13), is quite restrictive and it is not satisfied by a general extended stochastic transformation.

In [20] (Theorem 4.6 and Remark 4.7) it is proved that if we restrict to the family of transformation generating a *Doob h-transform* (which is a probabilistic requirement equivalent to the analytic conditions that there exists a function k such that $L(k) = 0$ and $H = \sigma^T \cdot \nabla k$) the converse of Theorem 23 also holds.

4.3 Weak Symmetries of Numerical Schemes for SDEs

An important family of iterated random maps of the form (10) is given by the discrete numerical approximation schemes for Brownian motion driven SDEs. In this section we analyze only the simplest numerical scheme for a given SDE: the Euler scheme. A more general analysis considering also the Milstein scheme is given in [2, 22].

Let (μ, σ) be a smooth SDE, then we can define the following Euler discrete approximation scheme

$$X^i_{t_\ell} = X_{t_{\ell-1}} + \mu^i(X_{t_{\ell-1}})(t_\ell - t_{\ell-1}) + \sigma^i_\alpha(X_{t_{\ell-1}})(W^\alpha_{t_\ell} - W^\alpha_{t_{\ell-1}}), \qquad (14)$$

where $\ell \in \mathbb{N}$, $\alpha = 1, \ldots, n$. Let us consider the semimartingale $Z = (Z^0, Z^\alpha) \in \mathbb{R}^{n+1}$ defined by

$$Z^0_{t_\ell} = t_\ell \quad Z^\alpha_{t_\ell} = W^\alpha_{t_\ell}, \qquad (15)$$

with $\alpha \in \{1, \ldots, n\}$. We have that Eq. (14) can be written as an iterated random map of the form (10) with Ψ given by

$$\Psi^i(x, z) = x^i + \mu^i(x)z^0 + \sigma^i_\alpha(x)z^\alpha,$$

with $x \in M$ and $z \in N$, and with the driving process given by the discrete time semimartingale $Z = (Z^0, Z^\alpha)$ given by the expression (15). Consider the action $\Xi_B(z) = (z^0, B^\beta_\alpha z^\alpha)$ (where $B \in O(n)$) of $O(n)$ on $\mathbb{R}^{n+1}$. We have the following important theorem.

Theorem 25 *The discrete semimartingale (15) admits $O(n)$ with action Ξ_B as gauge symmetry group (in the sense of Definition 14).*

Proof The proof can be found in [2] Theorem 7.2. □

Since the numerical scheme (14) can be rewritten as a random map Ψ, and $O(n)$ is a gauge symmetry group for Z with action Ξ_B, we can say that (Y, C) is a weak symmetry for the numerical scheme (14) if it is a weak symmetry for the random map Ψ. The determining equations for SDE (14) reads

$$Y^i(\Psi(x, z^0, z^\alpha)) - Y^j(x)\partial_{x^j}(\Psi^i)(x, z^0, z^\alpha) = -C^\alpha_\beta(x)z^\beta \partial_{z^\alpha}(\Psi^i)(x, z^0, z^\alpha) \quad (16)$$

with $x \in M$ and $z \in N$. There is an important relation between weak symmetries of the numerical scheme (14) and of the SDE (μ, σ).

Theorem 26 *Let $V = (Y, C, 0)$ be a weak infinitesimal symmetry of SDE (μ, σ). When $Y^i(x)$ are polynomials of at most degree one in $x^1, \ldots, x^m$ then (Y, C) is a weak symmetry of the Euler discretization scheme (14). Conversely if for a given $x_0 \in M$, $\text{span}\{\sigma_1(x_0), \ldots, \sigma_n(x_0)\} = \mathbb{R}^m$ then also the converse is true.*

Proof The proof can be found in [2] Theorem 7.3 (see also [22] where the Theorem is proven for $C = 0$). □

When we have a solvable Lie algebra of weak symmetries of the numerical scheme (14) we can find a weak (finite) stochastic transformation $T = (\Phi, B)$ putting Eq. (14) in triangular form. We can use this privileged coordinate system for integrating the SDE (μ, σ). This strategy was applied in [22] to linear scalar equations and some theoretical and practical advantages in terms of the forward error for long time integration have been put in evidence.

5 Other Results on the Lie Symmetry Analysis of Stochastic Systems

5.1 Random Symmetries

In this Section we try to summarize, very shortly and using our language introduced in the first part of the present paper, the idea of Random symmetries, first proposed by Gaeta and Spadaro in [33] and which has received further investigation in [27, 28, 31, 43, 44] (see also the review [30]). We use here some notations closer to the ones used in the first part of the present paper and different from the ones used in the articles cited above. For simplicity we consider here only autonomous equations and autonomous transformations (i.e. non depending explicitly on time t).

The main idea of random symmetries is to consider transformations $\varphi : M \times \mathbb{R}^n \to M$ depending both on the variable X of the process (X, W) and on the Brownian motion W. In other words one considers a triad $T_r = (\varphi, B_0, \eta_0)$ where $\varphi(\cdot, w)$ is a diffeomorphism for each $w \in \mathbb{R}^n$, $B_0 \in O(n)$ and $\eta_0 \in \mathbb{R}_+$ are constants. In this case one takes, in the language used before, an SDE (μ_r, σ_r), where $\mu_r : M \times \mathbb{R}^n \to \mathbb{R}^m$ and $\sigma_r : M \times \mathbb{R}^n \to \mathbb{R}^{m \times n}$ are smooth functions, generically depending on both $x \in M$ and $w \in \mathbb{R}^n$. In this case (X, W) is a solution of the SDE (μ_r, σ_r) if

$$dX_t^i = \mu_r^i(X_t, W_t)dt + \sigma_{r,\alpha}^i(X_t, W_t)dW_t^\alpha.$$

One can then also consider the random generator

$$L_r(f)(x, w) = \mu_r^i \partial_{x^i} f + \frac{1}{2} \sum_{\alpha=1}^n (\sigma_{r,\alpha}^i \sigma_{r,\alpha}^j \partial_{x^i x^j} f + 2\sigma_{r,\alpha}^i \partial_{x^i w^\alpha} f + \partial_{w^\alpha w^\alpha} f),$$

where $x \in M$, $w \in \mathbb{R}^n$ and $f \in C^2(M \times \mathbb{R}^n)$. Using our language, the action of the random transformation T_r on the processes (X, W) is given by

$$P_{T_r}(X)_t = \varphi\left(X_{\frac{t}{\eta_0}}, W_{\frac{t}{\eta_0}}\right)$$
$$P_{T_r}(W)_t = \sqrt{\eta_0} B_0 \cdot W_{\frac{t}{\eta_0}}.$$

Using Itô formula one can determine the action of T_r on the SDE (μ, σ), which is given by the following relations

$$E_{T_r}(\mu_r)(x, w) = \frac{1}{\eta_0} L_r(\varphi) \left(\varphi^{-1} \left(x, \frac{B_0^{-1}}{\sqrt{\eta_0}} \cdot w \right), \frac{B_0^{-1}}{\sqrt{\eta_0}} \cdot w \right)$$

$$E_{T_r}(\sigma_r)(x, w) = \frac{1}{\sqrt{\eta_0}} (\nabla_x \varphi \cdot \sigma_r \cdot B_0^{-1} + \nabla_w \varphi \cdot B_0^{-1}) \left(\varphi^{-1} \left(x, \frac{B_0^{-1}}{\sqrt{\eta_0}} \cdot w \right), \frac{B_0^{-1}}{\sqrt{\eta_0}} \cdot w \right).$$

Definition 27 We call *finite random symmetry* of the SDE (μ_r, σ_r) a random transformation T_r such that for any solution (X, W) to (μ_r, σ_r) also $P_{T_r}(X, W)$ is solution to (μ_r, σ_r).

Theorem 28 *A random transformation T_r is a symmetry of the SDE (μ_r, σ_r) if and only if $E_{T_r}(\mu_r, \sigma_r) = (\mu_r, \sigma_r)$.*

Proof The proof is a consequence of Itô lemma and uniqueness of the martingale representation for processes adapted to a Brownian motion filtration (see [27] for more details). □

We can introduce also infinitesimal random transformation (Y_r, C_0, τ_0), where $Y_r : M \times \mathbb{R}^n \to \mathbb{R}^m$, C_0 is an antisymmetric matrix and $\tau_0 \in \mathbb{R}$, and the related infinitesimal symmetries.

Theorem 29 *An infinitesimal random transformation $V = (Y_r, C_0, \tau_0)$ is a symmetry of the SDE (μ_r, σ_r) if and only if the following determining equations are satisfied*

$$Y_r(\mu) + \partial_{w^\alpha} \mu \cdot C_{0,\beta}^\alpha w^\beta - L_r(Y) = \tau_0 \mu$$

$$(Y_r(\sigma_r) + \partial_{w^\alpha} \sigma_r C_{0,\beta}^\alpha w^\beta) - \nabla_x Y_r \cdot \sigma_r - \nabla Y_r = \frac{1}{2} \tau_0 \sigma_r + \sigma_r \cdot C_0.$$

Proof The proof is provided in [27]. □

Remark 30 It is obviously possible to consider B_0 and τ_0 to be functions of x and w. Unfortunately for general transformations of this form the random transformation acts in a non-local way on the coefficients (μ_r, σ_r). This makes it difficult to write some explicit determining equations.

Also in this case in [28] a theorem of reduction and reconstruction has been derived. Using our language one can call *(infinitesimal) simple random transformation* an infinitesimal random transformation of the form $V_r = (Y_r, 0, 0)$.

Theorem 31 *Let $V_{r,1} = (Y_{r,1}, 0, 0), \ldots, V_{r,k} = (Y_{r,k}, 0, 0)$ be a solvable Lie algebra of symmetries of the SDE (μ_r, σ_r) then there exists a random transformation $T_r = (\varphi, I_n, 1)$ such that $E_{T_r}(\mu_r, \sigma_r)$ is in triangular form with respect the first k variables.*

Proof The proof can be found in [28]. □

5.2 Variational SDEs and Noether Theorem

In this section we want to treat a special form of SDEs, admitting a variational structure which arises as a stochastic perturbation of the standard Lagrangian mechanics (see [64]). In this case it is possible to generalize many properties of the Lagrangian mechanics (see [12, 45, 47, 58, 62, 64] and also [1, 11, 38]). We show here one of the simplest result concerning variational stochastic systems and symmetries, namely a stochastic version of Noether's theorem. For simplicity we present here the case involving only regular SDEs, for a weak version holding also for semimartingales see [12, 45]. See [15] for recent results.

Consider the following SDE with additive noise, over $\mathbb{R}^m$,

$$dX_t = b(X_t, t)dt + dW_t \qquad (17)$$

where $t \geq 0$, with initial condition $X_0 = x \in \mathbb{R}^m$ and smooth time dependent drift b. This kind of system can be viewed as a stochastic perturbation of a classical ODE of the form

$$\dot{X}_t = b(X_t, t),$$

with $t \geq 0$ and $X_0 = x$. Furthermore we have that

$$v(x,t) := \lim_{h \to 0} \mathbb{E}\left[\left.\frac{X_{t+h} - X_t}{h}\right| X_t = x\right] = b(x,t), \qquad (18)$$

where $\mathbb{E}[\cdot|X_t = x]$ denotes the conditional expectation given that $X_t = x$, with $x \in \mathbb{R}^m, t \in \mathbb{R}_+$. For this reason we can consider $v(x, t) = b(x, t)$ as a kind of generalized velocity for the system. Consider now the Lagrangian

$$L(v, x) = \frac{|v|^2}{2} + V(x),$$

where $x \in \mathbb{R}^m$ and $v \in T\mathbb{R}^m$, for some smooth function V bounded from below.

We define a special family of Eq. (17) for which

$$S[b] := \mathbb{E}\left[\int_0^T L(v(X_t, t), X_t)dt\right] \qquad (19)$$

is minimized. Putting

$$S(x,t) = \min_{b \in C^\infty(M, \mathbb{R}^m)} \left(\mathbb{E}\left[\int_0^t L(v(X_s, s), X_s)ds\right]\right)$$

we have that S solves the Hamilton-Jacobi-Bellman equation

$$\partial_t S + \frac{1}{2}\Delta S - \frac{1}{2}\frac{|\nabla S|^2}{2} + V = 0 \tag{20}$$

with final condition $S(x, T) = 0$ and we obtain that the minimizer of the action (19) is exactly

$$b(x, t) = v(x, t) = \nabla S(x, t). \tag{21}$$

We can introduce the concept of symmetry of the Lagrangian L and of the variational problem (17) with action (19).

Consider a vector field $Y : \mathbb{R}^m \times \mathbb{R}_+ \to \mathbb{R}^m$ on $\mathbb{R}^m$ (depending also on the time t) and a time vector field $T : \mathbb{R}_+ \to \mathbb{R}$. They generate a one parameter group of transformations of the form

$$x' = \Phi_a(x, t) \quad t' = f_a(t),$$

with $a \in \mathbb{R}$, $t \in \mathbb{R}_+$ and $x \in \mathbb{R}^m$. Using the expression of the velocity (18), we obtain that v is transformed in the following way

$$v'(x', t') = (\nabla \Phi_a \cdot v)(\Phi_{-a}(x', t'), f_{-a}(t')). \tag{22}$$

Definition 32 We say that the infinitesimal transformation (Y, T) leaves invariant the Lagrangian L if for each $a \in \mathbb{R}$

$$\frac{|v(\Phi_{-a}(x', t'), f_{-a}(t'))|^2}{2} + V(\Phi_{-a}(x', t')) = L(v'(x', t'), x')$$

where v' is given by the expression (22) and v satisfies Eq. (21) (and S satisfies Eq. (20)).

We introduce now the following expression

$$P(x, t) = v(x, t) \cdot \nabla Y(x, t) + T(t)\left(\frac{|v(x, t)|^2}{2} + V(x)\right).$$

The functional P can be view as a generalization to the stochastic case of the deterministic momentum associated with the vector field Y and time change T. In the stochastic case we cannot expect the quantity $P(X_t, t)$ to be a constant as a function of t. Indeed if $P(X_t, t)$ were a constant of motion, the law of the process $X_t^{x_0}$ (i.e. the solution to SDE (17) such that $X_0^x = x_0$) would be supported on the $m - 1$ dimensional manifold $P(x, t) = P(x_0, 0)$. But for processes with smooth drift and additive noise the support of the law of $X_t^{x_0}$ is the whole $\mathbb{R}^m$. Nevertheless also in the stochastic case the symmetry (Y, T) is associated with the conservation of $P(X_t, t)$, understood not in the pathwise sense, but rather in the mean sense. Indeed the following theorem of [62] holds.

Theorem 33 *Suppose that v is a solution of the variational problem (17) with action (19) (namely suppose that v satisfies the relation (21) with S satisfying Eq. (20)) and let (Y, T) be a symmetry of the Lagrangian L, then for any solution to the Eq. (17) we have that the process*

$$\mathfrak{P}_t := P(X_t, t)$$

is a martingale with respect to the filtration generated by the Brownian motion W in (17).

Proof The proof is given in [62]. □

5.3 Finite Dimensional Reduction of SPDEs

In principle, it is possible to develop a concept of Lie point symmetry for SPDEs generalizing what has been done for deterministic PDEs (see [56, 57]). Unfortunately this approach is not so fruitful for two reasons. The first one is that the definition of invariance of an SPDEs with respect to Lie point transformations is too restrictive: in fact the SPDEs with this kind of invariance are very few and not so useful in the applications (see [10] where this approach is applied to the Zakai equation). The second reason is that, assuming a notion of symmetry of an SPDE based on Lie point transformations, the use of the invariant solutions to an SPDE for reducing the considered SPDE to a finite dimensional SDE, turns out to be too restrictive since the dimension of the reduced SDE is fixed by the order of the considered equation and it is often too low for being interesting in the applications.

Instead of extending the notion of symmetry of PDEs to the SPDEs case and then using this property for reducing an SPDE to a finite dimensional SDE, we make these two steps at once facing directly the problem of reducing an SPDE to a finite dimensional SDE. We here consider an SPDE of the form

$$dU_t(x) = F_\alpha(x, U_t(x), \partial^\sigma(U_t(x))) \circ dS_t^\alpha \qquad (23)$$

where $x \in M \subset \mathbb{R}^m$ and $\alpha = 1, \ldots, r$, $U_t(x)$ is a semimartingale taking values in $\mathbb{R}$, $F_\alpha(x, u, u_\sigma)$ are smooth functions of the independent coordinates x^i, the dependent coordinate u and its derivatives u_σ (where $\sigma \in \mathbb{N}_0^m$ is a multi-index denoting the number of derivatives with respect the coordinates x^i), $S^1, \ldots, S^r$ are continuous semimartingales and $\circ$ denotes the Stratonovich integral (see, e.g., [37, 59]).

A finite dimensional solution to the SPDE (23) is a solution $U_t(x)$ to (23) for which there exists a function $K : M \times \mathbb{R}^n \to \mathbb{R}$ and a semimartingale B_t (adapted to the filtration of $S^1, \ldots, S^r$) taking values in $\mathbb{R}^n$ such that

$$U_t(x) = K(x, B_t). \qquad (24)$$

Generic SPDEs of the form (23) do not admit finite dimensional solutions, but it is possible, using the methods of Lie symmetry analysis of PDEs (in particular the notion of differential constraints see [16, 17]), to provide a sufficient condition such that solutions, of the form (24), exist.

We introduce the vector fields

$$D_{x^i} = \partial_{x^i} + \sum_\sigma u_{\sigma+1_i} \partial_{u_\sigma}$$

where $1_i = (\delta_{ij})_{j=1,\ldots,m} \in \mathbb{N}_0^m$. We define also $D^\sigma = D_{x^1}^{\sigma_1} \cdots D_{x^m}^{\sigma_m}$. We denote by $\mathfrak{G}$ the set of smooth real-valued functions of the independent coordinates x^i, the dependent coordinate u and of a finite number of its derivatives u_σ. If $F, G \in \mathfrak{G}$ we write

$$\{F, G\} = \sum_{\sigma \in \mathbb{N}_0} (D^\sigma F \cdot \partial_{u_\sigma} G - \partial_{u_\sigma} F \cdot D^\sigma G).$$

We have that the brackets $\{\cdot, \cdot\}$ gives to $\mathfrak{G}$ the structure of a (real) Lie algebra.

Definition 34 We say that $F_1, \ldots, F_r$ generates a finite dimensional Lie algebra if span$\{F_1, \ldots, F_r\}$ generates a finite dimensional Lie subalgebra of $\mathfrak{G}$ with respect to the Lie brackets $\{\cdot, \cdot\}$.

Let $F \in \mathfrak{G}$ and let $f \in C^\infty(M, \mathbb{R})$ be a smooth function. We then say that the evolution equation F has a unique solution with respect to the initial condition f if then there exist a $T > 0$ and a unique function $v \in C^\infty(M \times [0, T], \mathbb{R})$ such that

$$\partial_t v(x, t) = F(x, v(x, t), \partial_x^\sigma v(x, t))$$

and $v(x, 0) = f(x)$.

Thanks to the previous definitions we can obtain the following theorem.

Theorem 35 *Suppose that S_t^1 in (23) satisfies $S_t^1 = t$, and that the evolution equation F_1 has a solution with respect to the initial condition $f \in C^\infty(M, \mathbb{R})$, and that $F_2, \ldots, F_r$ are at most of first degree (i.e. they are independent of u_σ for $|\sigma| > 1$), and that $F_1, \ldots, F_r$ generates a finite dimensional Lie algebra. Then there exist a smooth function $K : M \times \mathbb{R}^n \to \mathbb{R}$, such that $K(x, 0) = f(x)$, and a stochastic process B_t, taking values in $\mathbb{R}^n$, defined till a stopping time $\tau > 0$ and such that $B_0 = 0$, such that $U_t(x) = K(x, B_t)$ is a solution to Eq. (23) for $t < \tau$.*

Proof The theorem in this formulation is proven in the particular case of Zakai equation and in [8, 9] general in [13, 17]. A related result using a different formulation and with different hypotheses is proven in [25]. □

Acknowledgements This research is funded by the DFG under Germany's Excellence Strategy - GZ 2047/1, Project-ID 390685813. The authors want to thank Stefania Ugolini and the Dipartimento di Matematica of Università degli Studi di Milano for the warm hospitality.

References

1. Albeverio, S., Cruzeiro, A.B., Holm, D. (eds.): Stochastic geometric mechanics, vol. 202 of Springer Proceedings in Mathematics & Statistics. Springer, Cham (2017). Papers from the Research Semester "Geometric Mechanics—Variational and Stochastic Methods" held at the Centre Interfacultaire Bernoulli (CIB), Ecole Polytechnique Fédérale de Lausanne, Lausanne, January–June (2015)
2. Albeverio, S., De Vecchi, F.C., Morando, P., Ugolini, S.: Weak symmetries of stochastic differential equations driven by semimartingales with jumps. Electron. J. Probab. 25 (2020)
3. Albeverio, S., De Vecchi, F.C., Morando, P., Ugolini, S.: Random transformations and invariance of semimartingales on lie groups. Random Oper. Stoch. Equ. (2021)
4. Bluman, G.W., Cheviakov, A.F., Anco, S.C.: Applications of symmetry methods to partial differential equations. In: Applied Mathematical Sciences, vol. 168. Springer, New York (2010)
5. Bluman, G.W., Kumei, S.: Symmetries and differential equations. In: Applied Mathematical Sciences, vol. 81. Springer, New York (1989)
6. Cohen de Lara, M.: A note on the symmetry group and perturbation algebra of a parabolic partial differential equation. J. Math. Phys. **32**(6), 1445–1449 (1991)
7. Cohen de Lara, M.: Geometric and symmetry properties of a nondegenerate diffusion process. Ann. Probab. **23**(4), 1557–1604 (1995)
8. De Lara, M.C.: Finite-dimensional filters. I. The Wei-Norman technique. SIAM J. Control Optim. **35**(3), 980–1001 (1997)
9. De Lara, M.C.: Finite-dimensional filters. II. Invariance group techniques. SIAM J. Control Optim. **35**(3), 1002–1029 (1997)
10. De Lara, M.C.: Reduction of the Zakai equation by invariance group techniques. Stochastic Process. Appl. **73**(1), 119–130 (1998)
11. Cruzeiro, A.B., Holm, D.D., Ratiu, T.S.: Momentum maps and stochastic clebsch action principles (2016). arXiv:1604.04554
12. Cruzeiro, A.B., Lassalle, R.: Weak calculus of variations for functionals of laws of semimartingales (2015). arXiv:1501.05134
13. De Vecchi, F.C.: Finite dimensional solutions to SPDEs and the geometry of infinite jet bundles (2017). arXiv:1712.08490
14. De Vecchi, F.C.: Lie symmetry analysis and geometrical methods for finite and infinite dimensional stochastic differential equations. Ph.D. thesis, Università degli Studi di Milano (2018)
15. De Vecchi, F.C., Mastrogiacomo, E., Turra, M., Ugolini, S.: Noether theorem in stochastic optimal control problems via contact symmetries. Mathematics **9**(9) (2021)
16. De Vecchi, Francesco C., Morando, Paola: Solvable structures for evolution PDEs admitting differential constraints. J. Geom. Phys. **124**, 170–179 (2018)
17. Francesco C. De Vecchi and Paola Morando. The geometry of differential constraints for a class of evolution PDEs. *J. Geom. Phys.*, 156:103771, 23, 2020
18. De Vecchi, F.C., Morando, P., Ugolini, S.: Reduction and reconstruction of stochastic differential equations via symmetries. J. Math. Phys. **57**(12), 123508, 22 (2016)
19. De Vecchi, F.C., Morando, P., Ugolini, S.: Symmetries of stochastic differential equations: a geometric approach. J. Math. Phys. **57**(6), 063504, 17 (2016)
20. De Vecchi, F.C., Morando, P., Ugolini, S.: Symmetries of stochastic differential equations using Girsanov transformations. J. Phys. A **53**(13), 135204, 31 (2020)
21. De Vecchi, F.C., Morando, P., Ugolini, S.: Reduction and reconstruction of sdes via Girsanov and quasi doob symmetries. J. Phys. A **54**(18), 185203 (2021)
22. De Vecchi, Francesco C., Romano, Andrea, Ugolini, Stefania: A symmetry-adapted numerical scheme for SDEs. J. Geom. Mech. **11**(3), 325–359 (2019)
23. Diaconis, Persi, Freedman, David: Iterated random functions. SIAM Rev. **41**(1), 45–76 (1999)
24. Ferrario, Benedetta: A note on a result of Liptser-Shiryaev. Stoch. Anal. Appl. **30**(6), 1019–1040 (2012)
25. Filipović, Damir, Teichmann, Josef: Existence of invariant manifolds for stochastic equations in infinite dimension. J. Funct. Anal. **197**(2), 398–432 (2003)

26. Fredericks, E., Mahomed, F.M.: Symmetries of first-order stochastic ordinary differential equations revisited. Math. Methods Appl. Sci. **30**(16), 2013–2025 (2007)
27. Gaeta, G.: W-symmetries of Ito stochastic differential equations. J. Math. Phys. **60**(5), 053501, 29 (2019)
28. Gaeta, G., Lunini, C.: Symmetry and integrability for stochastic differential equations. J. Nonlinear Math. Phys. **25**(2), 262–289 (2018)
29. Gaeta, Giuseppe: Lie-point symmetries and stochastic differential equations. II. J. Phys. A **33**(27), 4883–4902 (2000)
30. Gaeta, Giuseppe: Symmetry of stochastic non-variational differential equations. Phys. Rep. **686**, 1–62 (2017)
31. Gaeta, Giuseppe: Integration of the stochastic logistic equation via symmetry analysis. J. Nonlinear Math. Phys. **26**(3), 454–467 (2019)
32. Gaeta, G., Quintero, N.R.G.: Lie-point symmetries and stochastic differential equations. J. Phys. A **32**(48), 8485–8505 (1999)
33. Gaeta, G., Spadaro, F.: Random Lie-point symmetries of stochastic differential equations. J. Math. Phys. **58**(5), 053503, 20 (2017)
34. Glover, Joseph: Symmetry groups and translation invariant representations of Markov processes. Ann. Probab. **19**(2), 562–586 (1991)
35. Joseph Glover. Symmetry groups in Markov processes and potential theory. In *Functional analysis. V, Vol. I (Dubrovnik, 1997)*, volume 44 of *Various Publ. Ser. (Aarhus)*, pages 19–33. Univ. Aarhus, Aarhus, 1998
36. Hawkins, T.: Emergence of the theory of Lie groups. In: Sources and Studies in the History of Mathematics and Physical Sciences. Springer, New York (2000). An essay in the history of mathematics 1869–1926
37. He, S.W., Wang, J.G., Yan, J.A.: Semimartingale theory and stochastic calculus. In: Kexue Chubanshe (Science Press), Beijing; CRC Press, Boca Raton, FL (1992)
38. Holm, D.D.: Variational principles for stochastic fluid dynamics. Proc. A. **471**(2176), 20140963, 19 (2015)
39. Holm, D.D., Schmah, T., Stoica, C.: Geometric mechanics and symmetry, vol. 12 of Oxford Texts in Applied and Engineering Mathematics. Oxford University Press, Oxford: From finite to infinite dimensions. With solutions to selected exercises by David C. P, Ellis (2009)
40. Kosmann-Schwarzbach, Y.: The Noether theorems. In: Sources and Studies in the History of Mathematics and Physical Sciences. Springer, New York (2011). Invariance and conservation laws in the twentieth century, Translated, revised and augmented from the 2006 French edition by Bertram E. Schwarzbach
41. Kozlov, R.: The group classification of a scalar stochastic differential equation. J. Phys. A **43**(5), 055202, 13 (2010)
42. Kozlov, R.: Symmetries of systems of stochastic differential equations with diffusion matrices of full rank. J. Phys. A **43**(24), 245201, 16 (2010)
43. Kozlov, R.: Lie point symmetries of Stratonovich stochastic differential equations. J. Phys. A **51**(50), 505201, 15 (2018)
44. Kozlov, R.: Random Lie symmetries of Itô stochastic differential equations. J. Phys. A **51**(30), 305203, 22 (2018)
45. Lassalle, R., Zambrini, J.C.: A weak approach to the stochastic deformation of classical mechanics. J. Geom. Mech. **8**(2), 221–233 (2016)
46. Lázaro-Camí, Joan-Andreu., Ortega, Juan-Pablo.: Reduction, reconstruction, and skew-product decomposition of symmetric stochastic differential equations. Stoch. Dyn. **9**(1), 1–46 (2009)
47. Lescot, P., Zambrini, J.C.: Isovectors for the Hamilton-Jacobi-Bellman equation, formal stochastic differentials and first integrals in Euclidean quantum mechanics. In: Seminar on Stochastic Analysis, Random Fields and Applications IV, vol. 58 of Progress in Probability, pp. 187–202. Birkhäuser, Basel (2004)
48. Liao, Ming: Symmetry groups of Markov processes. Ann. Probab. **20**(2), 563–578 (1992)
49. Liao, M.: Invariant diffusion processes in Lie groups and stochastic flows. In: Stochastic analysis (Ithaca, NY, 1993), vol. 57 of Proceedings of Symposia in Pure Mathematics, pp. 575–591. American Mathematical Society, Providence, RI (1995)

50. Liao, Ming: Markov processes invariant under a Lie group action. Stochastic Process. Appl. **119**(4), 1357–1367 (2009)
51. Marsden, J.E., Ratiu, T.S.: Introduction to mechanics and symmetry, vol. 17 of Texts in Applied Mathematics. Springer, New York, 2nd ed. (1999). A basic exposition of classical mechanical systems
52. Misawa, Tetsuya: Conserved quantities and symmetry for stochastic dynamical systems. Phys. Lett. A **195**(3–4), 185–189 (1994)
53. Neuenschwander, Dwight E.: Emmy Noether's Wonderful Theorem. Johns Hopkins University Press, Baltimore, MD, updated edition (2017)
54. Noether, Emmy: Invariant variation problems. Transp. Theory Statist. Phys. **1**(3), 186–207 (1971)
55. Noether, E.: Gesammelte Abhandlungen/Collected papers. In: Springer Collected Works in Mathematics. Springer, Heidelberg (2013). Edited and with an introduction by Nathan Jacobson, With an introductory address by P. S. Alexandrov, Reprint of the 1983 edition
56. Olver, P.J.: Applications of Lie groups to differential equations, vol. 107 of Graduate Texts in Mathematics, 2nd ed. Springer, New York (1993)
57. Olver, Peter J.: Equivalence, Invariants, and Symmetry. Cambridge University Press, Cambridge (1995)
58. Privault, N., Zambrini, J.C.: Stochastic deformation of integrable dynamical systems and random time symmetry. J. Math. Phys. **51**(8), 082104, 19 (2010)
59. Rogers, L.C.G., Williams, D.: Diffusions, Markov processes, and martingales, vol. 2. Cambridge Mathematical Library. Cambridge University Press, Cambridge (2000). Itô calculus, Reprint of the second (1994) edition
60. Srihirun, B., Meleshko, S.V., Schulz, E.: On the definition of an admitted Lie group for stochastic differential equations with multi-Brownian motion. J. Phys. A **39**(45), 13951–13966 (2006)
61. Stephani, Hans: Differential Equations: Their Solution Using Symmetries. Cambridge University Press, Cambridge (1989)
62. Thieullen, M., Zambrini, J.C.: Symmetries in the stochastic calculus of variations. Probab. Theory Relat. Fields **107**(3), 401–427 (1997)
63. Ünal, Gazanfer: Symmetries of Itô and Stratonovich dynamical systems and their conserved quantities. Nonlinear Dynam. **32**(4), 417–426 (2003)
64. Zambrini, J.-C.: The research program of stochastic deformation (with a view toward geometric mechanics). In: Stochastic analysis: a series of lectures, vol. 68 of Progress in Probability, pp. 359–393. Birkhäuser/Springer, Basel (2015)

Markov Processes with Jumps on Manifolds and Lie Groups

David Applebaum and Ming Liao

Abstract We review some developments concerning Markov and Feller processes with jumps in geometric settings. These include stochastic differential equations in Markus canonical form, the Courrège theorem on Lie groups, and invariant Markov processes on manifolds under both transitive and more general Lie group actions.

Keywords Feller process · Lévy process · Stochastic differential equation · Manifold · Lie group · Symmetric space

1 Introduction

Stochastic differential geometry is a deep and beautiful subject. It is essentially the study of stochastic processes that take their values in manifolds, and so it naturally sits at the intersection of probability theory with differential geometry, but it also impacts on, and makes use of techniques from real, stochastic and functional analysis, dynamical systems and ergodic theory. If the manifold has the additional structure of being a Lie group, then more tools are available and the results are of considerable interest in their own right. By far the majority of work on the subject has arisen from studying Markov processes that arise as the solutions of stochastic differential equations (SDEs) on manifolds or Lie groups. From its beginnings with the pioneering work of Itô in 1950 [30] until the 1990s, the emphasis was on processes with continuous sample paths that are obtained by solving SDEs driven by Brownian motion. For accounts of this work, see [19, 20, 27] and references therein.

We dedicate this article to the memory of Hiroshi Kunita (1937–2019).

D. Applebaum (✉)
School of Mathematics and Statistics, University of Sheffield, Sheffield S3 7RH, UK
e-mail: D.Applebaum@sheffield.ac.uk

M. Liao
Department of Mathematics, Auburn University, Auburn, AL, USA
e-mail: liaomin@auburn.edu

More recently, there has been increasing interest in studying processes with jumps which are solutions of SDEs driven by Lévy processes. The material reviewed in this paper is wholly concerned with the jump case, including Lévy processes in Lie groups and manifolds, and generalisations.

The organisation of the paper is as follows. We begin with a short Sect. 2 that reviews the key definitions of Markov and Feller processes in a suitably general setting. In Sect. 3, we describe SDEs driven by Lévy processes on manifolds in Markus canonical form. If unique solutions exist, then they give rise to a Markov process. As an example we show how to obtain a Lévy process on a compact manifold by projection from the solution of an SDE on the frame bundle. When there are no jumps, this is precisely the celebrated Eels–Elworthy construction of Brownian motion on a manifold. The next section is more analytic. We outline the proof of a global version of the Courrège theorem in a Lie group, which gives a canonical form for a linear operator that satisfies the positive maximum principle (PMP). The probabilistic importance of this result is that the generators of all sufficiently rich Feller processes satisfy the PMP, and so their generators must be of this form. Indeed we see that the generators are characterised by a real-valued function, a vector-valued function, a matrix-valued function, and a kernel, that may be probabilistically interpreted as describing killing, drift, diffusion and jump intensity (respectively). We also describe how, when the group is compact, the generator may be represented by a pseudo-differential operator in the Ruzhansky–Turunen sense.

Sections 5 and 7 deal with Markov processes that are suitably invariant (i.e. their transition probabilities are invariant) under the action of a Lie group. In Sect. 5, we examine the case where the group acts transitively. In this case the manifold is a homogeneous space and the Markov process is, in fact, a Lévy process. To consider the non-transitive case, we need the notion of inhomogeneous Lévy process in a homogeneous space, i.e. a process that has independent, but not necessarily stationary increments. These are described more fully in the short Sect. 6. In Sect. 7, we consider the non-transitive case where we may effectively assume that the manifold M is the product of another manifold M_1 and a homogeneous space M_2. Then our process is the product of a radial part, that lives in M_1, and an angular part that lives in M_2. In fact the radial part is a Markov process, and the angular part is an inhomogeneous Lévy process. In the case of sample path continuity, some more detailed results are given.

We emphasise that this survey is by no means comprehensive. Our goal is the limited one of giving an introduction to the subject, and placing the spotlight on some key themes where there is active work going on. For more systematic study of Lévy processes on Lie groups and invariant Markov processes on manifolds, see [32, 33]. An important topic that is not dealt with here is the application of Malliavin calculus on Wiener–Poisson space to study regularity of transition densities for jump-diffusions. The recently published monograph [31] presents a systematic account of key results in this area.

Notation. If E is a locally compact Hausdorff space, the Borel σ-algebra of E is denoted as $\mathcal{B}(E)$. We denote by $\mathcal{F}(E)$, the space of all real-valued functions on E, and $C_0(E)$ is the Banach space (with respect to the supremum norm) of all real-valued, continuous functions on E that vanish at infinity. If M is a finite-dimensional real C^∞-manifold, then $C_c^\infty(M)$ is the dense linear manifold in $C_0(M)$ comprising all smooth functions with compact support.

The trace of a real or complex $d \times d$ matrix A is written tr(A). We denote as B_1, the open ball of radius 1 in $\mathbb{R}^d$ that is centred on the origin.

2 Markov and Feller Processes

Let $(\Omega, \mathcal{F}, P)$ be a probability space and $(E, \mathcal{E})$ be a measurable space. An *E-valued stochastic process* is a family $X := (X_t, t \geq 0)$ of random variables defined on Ω and taking values in E (so for all $t \geq 0$, X_t is $\mathcal{F} - \mathcal{E}$ measurable). Now suppose that $\mathcal{F}$ is equipped with a filtration $(\mathcal{F}_t, t \geq 0)$. An adapted E-valued stochastic process X is a *Markov process* if for all bounded measurable functions $f : E \to \mathbb{R}$, and all $0 \leq s \leq t < \infty$,

$$\mathbb{E}(f(X_t)|\mathcal{F}_s) = \mathbb{E}(f(X_t)|X_s) \quad \text{(a.s.)}$$

We then obtain a family of linear operators $(T_{s,t}, 0 \leq s \leq t < \infty)$ (in fact, these are also contractions and positivity preserving) on the Banach space $B_\mathcal{E}(E)$ of all bounded measurable real-valued functions on E (equipped with the supremum norm) by the prescription

$$T_{s,t} f(x) = \mathbb{E}(f(X_t)|X_s = x).$$

From now on all Markov processes that are considered will be *homogeneous*, in that $T_{s,t} = T_{0,t-s} =: T_{t-s}$, unless otherwise stated. Then the family $(T_t, t \geq 0)$ is an algebraic operator (contraction) semigroup on $B_\mathcal{E}(E)$, in that for all $s, t \geq 0$:

$$T_{s+t} = T_s T_t, \text{ and } T_0 = I.$$

From now on, we will assume that E is a locally compact, second countable Hausdorff space and $\mathcal{E}$ is its Borel σ-algebra For each $t \geq 0$, $B \in \mathcal{E}, x \in E$ we define the transition probability by

$$p_t(x, B) = P(X_t \in B | X_0 = x) = \mathbb{E}(\mathbf{1}_{X_t^{-1}(B)} | X_0 = x).$$

We then have the representation

$$T_t f(x) = \int_E f(y) p_t(x, dy), \tag{2.1}$$

for all $t \geq 0$, $f \in B_{\mathcal{E}}(E)$, $x \in E$, where we have taken a regular version[1] of the transition probability. If the mappings $x \to p_t(x, B)$ are measurable, we have the Chapman–Kolmogorov equations

$$p_{s+t}(x, B) = \int_E p_s(y, B) p_t(x, dy).$$

We would like to write down a differential equation for the transition probabilities which would enable us to extract information about these. This should be of the form of *Kolmogorov's forward equation*

$$\frac{\partial p_t(x, B)}{\partial t} = A^\dagger p_t(x, B), \qquad (2.2)$$

where $A^\dagger$ is the formal adjoint of a linear operator A acting on a suitable space of functions. We say that our process X is a *Feller process* if $(T_t, t \geq 0)$ is a *Feller semigroup*, i.e.

1. $T_t(C_0(E)) \subseteq C_0(E)$ for all $t \geq 0$,
2. $\lim_{t \to 0} \|T_t f - f\| = 0$ for all $f \in C_0(E)$.

It then follows that $(T_t, t \geq 0)$ is a strongly continuous contraction semigroup on $C_0(E)$. Hence it has an *infinitesimal generator* A defined on a dense subspace Dom(A) of $C_0(E)$ so that for all $f \in \text{Dom}(A)$,

$$\lim_{t \to 0} \left\| \frac{T_t f - f}{t} - Af \right\| = 0.$$

It is precisely this operator that enables us to give meaning to (2.2).

3 Stochastic Differential Equations on Manifolds

There are two ways of making sense of stochastic differential equations (SDEs) on manifolds. The first dates back to Itô [30], and is nicely described in Chap. 5 of Ikeda–Watanabe [29]. It involves solving the equation in local co-ordinates in each chart and then showing that the solutions transform geometrically on overlaps. Another method, which can be found in Elworthy [20], involves using the Whitney or Nash embedding theorem to embed the manifold into a Euclidean space of larger dimension. In that larger space, we must then show that if the initial data lie on the embedded manifold, then so does the solution for all later times. If the driving noise is a continuous semimartingale we must use the Stratonovitch differential to set up the SDE, and for discontinuous semimartingales, the more general Marcus canonical

---

[1] Our assumptions on $(E, \mathcal{E})$ ensure that such a version exists.

form. However we cannot expect to get Markov processes as solutions with such great generality. Nonetheless, by the argument of Sect. 6.4.2 in [2] pp. 387–8, we find that if global solutions exist, then they yield Markov processes when the driving noise is a Lévy process. To be more specific, we follow [10]. Let M be a manifold of dimension d, and consider an $\mathbb{R}^d$-valued Lévy process $L = (L(t), t \geq 0)$ having Lévy measure ν and Lévy–Itô decomposition (see Chap. 2 of [2] for background and explanation of the notation)

$$L_i(t) = b_i t + \sum_{j=1}^{m} \sigma_{ij} B_j(t) + \int_{B_1 \setminus \{0\}} y_i \tilde{N}(t, dy) + \int_{B_1^c} y_i N(t, dy),$$

for all $i = 1, \ldots, d, t \geq 0$.

Let $Y_1, \ldots, Y_d$ be C^∞ vector fields which have the properties that

(A1) All finite linear combinations of $Y_1, \ldots, Y_d$ are complete,
(A2) Each Y_j has bounded derivatives to all orders in every co-ordinate system (obtained by smoothly embedding M into a Euclidean space).

Now consider the following SDE

$$dX_t = \sum_{i=1}^{d} Y_i(X_{t-}) \diamond dL(t) \tag{3.1}$$

with initial condition $X_0 = p$ (a.s.) (where $p \in M$). The $\diamond$ stands for the Markus canonical integral, so in a local co-ordinate system containing p we have the symbolic form

$$X_t = p + \sum_{i=1}^{d} b_i \int_0^t Y_i(X_{s-}) ds + \sum_{i=1}^{d} \sum_{j=1}^{m} \sigma_{ij} \int_0^t Y_i(X_{s-}) \circ dB_j(s)$$

$$+ \int_0^t \int_{B_1 \setminus \{0\}} \left[\exp\left(\sum_{j=1}^{d} y_j Y_j \right) (X_{s-}) - X_{s-} \right] \tilde{N}(t, dy)$$

$$+ \int_0^t \int_{B_1^c} \left[\exp\left(\sum_{j=1}^{d} y_j Y_j \right) (X_{s-}) - X_{s-} \right] N(t, dy)$$

$$+ \int_0^t \int_{B_1 \setminus \{0\}} \left[\exp\left(\sum_{j=1}^{d} y_j Y_j \right) (X_{s-}) - X_{s-} - \sum_{j=1}^{d} y_j Y_j(X(s-)) \right] \nu(dy) ds.$$

where exp is the exponential mapping from complete vector fields to diffeomorphisms, and $\circ$ is the Stratonovitch differential. Under the stated conditions, we do indeed obtain a global solution (that is in fact a stochastic flow of diffeomorphisms) and which, as discussed above will be a Markov process. In fact the conditions are always satisfied if the manifold is compact. We might also ask about the Feller property. In general this is not so easy to answer; for SDEs driven by Lévy processes in Euclidean space, it is sufficient for coefficients to be bounded as well as suitably Lipschitz continuous (see Sect. 6.7 of [2]). We do have the following result for manifolds.

Theorem 3.1 *If M is compact then the solution to (3.1) is a Feller process. If A is the infinitesimal generator of the transition semigroup, then $C^2(M) \subseteq \text{Dom}(A)$ and for all $f \in C^2(M)$, $x \in M$,*

$$Af(x) = \sum_{i=1}^{d} b_i Y_i f(x) + \frac{1}{2} \sum_{i=1}^{d} \sum_{j=1}^{m} a_{ij} Y_i Y_j f(x)$$
$$+ \int_{\mathbb{R}^d \setminus \{0\}} \left[f\left(\exp\left(\sum_{i=1}^{d} y_i Y_i\right) x\right) - f(x) - \sum_{i=1}^{d} y_i Y_i f(x) \mathbf{1}_{B_1(0)}(y) \right] \nu(dy),$$
(3.2)

where the matrix $a := \sigma \sigma^T$.

Proof We just sketch this as it is along similar lines to that of Theorem 6.7.4 in [2] pp. 402–3. Let Φ_t be the solution flow that takes each $x \in M$ to the solution of (3.1) at time t with initial condition $X_0 = x$ (a.s.). Then we have $T_t f(x) = \mathbb{E}(f(\Phi_t(x)))$ for each $t \geq 0$, $f \in C(M)$. Both (A1) and (A2) hold and so the mapping Φ_t is continuous. Hence $T_t : C(M) \to C(M)$ by dominated convergence. Using Itô's formula we deduce that for all $f \in C^2(M)$,

$$T_t f(x) - f(x) = \int_0^t T_s A f(x) ds,$$

and the result follows easily from here. □

Another interesting example of a class of Feller processes on compact Riemannian manifolds are the isotropic Lévy processes which are obtained by first solving the equation

$$dR_t = \sum_{j=1}^{d} H_j(R_{t-}) \diamond dL_j(t), \qquad (3.3)$$

on $O(M)$, the bundle of orthonormal frames over M. Here $H_1, \ldots, H_d$ are horizontal vector fields (with respect to the Riemannian connection on M), and L is an isotropic Lévy process on $\mathbb{R}^d$ (i.e. its laws are $O(d)$-invariant). The required isotropic process is given by $X_t = \pi(R_t)$ where $\pi : O(M) \to M$ is the canonical surjection. The generator of the Feller process X is the sum of a non-negative multiple of the Laplace–Beltrami operator and an integral superposition of jumps along geodesics weighted by the Lévy measure of L. For details see [8]. If L is a standard Brownian motion so that (3.3) reduces to a Stratonovitch equation, then X is Brownian motion on M whose generator is the Laplace–Beltrami operator.

Now return to the SDE (3.1). In [9], it is shown the this equation has a unique solution (which will be a Markov process) if (A1) and (A2) are replaced by the single condition that the vector fields $Y_1, \ldots, Y_d$ generate a finite-dimensional Lie algebra. In general this is quite a strong assumption; as we are essentially saying that solution flow only explores a "small" finite-dimensional part of the "huge" infinite-dimensional diffeomorphism group of M. But one case where it occurs naturally is if M is a Lie group, which we now denote as G, and if $Y_1, \ldots, Y_d$ are assumed to be left-invariant vector fields which form a basis for the Lie algebra $\mathfrak{g}$ of G. Before proceeding further, we will define a Lévy process on a Lie group. This is precisely as in the well-known Euclidean space, i.e. a càdlàg process $X = (X_t, t \geq 0)$, that is stochastically continuous, satisfies $X_0 = e$ (a.s.), where e is the neutral element of G and has stationary and independent increments, where the increment between times s and t with $s \leq t$, is understood to be $X_s^{-1} X_t$. Since the pioneering work of Hunt in 1956 [28], it has been known that Lévy processes are Feller processes (take the filtration to be the natural one coming from the process). The associated Feller semigroup is defined for $f \in C_0(G), \sigma \in G, t \geq 0$ by

$$T_t f(\sigma) = \mathbb{E}(f(\sigma X_t)) = \int_G f(\sigma \tau) \rho_t(d\tau),$$

where ρ_t is the law of X_t. In fact $(\rho_t, t \geq 0)$ is a weakly continuous, convolution semigroup of probability measures on $(G, \mathcal{B}(G))$. Hunt also showed that the generator A has the following canonical representation on[2] $C_c^\infty(G) \subseteq \mathrm{Dom}(A)$: For each $\sigma \in G, f \in C_c^\infty(G)$,

$$Af(\sigma) = \sum_{i=1}^d b_i Y_i f(\sigma) + \frac{1}{2} \sum_{i,j=1}^d a_{ij} Y_i Y_j f(\sigma)$$
$$+ \int_{G \setminus \{e\}} \left(f(\sigma \tau) - f(\sigma) - \sum_{i=1}^d x_i(\tau) Y_i f(\sigma) \right) \nu(d\tau),$$
(3.4)

where $b = (b_1, \ldots, b_d) \in \mathbb{R}^d$, $a = (a_{ij})$ is a non-negative definite, symmetric $d \times d$ real-valued matrix and ν is a Lévy measure on G, i.e. a Borel measure on $G \setminus \{e\}$ which is such that for every canonical co-ordinate neighbourhood U of e we have

$$\int_{U \setminus \{e\}} \left(\sum_{i=1}^d x_i^2(g) \right) \nu(dg) < \infty, \text{ and } \nu(U^c) < \infty.$$

---

[2] In fact he obtained the representation on a larger domain, but we will not need that here.

Here $x_i \in C_c^\infty(G)$ for $i = 1, \ldots, d$ are such that $(x_1, \ldots, x_d)$ are canonical coordinates for G in the neighbourhood U of e. The triple (b, a, ν) are called the characteristics of X, and they uniquely determine $(\rho_t, t \geq 0)$.

In the case where the exponential map from $\mathfrak{g}$ to G is onto (e.g. if G is compact and connected, or simply connected and nilpotent), then the Lévy process X arises as the unique solution to the SDE (3.1), provided the characteristics of the driving Lévy process L are chosen so as to match those in (3.4). More generally, X is written as a more general type of stochastic integral equation (for details see [9, 32]).

Remark. A more general form of SDE on manifold than (3.1) was developed by Cohen [15]. In this case, the driving noise itself takes values in a manifold, which is in general different to that in which the solution flow takes values. After some years of neglect, this theory has recently found some new applications in work of [1].

4 The Positive Maximum Principle and Courrège Theory

Let E be a locally compact Hausdorff space and A be a linear mapping with domain $\mathrm{Dom}(A) \subseteq C_0(E)$ and range $\mathrm{Ran}(A) \subseteq \mathcal{F}(E)$. Let D_A be a linear subspace of $\mathrm{Dom}(A)$. We say that A satisfies the positive maximum principle (PMP) with respect to D_A if $f \in D_A$ and $f(y) = \sup_{x \in E} f(x) \geq 0 \Rightarrow Af(y) \leq 0$. It is usual in the literature (see e.g. [21]) to assume that A maps continuously i.e. that $\mathrm{Ran}(A) \subseteq C_0(E)$. Then one can prove powerful results—e.g. that A is dissipative, and hence if D_A is dense in $C_0(E)$, then A is closeable (see [21], Chap. 4, Lemma 2.1 (p. 165) and Chap. 1, Lemma 2.11 (p. 10)). We say that A has the full PMP if $D_A = \mathrm{Dom}(A)$.

The connection with our work is through the following well-known result.

Proposition 4.1 *If A is the generator of a Feller semigroup in $C_0(E)$ then it has the full PMP.*

Proof Using (2.1) we have for all $f \in \mathrm{Dom}(A)$ with $f(y) = \sup_{x \in E} f(x)$,

$$Af(y) = \lim_{t \to 0} \frac{T_t f(y) - f(y)}{t}$$
$$= \lim_{t \to 0} \frac{1}{t} \int_E (f(x) - f(y)) p_t(y, dx) \leq 0.$$

$\square$

There are stronger results than this. The celebrated *Hille–Yosida–Ray theorem* states that a linear operator A is the generator of a positivity preserving contraction semigroup $(T_t, t \geq 0)$ if and only if A is densely defined, closed, satisfies the full PMP and $\lambda I - A$ is onto $C_0(E)$ for all $\lambda > 0$. If such a semigroup is also *conservative*, i.e. it has an extension to the space of bounded measurable functions on E such that $T_t 1 = 1$ for all $t \geq 0$, then we may effectively use Kolomogorov's construction to obtain a Feller process for which $(T_t, t \geq 0)$ is the transition semigroup. For details see [14] pp. 13–238 and [21] pp. 165–73.

4.1 The Courrège Theorem on Euclidean Space and Manifolds

In this subsection we take $D_A = C_c^\infty(\mathbb{R}^d)$ and consider linear operators A that satisfy the PMP. The following key classification result was first published by Courrège [16]. In the following statement, we give the form (4.1) as can be found in Hoh [26].

Theorem 4.2 (Courrège theorem) *Let A be a linear operator from $C_c^\infty(\mathbb{R}^d)$ to $\mathbb{F}(\mathbb{R}^d)$. Then A satisfies the PMP if and only if there exists a unique quadruple $(a(\cdot), b(\cdot), c(\cdot), \nu(\cdot))$ wherein for all $x \in \mathbb{R}^d$, $a(x)$ is a $d \times d$ non-negative definite symmetric matrix, $b(x)$ is a vector in $\mathbb{R}^d$, $c(x)$ is a non-negative constant and $\nu(x, \cdot)$ is a Lévy measure on $\mathbb{R}^d$, so that for all $f \in C_c^\infty(\mathbb{R}^d)$*

$$Af(x) = -c(x)f(x) + \sum_{i=1}^d b_i(x)\partial_i f(x) + \sum_{i,j=1}^d a_{ij}(x)\partial_i \partial_j f(x)$$

$$+ \int_{\mathbb{R}^d \setminus \{0\}} \left(f(x+y) - f(x) - \sum_{i=1}^d y_i \partial_i f(x) \mathbf{1}_{B_1}(y) \right) \nu(x, dy) \quad (4.1)$$

In the same paper, Courrège proved that A is a pseudo-differential operator:

$$Af(x) = \int_{\mathbb{R}^d} e^{ix \cdot y} \eta(x, y) \widehat{f}(y) dy,$$

where $\widehat{f}$ denotes the usual Fourier transform: $\widehat{f}(y) = \frac{1}{(2\pi)^{d/2}} \int_{\mathbb{R}^d} e^{-ix \cdot y} f(x) dx$, and η is the symbol of the operator where

$$\eta(x, y) = -c(x) + ib(x) \cdot y - a(x)y \cdot y$$
$$+ \int_{\mathbb{R}^d \setminus \{0\}} (e^{ix \cdot y} - 1 - ix \cdot y \mathbf{1}_{B_1}(y)) \nu(x, dy), \quad (4.2)$$

so that formally:

$$\eta(x, y) = e^{-ix \cdot y} A(e^{ix \cdot y}). \quad (4.3)$$

If A is the generator of a Lévy process in $\mathbb{R}^d$, then $c = 0$, b, a and ν are independent of the value of $x \in \mathbb{R}^d$, (4.1) is the Euclidean version of (3.4) and η is the characteristic exponent whose form is that of the classical Lévy–Khintchine formula (see e.g. Chap. 3 of [2] and also Sect. 5 below). The general result (4.2) is important as it is a valuable source of probabilistic information about rich Feller processes, i.e. those for which $C_c^\infty(\mathbb{R}^d) \subseteq \text{Dom}(A)$, and so come under the auspices of this theory. See [14] and references therein for details.

The Courrège theorem has been generalised to manifolds, first by Courrège in the compact case [17] and then by Courrège, Bony and Prioret [13] in the general case. These authors succeeded in showing that if $A : C_c^\infty(M) \to \mathcal{F}(M)$ satisfies the PMP, then it has a decomposition of similar form to (4.1) relative to a local chart at a point. So they were able to describe the form of the operator in local co-ordinate systems. If the manifold is a Lie group or a symmetric space, we can obtain a global formalism.

4.2 The Courrège Theorem on Lie Groups

We summarise some of the main results of [11]. Let G be a Lie group with Lie algebra $\mathfrak{g}$. Here the strategy is to imitate the approach in [26], but with the vector fields $\{\partial_1, \ldots, \partial_d\}$ in Euclidean space replaced by the Lie algebra basis $\{Y_1, \ldots, Y_d\}$ for $\mathfrak{g}$. We say that a linear functional $T : C_c^\infty(G) \to \mathbb{R}$ satisfies the *positive maximum principle* (PMP), if whenever $f \in C_c^\infty(G)$ with $f(e) = \sup_{\tau \in G} f(\tau) \geq 0$ then $Tf \leq 0$. Then the linear operator A satisfies the PMP if and only if each of the linear functionals A_g satisfy the PMP, where $A_g f := A(L_{g^{-1}} f)(g)$ for each $g \in G$, $f \in C_c^\infty(G)$. Here L_g is the usual left translation defined by $L_g f(\sigma) = f(g\sigma)$ for $\sigma \in G$. We can recover the action of A from that of the $A_g's$ by $Af(g) = A_g L_g f$. So the problem is now reduced to studying the PMP for linear functionals T. Now it can be shown that if T satisfies the PMP, then it is a distribution of order 2. This is not a distribution in the usual sense. It is defined in the same way, but with the role of each partial derivative ∂_i replaced by Y_i for $i = 1, \ldots, d$.

We can then prove the first important result

Theorem 4.3 *Let* $T : C_c^\infty(G) \to \mathbb{R}$ *be a linear functional satisfying the PMP. Then there exists* $c \geq 0$, $b = (b_1, \ldots, b_d) \in \mathbb{R}^d$, *a non-negative definite symmetric* $d \times d$ *real-valued matrix* $a = (a_{ij})$, *and a Lévy measure* μ *on* G *such that for all* $f \in C_c^\infty(G)$,

$$Tf = -cf(e) + \sum_{i=1}^d b_i Y_i f(e) + \sum_{i,j=1}^d a_{ij} Y_i Y_j f(e)$$
$$+ \int_{G\{e\}} \left(f(g) - f(e) - \sum_{i=1}^d x_i(g) Y_i f(e) \right) \mu(dg), \tag{4.4}$$

The proof involves using a positivity argument (essentially the Riesz lemma) to show that there exists a Borel measure μ on $G \setminus \{e\}$ so that for all $f \in C_c^\infty(G \setminus \{e\})$

$$Tf = \int_{G\setminus\{e\}} f(g)\mu(dg). \qquad (4.5)$$

It then turns out that μ is a Lévy measure. Next we introduce a linear functional $S : C_c^\infty(G) \to \mathbb{R}$, by

$$Sf := \int_{G\setminus\{e\}} \left[f(g) - f(e) - \sum_{i=1}^d x_i(g) Y_i f(e) \right] \mu(dg).$$

Then S satisfies the PMP and so is a distribution of order 2. Hence so is $P = T - S$. But $\operatorname{supp}(P) \subseteq \{e\}$ and so P must take the form

$$Pf = -cf(e) + \sum_{i=1}^d b_i Y_i f(e) + \sum_{i,j=1}^d a_{ij} Y_i Y_j f(e).$$

It remains to prove that the matrix (a_{ij}) is positive definite and that $c \geq 0$. For this we refer the reader to the paper [11]. Once Theorem 4.3 is established, we can use the "left translation trick" described above to get the main result:

Theorem 4.4 *The mapping $A : C_c^\infty(G) \to \mathcal{F}(G)$ satisfies the PMP if and only if there exist functions c, b_i, a_{jk} ($1 \leq i, j, k \leq d$) from G to $\mathbb{R}$, wherein c is non-negative, and the matrix $a(\sigma) := (a_{jk}(\sigma))$ is non-negative definite and symmetric for all $\sigma \in G$, and a Lévy kernel[3] μ, such that for all $f \in C_c^\infty(G), \sigma \in G$,*

$$Af(\sigma) = -c(\sigma)f(\sigma) + \sum_{i=1}^d b_i(\sigma) Y_i f(\sigma) + \sum_{j,k=1}^d a_{jk}(\sigma) Y_j Y_k f(\sigma)$$

$$+ \int_{G\setminus\{e\}} \left(f(\sigma\tau) - f(\sigma) - \sum_{i=1}^d x_i(\tau) Y_i f(\sigma) \right) \mu(\sigma, d\tau). \qquad (4.6)$$

In [11] sufficient conditions are imposed on the coefficients to ensure that $A : C_c^\infty(G) \to C_0(G)$, and we will assume that these hold from now on.

We can represent linear operators satisfying the PMP as pseudo-differential operators when G is compact. To carry this out we need the unitary dual $\widehat{G}$ comprising equivalence classes (with respect to unitary equivalence) of all irreducible representations of $\widehat{G}$ in some complex (finite-dimensional) Hilbert space. So if $\pi \in \widehat{G}$, then for each $g \in G$, $\pi(g)$ is a unitary matrix acting in a space V_π which has dimension d_π. The Fourier transform $\widehat{f}$ of $f \in C_c^\infty(G)$ is the matrix-valued function on $\widehat{G}$ defined by

$$\widehat{f}(\pi) = \int_G f(g) \pi(g^{-1}) dg,$$

---

[3] We say that μ is a *Lévy kernel* if $\mu(\sigma, \cdot)$ is a Lévy measure for all $\sigma \in G$.

where integration is with respect to normalised Haar measure on G. The Ruzhansky–Turunen theory of pseudo-differential operators [35], starts from the Fourier inversion formula (just as in the classical case):

$$f(g) = \sum_{\pi \in \widehat{G}} d_\pi \operatorname{tr}(\widehat{f}(\pi)\pi(g)).$$

Then we say that $A : C_c^\infty(G) \to C_0(G)$ is a pseudo-differential operator with matrix valued symbol $j_A(\sigma, \pi)$ acting in V_π for each $\sigma \in G, \pi \in \widehat{G}$ if

$$Af(g) = \sum_{\pi \in \widehat{G}} d_\pi \operatorname{tr}(j_A(\sigma, \pi)\widehat{f}(\pi)\pi(g)).$$

In [3] it was shown that the generators of Lévy processes on compact Lie groups are pseudo differential operators, and this was further extended to some classes of Feller processes in [4]. Now we have the more general result:

Proposition 4.5 *If $A : C_c^\infty(G) \to C_0(G)$ satisfies the PMP then it is a pseudo-differential operator with symbol*

$$j_A(\sigma, \pi) = -c(\sigma)I_\pi + \sum_{i=1}^d b_i(\sigma)d\pi(Y_i) + \sum_{j,k=1}^d a_{jk}(\sigma)d\pi(Y_j)d\pi(Y_k)$$
$$+ \int_{G\setminus\{e\}} \left(\pi(\tau) - I_\pi - \sum_{i=1}^d x_i(\tau)d\pi(Y_i)\right) \mu(\sigma, d\tau), \quad (4.7)$$

for each $\pi \in \widehat{G}, \sigma \in G$.

Here if $Y \in \mathfrak{g}$, $d\pi(Y)$ is the skew-hermitian matrix acting in V_π which is uniquely defined by

$$\pi(\exp(tY)) = e^{td\pi(Y)},$$

for all $t \geq 0$. In the case where $c = 0$ and the characteristics b, a and μ are independent of $\sigma \in G$, (4.7) coincides with the form of the Lévy–Khinchine formula on compact Lie groups (see Sect. 5.5 in [6]).

Note that the analogue of (4.3) is

$$j_A(\sigma, \pi) = \pi(\sigma^{-1})A\pi(\sigma),$$

where $A\pi(\sigma)$ is the matrix $(A\pi_{ij}(\sigma))$. We recall that here G is compact and $\pi_{ij}(\cdot) \in C^\infty(G)$ for all $i, j = 1, \ldots d_\pi$, so that in this identity, the right hand side is a well-defined product of $d_\pi \times d_\pi$ matrices.

In [12] these ideas are extended to study linear operators satisfying the PMP on a symmetric space M. Observe that M may be identified with the homogeneous space G/K, where G is the isometry group of M (which turns out to be a Lie group), and

K is the compact subgroup of G comprising isometries leaving some given point fixed.[4] This enables us to use group theoretic techniques as described above, but with additional symmetries arising from the action of K. A feature of this theory is that, by using the properties of spherical functions, we can study a class of linear operators on M that satisfy the PMP and are pseudo-differential operators with scalar-valued symbols.

5 Invariant Markov Processes

Let M be a manifold under the action of a Lie group G, and let $X = (X_t; t \geq 0)$ be a Markov process in M with transition semigroup T_t as defined in Sect. 2. For simplicity, we write X_t for the process from now on. It should be clear from the context whether X_t is the whole process or just the random variable for a fixed time t. The Markov process X_t will be called invariant under the action of G, or G-invariant for short, if for any bounded Borel function f on M and $g \in G$,

$$T_t(f \circ g)(x) = (T_t f)(gx), \quad x \in M. \tag{5.1}$$

It is easy to show, see [33, Proposition 1.1], that the G-invariance of the Markov process X_t may also be characterized probabilistically as follows: We may think of a Markov process X_t as a family of processes, one for each starting point $x \in M$, all governed probabilistically by the same transition semigroup. If we denote X_t^z for the Markov process starting from $z \in M$, then X_t is G-invariant if and only if for any $g \in G$, gX_t^z and X_t^{gz} are equal in distribution as processes.

The G-invariance may also be defined for inhomogeneous Markov processes if T_t in (5.1) is replaced by $T_{s,t}$.

When $M = \mathbb{R}^d$ and $G = \mathbb{R}^d$ acts as translations on $\mathbb{R}^d$, it is well known that a càdlàg Markov process X_t in $\mathbb{R}^d$ is G-invariant, or translation-invariant, if and only if it has independent and stationary increments $X_t - X_s$ for $s < t$, that is, if it is a Lévy process in $\mathbb{R}^d$. The celebrated Lévy-Khintchine formula gives the characteristic function of each random variable within a Lévy process X_t:

$$E(e^{iX_t \cdot y}) = e^{t\psi(y)}$$

for any $y = (y_1, \ldots, y_d) \in \mathbb{R}^d$ by

$$\psi(y) = i \sum_{j=1}^{d} b_j y_j + \frac{1}{2} \sum_{j,k=1}^{d} a_{jk} y_j y_k + \int_{\mathbb{R}^d \setminus \{0\}} (e^{iz \cdot y} - 1 - iz \cdot y 1_{B_1}(z)) \nu(dz).$$

---

[4] K is independent of the choice of point chosen, up to isomorphism.

It provides a useful representation for a Lévy process in terms of a triple (b, a, ν), comprising a drift vector b, a covariance matrix $a = \{a_{ij}\}$ and a Lévy measure ν in the sense that its probability distribution is determined by the triple, and to any such triple, there is an associated Lévy process, unique in distribution, which also determines the triple. Note that (as we would expect) ψ coincides with η, as appears in (4.2), when the latter function is constant in the x-variable.

In the rest of paper, we will present some results on the more general invariant Markov processes under actions of Lie groups. The reader is referred to [33] for more details.

Because a Lie group G acts on itself by left translations $l_g : G \to G, x \mapsto gx$, for $g \in G$, we may first consider a Markov process X_t in G invariant under this G-action. This is a direct extension of a translation-invariant Markov process in $\mathbb{R}^d$ to a Lie group, and may be identified with a Lévy process in G as defined in Sect. 3, which is characterized by independent and stationary increments of the form $X_s^{-1} X_t$ for $s < t$. As already mentioned there, a Lévy process X_t in G is a Feller process, and its generator L has an explicit expression given by Hunt's formula (3.4), expressed in terms of a drift vector b, a covariance matrix a and a Lévy measure ν. Thus, a Lévy process in a Lie group is represented by a triple (b, a, ν) just as for a Lévy process in $\mathbb{R}^d$.

More generally, we may consider an invariant Markov process X_t in a manifold under the transitive action of a Lie group G. Here a transitive action means that for any two points x and y in M, there is $g \in G$ such that $gx = y$. Fix a point $o \in M$ and let $K = \{g \in G; go = o\}$. Then K is a closed subgroup G, called the isotropy subgroup at point o. The manifold M may be identified with the homogeneous space G/K, the space of left cosets $gK, g \in G$, via the map $gK \mapsto go$, and the G-action on M corresponds to the natural G-action on G/K given by $(g, hK) \mapsto ghK$ for $g, h \in G$. Thus, a G-invariant Markov process in M may be naturally identified with a G-invariant Markov process in the homogeneous space G/K.

A G-invariant Markov processes in G/K may also be characterised by independent and stationary increments as for a Markov process in G invariant under left translations, but to state this precisely will require a little preparation.

Let $M = G/K$ and let $\pi : G \to M$ be the natural projection $g \mapsto gK$. A Borel map $S : M \to G$ will be called a section map if

$$\pi \circ S = \mathrm{id}_M \text{ (the identity map on } M\text{).}$$

A section map always exists, but is not unique. It may not be smooth on M, but can be chosen to be smooth near any point in M.

In general, there is no natural product structure on $M = G/K$, but we may define $xy = S(x)y$ with the choice of a section map S. Although this definition depends on the choice of S, if Z is a random variable in M that has a K-invariant distribution, that is, if kZ is equal to Z in distribution for any $k \in K$, then the distribution of xZ does not depend on the choice of S. The increment of a process X_t in $M = G/K$ over the time interval $(s, t]$ is defined by $X_s^{-1} X_t = S(X_s)^{-1} X_t$, and its distribution does not depends on S when X_t is a G-invariant Markov process.

Markov Processes with Jumps on Manifolds and Lie Groups 39

Proposition 5.1 *A process X_t in $M = G/K$ with natural filtration $\{\mathcal{F}_t\}$ is a G-invariant Markov process if and only if it has independent and stationary increments in the sense that for any $s < t$ and any section map S, $X_{s,t} = S(X_s)^{-1} X_t$ is independent of $\mathcal{F}_s$ and its distribution depends only on $t - s$, not on S.*

The easy proof of the above proposition may be found in [33, Theorem 1.14]. Because of this result, a G-invariant Markov process in $M = G/K$ will be called a Lévy process.

When K is compact, a Lévy process in $M = G/K$ is a Feller process and its generator is determined by Hunt [28] to have essentially the same form as the generator of a Lévy process in G. To state this precisely will require some additional preparation.

For any $g \in G$, the conjugation map $G \to G$, given by $x \mapsto gxg^{-1}$, fixes the identity element e of G. Its differential map at e is a linear bijection from the Lie algebra $\mathfrak{g}$ of G to $\mathfrak{g}$, denoted as $\mathrm{Ad}(g)$. Let $\mathfrak{p}$ be an $\mathrm{Ad}(K)$-invariant subspace of $\mathfrak{g}$, complementary to the Lie algebra $\mathfrak{k}$ of K. The linear space $\mathfrak{p}$ may be identified with the tangent space $T_o M$ of M at o via the natural projection $\pi \colon G \to M$. Recall that we have chosen a basis $\{Y_1, \ldots, Y_d\}$ of $\mathfrak{g}$. We may assume $Y_1, \ldots, Y_n$ form a basis of $\mathfrak{p}$ and $Y_{n+1}, \ldots, Y_d$ form a basis of $\mathfrak{k}$, where $n = \dim(M)$. Recall the coordinate functions $x_1, \ldots, x_d \in C_c^\infty(G)$ associated to the basis $\{Y_1, \ldots, Y_d\}$ of $\mathfrak{g}$ as introduced in Sect. 3. They may be chosen to satisfy $g = \exp[\sum_{i=1}^n x_i(g) Y_i] \exp[\sum_{i=n+1}^d x_i(g) Y_i]$ for g in a neighborhood of e, where $\exp \colon \mathfrak{g} \to G$ is the Lie group exponential map. We now define coordinate functions $y_1, \ldots, y_n$ on M associated to the basis $\{Y_1, \ldots, Y_n\}$ of $\mathfrak{p}$ to be functions in $C_c^\infty(M)$ such that $\sigma = \pi\{\exp[\sum_{i=1}^n y_i(\sigma) Y_i]\}$ for σ in a neighborhood of o. They may be chosen to satisfy $\sum_{i=1}^n y_i(\sigma) \mathrm{Ad}(K) Y_i = \sum_{i=1}^n y_i(k\sigma) Y_i$ for all $\sigma \in M$ and $k \in K$ (see [33, Sect. 3.1]). Note that $x_i = y_i \circ \pi$ on a neighborhood of e for $1 \leq i \leq n$.

For $k \in K$, let $[\mathrm{Ad}(K)]$ denote the matrix representing $\mathrm{Ad}(K)$; $\mathfrak{p} \to \mathfrak{p}$ under the basis of $\{Y_1, \ldots, Y_n\}$, that is, $\mathrm{Ad}(K)\xi_j = \sum_{i=1}^n [\mathrm{Ad}(K)]_{ij} Y_i$. A vector $b = (b_1, \ldots, b_n) \in \mathbb{R}^n$ will called $\mathrm{Ad}(K)$-invariant if $[\mathrm{Ad}(K)]b = b$ for $k \in K$ and an $n \times n$ non-negative definite real symmetric matrix a will be called $\mathrm{Ad}(K)$-invariant if $[\mathrm{Ad}(K)]a[\mathrm{Ad}(K)]' = a$ for $k \in K$, where the prime denotes the matrix transpose. A measure μ on M is called K-invariant if $k\mu = \mu$ for $k \in K$, where $k\mu$ is the measure $k\mu(B) = \mu(k^{-1}(B))$. Note that for a K-invariant measure μ, the value of the integral

$$\int_M f(xy) \mu(dy) = \int f(S(x)y) \mu(dy), \quad x \in M,$$

does not depend on the choice of the section map S. A Lévy measure on M is defined in the same way as a Lévy measure on G, given in Sect. 3, but using the coordinate functions on M and a neighborhood U of o, and with the additional requirement that it is K-invariant.

The following result is Hunt's generator formula on $M = G/K$. The present form is taken from [33]. By this result, a Lévy process in $M = G/K$ is represented, in distribution, by a triple (b, a, ν) comprising an $\mathrm{Ad}(K)$-invariant vector b, an $\mathrm{Ad}(K)$-invariant matrix a and a Lévy measure ν on M, just as for a Lévy process in $\mathbb{R}^d$.

Theorem 5.2 *A Lévy process in $M = G/K$ is a Feller process. Let A be its generator. Then the domain of A contains $C_c^\infty(M)$, and there is a unique triple (b, a, ν) as above such that A restricted to $C_c^\infty(M)$ is given by (3.4) with d replaced by n, x_i by y_i and $G\setminus\{e\}$ by $M\setminus\{o\}$. Moreover, given any triple (b, a, ν), there is a Lévy process in M starting at o, unique in distribution, such that its generator has the above expression.*

Because the action of K on M fixes the point o, it induces a linear action of K on the tangent space $T_o M$. The homogeneous space $M = G/K$ is called irreducible if $T_o M$ has no proper subspace that is invariant under this K-action. For example, the sphere $S^{n-1} = O(n)/O(n-1)$ is irreducible as a homogeneous space of the orthogonal group $O(n)$ on $\mathbb{R}^n$.

On an irreducible homogeneous space $M = G/K$, it can be shown (see [33, Sect. 3.2]) that under a suitable choice of the basis $\{Y_1, \ldots, Y_n\}$ of $\mathfrak{p}$, any $\mathrm{Ad}(K)$-invariant matrix is proportional to the identity matrix I, and when $\dim(M) > 1$, there is no nonzero $\mathrm{Ad}(K)$-invariant vector. In this case, Hunt's generator formula (3.4) on M takes the following simpler form: For $f \in C_c^\infty(M)$,

$$Af(\sigma) = \frac{\alpha}{2} \sum_{i=1}^n Y_i Y_i f(\sigma) + \int_G [f(\sigma\tau) - f(\sigma)] \nu(d\tau), \qquad (5.2)$$

where the integral is understood as the principal value, that is, as the limit of $\int_{U^c} [f(\sigma\tau) - f(\sigma)] \nu(d\tau)$ when the K-invariant neighborhood U of o shrinks to o. Therefore, on an irreducible homogeneous space $M = G/K$ with $\dim(M) > 1$, a Lévy process is represented, in distribution, by a pair (α, ν) comprising a real number $\alpha \geq 0$ and a Lévy measure ν on M.

We note that the choice of the basis $\{Y_1, \ldots, Y_n\}$ of $\mathfrak{p}$ means that it is chosen to be orthonormal under an $\mathrm{Ad}(K)$-invariant inner product on $\mathfrak{p}$.

Recall that the Lévy-Khintchine formula gives the characteristic function of a Lévy process in $\mathbb{R}^d$. Such a formula may be extended to a general Lie group or a homogeneous space. See [6, Sect. 5.5] and [25] for its extension to a compact Lie group in terms of group representations, and [7] for the extension to a general Lie group. In these generalisations, the role of the characteristic exponent is played by a matrix-valued function on the unitary dual (in the compact case), or a function whose values are linear operators in a Hilbert space (in the general case). On a symmetric space, which is a special type of homogeneous space, Gangolli [24] obtained a Lévy-Khintchine type formula, in terms of the spherical transform, that closely resembles the classical form in that the characteristic exponent is scalar-valued. A simpler approach to Gangolli's result based on Hunt's generator formula may be found in [5, 34]. See also [33, Chap. 5] for a more systematic discussion.

We have considered invariant Markov processes under a transitive action of a Lie group, and have identified such processes with Lévy processes, which are characterized by independent and stationary increments. We will next consider the non-transitive action of a Lie group G. In this case, the state space M is a collection of G-orbits, and an invariant Markov process may be decomposed into a "radial" part, that is transversal to G-orbits, and an "angular" part, that lives in a G-orbit and is a G-invariant inhomogeneous Markov process. Such a process may be characterized by independent, but not necessarily stationary, increments, and so will be called an inhomogeneous Lévy process. Before we discuss the decomposition of an invariant Markov process under a non-transitive action, we will first briefly discuss inhomogeneous Lévy processes in Lie groups and homogeneous spaces.

6 Inhomogeneous Lévy Processes

Let G be a Lie group and K be a compact subgroup as before. A càdlàg inhomogeneous Markov process in G or in $M = G/K$, that is invariant under the left translation in G or G-invariant in M, will be called an inhomogeneous Lévy process. This definition is justified by the following easy proposition (see [33, Theorem 1.24]).

Proposition 6.1 *Let X_t be a càdlàg process in G or in $M = G/K$ with natural filtration $\{\mathcal{F}_t\}$. Then X_t is an inhomogeneous Lévy process if and only if it has independent increments in the sense that for any $s < t$, $X_s^{-1} X_t$ is independent of $\mathcal{F}_s$, where $X_s^{-1} X_t$ is understood as $S(X_s)^{-1} X_t$ on M for any section map S and its distribution is assumed not to depend on S.*

Recall that a Lévy process in G is a Feller process and the domain of its generator A contains $C_c^\infty(G)$. By a standard result in Markov process theory (see Proposition 1.7 in [21, Chap. 4]), for any $f \in C_c^\infty(G)$,

$$f(X_t) - \int_0^t Af(X_s)\,ds$$

is a martingale under the natural filtration of X_t, where the generator A, given by (3.4), is expressed in terms of a triple (b, a, ν). This martingale property in fact provides a complete representation of the Lévy process X_t, in distribution, in the sense that given any triple (b, a, ν), there is a Lévy process, unique in distribution, having this martingale property.

A càdlàg process X_t is called stochastically continuous if $P(X_{t-} = X_t) = 1$ for any $t > 0$. A Lévy process is stochastically continuous, but an inhomogeneous Lévy process may not be. Feinsilver [22] obtained a martingale representation for a stochastically continuous inhomogeneous Lévy process in a Lie group G by a time-dependent triple. Such a representation holds also for inhomogeneous Lévy processes in a homogeneous space G/K, even for non-stochastically continuous ones, see [33]. The general representation is rather complicated, so we will not discuss it here, but will present this representation on an irreducible G/K where it takes a very simple form.

A family of Lévy measures $\nu(t, \cdot)$, indexed by time $t \geq 0$, will be called a Lévy measure function if it is nondecreasing and continuous under weak convergence. It will be called an extended Lévy measure function if it may not be continuous, but is càdlàg, and $\nu(t, M) - \nu(t-, M) \leq 1$ for any $t > 0$. The extended Lévy measure function as defined in [33] requires an additional assumption, but on an irreducible $M = G/K$, it reduces to the above definition. The following result is [33, Theorem 8.16], which may be regarded as an extension of Hunt's generator formula (5.2) for Lévy processes on an irreducible G/K to inhomogeneous Lévy processes. We will assume the basis $\{Y_1, \ldots, Y_n\}$ of $\mathfrak{p}$ is chosen as described above in relation to (5.2).

Theorem 6.2 *Let $M = G/K$ be irreducible with $\dim(M) > 1$. For any inhomogeneous Lévy process X_t in M, there is a unique pair (a, ν) comprising a continuous nondecreasing function $a(t)$ with $a(0) = 0$ and an extended Lévy measure function $\nu(t, \cdot)$ such that for any $f \in C_c^\infty(M)$,*

$$f(X_t) - \int_0^t \frac{1}{2} \sum_{i=1}^n Y_i Y_i f(X_s) da(s) - \int_0^t \int_M [f(X_s \tau) - f(X_s)] \nu(ds\, d\tau) \quad (6.1)$$

is a martingale under the natural filtration of X_t, where the integral $\int_M [\cdots]$ is understood as the principal value as in (5.2). Moreover, X_t is stochastically continuous if and only if ν is a Lévy measure function.

Conversely, given any pair (a, ν) as above, there is an inhomogeneous Lévy process X_t in M with $X_0 = o$, unique in distribution, that has the above martingale property. Moreover, if ν is a Lévy measure function, then the uniqueness holds among all càdlàg processes in M.

7 Decomposition of Invariant Markov Processes

Let a Lie group G act on a manifold M. In this section we will not assume that the action is transitive; then M is a collection of G-orbits. Because G acts transitively on each G-orbit, any G-orbit is a homogeneous space G/K for some closed subgroup K. By the Principal Orbit Type Theorem [18], there is a subgroup K such that the G-orbits of type G/K fill up an open dense subset of M. Thus, by removing a small subset if necessary, we may assume M is a union of G-orbits of type G/K. Then it is reasonable to assume there is a submanifold M_1 of M that is transversal to G-orbits in the sense that it intersects each G-orbit at exactly one point, and

$$M = M_1 \times M_2, \quad M_2 = G/K. \quad (7.1)$$

The above will be assumed with K being compact. Note that G acts on M through its natural action on $M_2 = G/K$, and it acts trivially on M_1.

Markov Processes with Jumps on Manifolds and Lie Groups

For example, the orthogonal group $O(n)$ acts non-transitively on $\mathbb{R}^n$. Its orbits are spheres centered at the origin of type $O(n)/O(n-1)$, except the origin is an orbit by itself. After removing the origin, $\mathbb{R}^n$ becomes $M_1 \times M_2$, where M_1 is a half line from the origin and M_2 is the unit sphere. This is just the usual spherical polar decomposition.

Let X_t be a G-invariant Markov in M, and let $X_t = (Y_t, Z_t)$, where $Y_t \in M_1$ and $Z_t \in M_2$, be the decomposition (7.1). Borrowing the terms from the polar decomposition of $\mathbb{R}^2$, we will call Y_t the radial part and Z_t the angular part of the process X_t, noting that both may be multi-dimensional. It is easy to show that Y_t is a Markov process in M_1, and with some effort, it can be shown that Z_t is an inhomogeneous Lévy process in $M_2 = G/K$ under the conditional distribution given Y_t, but to state this more precisely requires some preparation. See Sects. 1.5 and 1.6 of [33] for more details.

Let $D(M)$ be the space of càdlàg paths in M equipped with the σ-algebra $\mathcal{F}$ generated by the coordinate maps $\omega \mapsto \omega(t), t \geq 0$, for $\omega \in D(M)$. The distribution of the process X_t with $X_0 = x$ induces a probability measure P_x on $D(M)$. We may regard X_t as the coordinate process on $D(M)$ by setting $X_t(\omega) = \omega(t)$. Similarly, Y_t and Z_t are regarded as coordinate processes on $D(M_1)$ and $D(M_2)$ (respectively). The product $M = M_1 \times M_2$ induces a product $D(M) = D(M_1) \times D(M_2)$. Let $J_1: D(M) \to D(M_1)$ and $J_2: D(M) \to D(M_2)$ be the natural projections, and let $\mathcal{F}^1$ and $\mathcal{F}^2$ be respectively the σ-algebras on $D(M_1)$ and $D(M_2)$ generated by the coordinate maps. We may regard $\mathcal{F}^1$ as a σ-algebra on $D(M)$ by identifying it with $J_1^{-1}(\mathcal{F}^1) = \{J_1^{-1}(B); B \in \mathcal{F}^1\}$.

Let $x = (y, z) \in M = M_1 \times M_2$. By the G-invariance of the Markov process X_t, $Q_y = P_x \circ J_1^{-1}$ is a probability measure on $D(M_1)$ depending only on y, and is in fact the distribution of the radial process Y_t.

Because $(D(M), \mathcal{F})$ is a standard Borel space, there is a regular conditional distribution $P_x^{y(\cdot)}(\cdot)$ of P_x given $\mathcal{F}^1$, that is, for any $F \in \mathcal{F}$,

$$P_x(F \mid \mathcal{F}^1)(y(\cdot)) = P_x^{y(\cdot)}(F) \quad \text{for } Q_y\text{-almost all } y(\cdot) \in D(M_1),$$

where $P_x^{y(\cdot)}(\cdot)$ is a probability kernel from $D(M_1)$ to $D(M)$ in the sense that it is a probability measure on $D(M)$ for each fixed $y(\cdot) \in D(M_1)$ and $P_x^{y(\cdot)}(F)$ is $\mathcal{F}^1$-measurable in $y(\cdot)$ for each $F \in \mathcal{F}$.

Let $R_z^{y(\cdot)} = P_x^{y(\cdot)} \circ J_2^{-1}$. Then $R_z^{y(\cdot)}$ is a probability kernel from $D(M_1)$ to $D(M_2)$, and for $F_1 \in \mathcal{F}^1$ and $F_2 \in \mathcal{F}^2$,

$$P_x(F_1 \cap F_2) = \int_{F_1} R_z^{y(\cdot)}(F_2) Q_y(dy(\cdot)).$$

Therefore, $R_z^{y(\cdot)}$ is the conditional distribution of the angular process Z_t given the radial process Y_t. The following result is Theorems 1.31 and 1.39 in [33].

Theorem 7.1 *For $y \in M_1$ and $z \in M_2$, the radial process Y_t is a Markov process in M_1 under Q_y, and for Q_y-almost all $y(\cdot) \in D(M_1)$, the angular process Z_t is an inhomogenous Lévy process in $M_2 = G/K$ under $R_z^{y(\cdot)}$.*

We will now consider a continuous G-invariant Markov process X_t in $M = M_1 \times M_2$ and assume $M_2 = G/K$ is irreducible. An example is a Brownian motion in $\mathbb{R}^n$, which is invariant under the orthogonal group $O(n)$, and in that case, the associated decomposition $X_t = (Y_t, Z_t)$ is just the usual spherical polar decomposition.

It is well known that the radial process Y_t and the angular process Z_t are not independent, but they are independent up to a random time change. More precisely, there is a spherical Brownian motion W_t in the unit sphere S^{n-1}, independent of Y_t, and a time change process $a(t)$ adapted to the natural filtration of Y_t, such that $Z_t = w_{a(t)}$. This is called the skew-product decomposition of the Brownian motion. Here the time change process $a(t)$ is a real-valued nondecreasing process with $a(0) = 0$.

This skew-product is generalized by Galmarino [23] to any continuous $O(n)$-invariant Markov process in $\mathbb{R}^n$. Using the decomposition of invariant Markov processes (Theorem 7.1) and the representation of inhomogeneous Lévy processes in irreducible homogeneous spaces (Theorem 6.2), we may obtain the skew-product on a more general space by a conceptually more transparent proof. See [33, Sect. 9.2] for more details. Note that the time change process $a(t)$ in the following theorem is the same $a(t)$ as in Theorem 6.2.

Theorem 7.2 *Let X_t be a continuous G-invariant Markov process in $M = M_1 \times M_2$ and assume $M_2 = G/K$ is irreducible with $\dim(M_2) > 1$. Then there are a Riemannian Brownian motion W_t in M_2 under a G-invariant Riemannian metric, independent of the radial process Y_t, and a time change process $a(t)$ that is adapted to the natural filtration of Y_t, such that $Z_t = W_{a(t)}$.*

Acknowledgements We thank the referee for some helpful suggestions.

References

1. Albeverio, S., De Vecchi, F.C., Morando, P., Ugolini, S.: Random transformations and invariance of semimartingales on Lie groups, Random Operators and Stochastic Equations (2021)
2. Applebaum, D.: Lévy Processes and Stochastic Calculus, 2nd edn. Cambridge University Press, Cambridge (2009)
3. Applebaum, D.: Infinitely divisible central probability measures on compact Lie groups - regularity, semigroups and transition kernels. Annals Prob. **39**, 2474–96 (2011)
4. Applebaum, D.: Pseudo differential operators and Markov semigroups on compact Lie groups. J. Math Anal. App. **384**, 331–48 (2011)
5. Applebaum, D.: Compound Poisson processes and Lévy processes in groups and symmetric spaces. J. Theor. Probab. **13**, 383–425 (2000)
6. Applebaum, D.: Probability on Compact Lie groups. Springer, Berlin (2014)
7. Applebaum, D.: Operator-valued stochastic differential equations arising from unitary group representations. J. Theor. Probab. **14**, 61–76 (2001)

8. Applebaum, D., Estrade, A.: Isotropic Lévy processes on Riemannian manifolds. Ann. Prob **28**, 188–84 (2000)
9. Applebaum, D., Kunita, H.: Lévy flows on manifolds and Lévy processes on Lie groups. J. Math Kyoto Univ. **33**, 1103–23 (1993)
10. Applebaum, D., Tang, F.: Stochastic flows of diffeomorphisms driven by infinite-dimensional semimartingales with jumps. Stoch. Proc. Appl. **92**, 219–36 (2001)
11. Applebaum, D., Le Ngan, T.: The positive maximum principle on Lie groups. J. London Math. Soc. **101**, 136–55 (2020)
12. Applebaum, D., Le Ngan, T.: The positive maximum principle on symmetric spaces. Positivity **24**, 1519–1533 (2020). https://doi.org/10.1007/s11117-020-00746-w
13. Bony, J.M., Courrège, P., Priouret, P.: Semi-groupes de Feller sur une variété à bord compacte et problèmes aux limites intégro- différentiels du second-ordre donnant lieu au principe du maximum. Ann. Inst. Fourier, Grenoble **18**, 369–521 (1968)
14. Böttcher, B., Schilling, R., Wang, J.: Lévy Matters III, Lévy Type Processes, Construction, Approximation and Sample Path Properties. Lecture Notes in Mathematics, vol. 2099. Springer International Publishing, Switzerland (2013)
15. Cohen, S.: Géometrie différentielle stochastique avec sauts I. Stoch. Stoch. Rep. **56**, 179–203 (1996)
16. Courrège, P.: Sur la forme intégro-différentielle des opérateurs de C_k^∞ dans C satifaisant au principe du maximum. Sém. Théorie du Potentiel exposé **2**, 38(1965/66)
17. Courrège, P.: Sur la forme intégro-différentielle du générateur infinitésimal d'un semi-group de Feller sur une variété. Séminaire Brelot-Choquet-Deny. Théorie du Potentiel, tome **10**(1), 1–48 (1965–66) exp. no. 3
18. Tom Dieck, T.: Transformation groups. de Gruyter, Berlin (1987)
19. Elworthy, K.D.: Stochastic Differential Equations on Manifolds. Cambridge University Press, Cambridge (1983)
20. Elworthy, K.D.: Geometric aspects of diffusions on manifolds in École d'Été de Probabilitès de Saint-Flour XV-XVII, 1985–87, 277–425. Lecture Notes in Math, vol. 1362. Springer, Berlin (1988)
21. Ethier, S.N., Kurtz, T.G.: Markov Processes, Characterisation and Convergence. Wiley, New York (1986)
22. Feinsilver, P.: Processes with independent increments on a Lie group. Trans. Amer. Math. Soc. **242**, 73–121 (1978)
23. Galmarino, A.R.: Representation of an isotropic diffusion as a skew product. Z. Wahrscheinlichkeitstheor. Verw. Geb. **1**, 359–378 (1963)
24. Gangolli, R.: Isotropic infinitely divisible measures on symmetric spaces. Acta Math. **111**, 213–246 (1964)
25. Heyer, H.: Infinitely divisible probability measures on compact groups. In: Lectures on Operator Algebras. Lecture Notes in Mathematics, vol. 247, pp. 55-249. Springer, Berlin (1972)
26. Hoh, W.: Pseudo Differential Operators Generating Markov Processes. Habilitationschrift Universität Bielefeld (1998). http://citeseerx.ist.psu.edu/viewdoc/download?doi=10.1.1.465.4876&rep=rep1&type=pdf
27. Hsu, E.P.: Stochastic Analysis on Manifolds. American Mathematical Society, Providence (2002)
28. Hunt, G.A.: Semigroups of measures on Lie groups. Trans. Amer. Math. Soc. **81**, 264–93 (1956)
29. Ikeda, N., Watanabe, S.: Stochastic Differential Equations and Diffusion Processes, 2nd edn. North Holland–Kodansha (1989)
30. Itô, K.: Stochastic differential equations on differentiable manifolds. Nagoya Math. J. **1**, 35–47 (1950)
31. Kunita, H.: Stochastic Flows and Jump Diffusions. Springer, Berlin (2019)
32. Liao, M.: Lévy Processes in Lie Groups. Cambridge University Press, Cambridge (2004)
33. Liao, M.: Invariant Markov Processes Under Lie Group Actions. Springer, Berlin (2018)

34. Liao, M., Wang, L.: Lévy-Khinchin formula and existence of densities for convolution semigroups on symmetric spaces. Potential Anal. **27**, 133–150 (2007)
35. Ruzhansky, M., Turunen, V.: Pseudo-differential Operators and Symmetries: Background Analysis and Advanced Topics. Birkhäuser, Basel (2010)

Asymptotic Expansion for a Black–Scholes Model with Small Noise Stochastic Jump-Diffusion Interest Rate

Francesco Cordoni and Luca Di Persio

Abstract In the present paper we study the asymptotic expansion for a Black–Scholes model with small noise stochastic jump-diffusion interest rate. In particular, we consider the case when the small perturbation is due to a general, but small, noise of Lévy type. Moreover, we provide explicit expressions for the involved expansion coefficients as well as accurate estimates on the remainders.

Keywords Mathematical finance · Asymptotic expansions · Stochastic interest rate models · Corrections for the Black–Scholes type models · Jump-diffusion models

MSC 2020: 35R60 · 60H15 · 65C30 · 91B24 · 91B25 · 91B30 · 91B70 · 91G10 · 91G80

1 Introduction

In the present paper we provide a correction around the well-known Black–Scholes (BS) formula for option pricing under the assumptions that the interest rate is stochastic. The latter description has gained particular attention after the 2008 worldwide credit crunch. In fact, starting from that date, several critiques to the standard BS model have been posed, mainly because of its too tight (deterministic) assumptions on the coefficients. Consequently, a series of alternatives to the classic BS model has been proposed, particularly allowing for stochastic volatility models in incomplete markets as well as to consider arbitrage and non constant interest rates.

Unfortunately, even if such generalizations provide more realistic settings, conversely to the BS model, they lack of a closed formula for option pricing. Therefore,

---

F. Cordoni · L. Di Persio (✉)
Department of Computer Science, University of Verona, Strada le Grazie, 15, 37134 Verona, Italy
e-mail: luca.dipersio@univr.it

F. Cordoni
e-mail: francesogiuseppe.cordoni@univr.it

numerical simulations are needed, often implying non trivial algorithmic procedures with associated significant computational costs.

One way to avoid such big numerical efforts is represented by ad hoc asymptotic expansion techniques, see, e.g., [9, 14–16, 20, 22, 25–27, 29, 33, 40, 42, 44, 46, 49–51]. Moreover, recent applications have been provided specifically w.r.t. the *financial* framework, as, e.g., in [12, 13, 31, 32, 37, 45, 52].

Following latter idea, we shall provide small noise asymptotic expansions for a BS model subjected to a small noise interest rate of jump-diffusion type. Our approach, inspired by the *heuristic* idea written in [28, Sect. 6.2], is based on the mathematically rigorous results derived in [6], see also [1, 22]. Let us recall that possible extensions to certain class of SPDEs are also possible, see, e.g., [2–4].

In particular, we shall consider an asset evolving according to a geometric Brownian motion also allowing for an interest rate characterized by a jump-diffusion noise term of *small* amplitude $\epsilon > 0$, then considering both a formal, based on [28, Sect. 6.2], resp. asymptotic, based on [6], expansion in ϵ. Indeed, considering *small* perturbations, we can approximate the considered model by a finite recursive system of N linear equations with random coefficients, hence exploiting its solutions by providing a formal, resp. an asymptotic, *smooth functions* approximation of the original solution. Analogously, within the usual *risk neutral* setting, we derive the corresponding option price approximation.

The paper is organized as follows: in Sect. 2 the general asymptotic expansions and related main results are presented; in Sect. 3 we apply the results obtained in Sect. 2 to provide corrections up to the first order for pricing plain vanilla options in a Black–Scholes model characterized by a small noise stochastic interest rate.

2 The Asymptotic Expansion

2.1 The General Setting

Let us consider the following stochastic differential equation (SDE), indexed by a parameter $\epsilon \geq 0$

$$\begin{cases} dX_t^\epsilon = \mu^\epsilon \left(X_t^\epsilon\right) dt + \sigma^\epsilon \left(X_t^\epsilon\right) dL_t \,, \\ X_0^\epsilon = x_0^\epsilon \in \mathbb{R}, \quad t \in [0, \infty) \end{cases} ; \qquad (1)$$

where L_t, $t \in [0, \infty)$, is a real-valued Lévy process of jump diffusion type, subjected to some restrictions which will be later specified, while $\mu^\epsilon : \mathbb{R}^d \to \mathbb{R}$, $\sigma^\epsilon : \mathbb{R}^d \to \mathbb{R}^{d \times d}$ are Borel measurable functions, for any $\epsilon \geq 0$ satisfying some additional technical conditions which guarantee existence and uniqueness of strong solutions, namely they are locally Lipschitz and with sub-linear growth at infinity, see, e.g., [10, 11, 30, 35, 43, 47]. If the Lévy process L_t has a jump component, then X_t^ϵ in Eq. (1) has to be understood as $X_{t-}^\epsilon := \lim_{s \uparrow \uparrow t} X_s^\epsilon$, see, e.g., [43] for details.

Hypothesis 2.1 *Let us assume that:*

(i) $\mu^\epsilon, \sigma^\epsilon \in C^{k+1}(\mathbb{R})$ *in the space variable, for any fixed value* $\epsilon \geq 0$ *and* $\forall k \in \mathbb{N}_0 := \mathbb{N} \cup \{0\}$;

(ii) *the maps* $\epsilon \mapsto \alpha^\epsilon(x)$, *where* $\alpha = \mu, \sigma$, *are elements of* $C^M(I)$ *in* ϵ, *for some* $M \in \mathbb{N}$, *for every fixed* $x \in \mathbb{R}$, *having defined* $I := [0, \epsilon_0]$, *while* $\epsilon_0 > 0$.

We show that under Hypothesis 2.1 and standard smoothness conditions on μ^ϵ and σ^ϵ to guarantee the well posedness of the random coefficients X_t^i, $i = 0, 1, \ldots, N$ in (2) below, a solution X_t^ϵ to Eq. (1) can be represented as a power series w.r.t. the parameter ϵ, namely:

$$X_t^\epsilon = X_t^0 + \epsilon X_t^1 + \epsilon^2 X_t^2 + \cdots + \epsilon^N X_t^N + R_N(t, \epsilon), \tag{2}$$

where $X^i : [0, \infty) \to \mathbb{R}, i = 0, \ldots, N$, are continuous functions, while $|R_N(t, \epsilon)| \leq C_N(t)\epsilon^{N+1}, \forall N \in \mathbb{N}$ and $\epsilon \geq 0$, for some $C_N(t)$ independent of ϵ, but in general dependent of randomness through $X_t^0, X_t^1, \ldots, X_t^N$.

For $n \in \mathbb{N}$, the functions X_t^i are determined recursively as solutions of random differential equations in terms of the X_t^j, $j \leq i - 1, \forall i \in \{1, \ldots, N\}$, see, e.g., [1, 6], for further details.

In particular, we have the following fundamental result:

Proposition 2.1 *Let* $x(\epsilon)$ *be as:*

$$x(\epsilon) = \sum_{j=0}^{N} \epsilon^j x_j + R_N^x(\epsilon), \quad N \in \mathbb{N}_0, \quad x_j \in \mathbb{R}, j = 0, 1, \ldots, N, \tag{3}$$

and f_ϵ *be as:*

$$f_\epsilon(x) = \sum_{j=0}^{K} f_j(x)\epsilon^j + R_K^{f_\epsilon}(\epsilon, x), \tag{4}$$

with $f_j \in C^{M+1}, j = 0, \ldots, K$. *Then*

$$f_\epsilon(x(\epsilon)) = \sum_{k=0}^{K+M} \epsilon^k [f_\epsilon(x(\epsilon))]_k + R_{K+M}(\epsilon),$$

with $|R_{K+M}(\epsilon)| \leq C_{K+M}\epsilon^{K+M+1}$, *for some constant* $C_{K+M} \geq 0$ *independent of both* ϵ *and* $0 \leq \epsilon \leq \epsilon_0$, *the coefficients* $[f_\epsilon(x(\epsilon))]_k$, *being defined by:*

$[f_\epsilon(x(\epsilon))]_0 = f_0(x_0);$

$[f_\epsilon(x(\epsilon))]_1 = Df_0(x_0)x_1 + f_1(x_0);$

$[f_\epsilon(x(\epsilon))]_2 = Df_0(x_0)x_2 + \frac{1}{2}D^2 f_0(x_0)x_1^2 + Df_1(x_0)x_1 + f_2(x_0);$

$[f_\epsilon(x(\epsilon))]_3 = Df_0(x_0)x_3 + \frac{1}{6}D^3 f_0(x_0)x_1^3 + Df_1(x_0)x_2 + Df_2(x_0)x_1 + D^2 f_1(x_0)x_1^2 + f_3(x_0).$

The general case has the following form

$$[f_\epsilon(x(\epsilon))]_k = Df_0(x_0)x_k + \frac{1}{k!}D^k f_0(x_0)x_1^k + f_k(x_0) + B_k^f(x_0, x_1, \ldots, x_{k-1}), \quad k = 1, \ldots, K+M \tag{5}$$

where B_k^f is a real function depending on $(x_0, x_1, \ldots, x_{k-1})$ only.

Remark 1 Let us note that $[f_\epsilon(x(\epsilon))]_k$ depends linearly on x_k, but not linearly in the inhomogeneity involving the coefficients x_j, $0 \le j \le k-1$ in (3). If $x(\epsilon)$ satisfies (3) and both μ^ϵ and σ^ϵ have the properties of the function f_ϵ, then the coefficients $\mu^\epsilon(x(\epsilon))$ and $\sigma^\epsilon(x(\epsilon))$ on the right hand side of (1) can be rewritten in powers of ϵ, for $0 \le \epsilon \le \epsilon_0$, as follows

$$\mu^\epsilon(x(\epsilon)) = \sum_{k=0}^{K_\mu + M_\mu} [\mu^\epsilon(x(\epsilon))]_k \epsilon^k + R^\mu_{K_\mu + M_\mu}(\epsilon);$$

$$\sigma^\epsilon(x(\epsilon)) = \sum_{k=0}^{K_\sigma + M_\sigma} [\sigma^\epsilon(x(\epsilon))]_k \epsilon^k + R^\sigma_{K_\sigma + M_\sigma}(\epsilon);$$

where the natural numbers K_α and M_α, resp. $\alpha = \mu, \sigma$, depend on the functions μ^ϵ, resp. σ^ϵ, while

$$|R^\alpha_{K_\alpha + M_\alpha}(\epsilon)| \le C_{K_\alpha + M_\alpha} \epsilon^{K_\alpha + M_\alpha + 1},$$

for some constants $C_{K_\alpha + M_\alpha}$ depending on C_j, $j = 0, \ldots, K_\alpha + M_\alpha$, but independently of both ϵ and $0 \le \epsilon \le \epsilon_0$.

2.2 The Asymptotic Character of the Expansion of the Solution X_t^ϵ of the SDE in Powers of ϵ

Theorem 2.2 *Let us assume that the coefficients α^ϵ, $\alpha = \mu, \sigma$ of the SDE (1) belong to $C^{K_\alpha}(I)$ as functions of ϵ, $\epsilon \in I = [0, \epsilon_0]$, $\epsilon_0 > 0$, and are in $C^{M_\alpha}(\mathbb{R})$ as functions of x. Moreover, let us assume that α^ϵ are such that there exists a solution X_t^ϵ in the probabilistic strong, resp. weak, sense for (1), and that the recursive system of random differential equations*

$$dX_t^j = [\mu^\epsilon(X_t^\epsilon)]_j dt + [\sigma^\epsilon(X_t^\epsilon)]_j dL_t, \quad j = 0, 1, \ldots, N, \ t \ge 0,$$

has a unique solution. Then there exists a sequence $\epsilon_n \in (0, \epsilon_0]$, $\epsilon_0 > 0$, $\epsilon_n \downarrow 0$ as $n \to \infty$ such that $X_t^{\epsilon_n}$ has an asymptotic expansion in powers of ϵ_n, up to order N, in the following sense:

$$X_t^{\epsilon_n} = X_t^0 + \epsilon_n X_t^1 + \cdots + \epsilon_n^N X_t^N + R_N(\epsilon_n, t),$$

with

$$st - lim_{\epsilon_n \downarrow 0} \frac{\sup_{s \in [0,t]} |R_N(\epsilon_n, s)|}{\epsilon_n^{N+1}} \le C_{N+1},$$

for some deterministic $C_{N+1} \ge 0$, independent of $\epsilon \in I$, where $st - lim$ stands for the limit in probability.

Proof See [1, 6] for details. □

3 The Black–Scholes Model with Stochastic Interest Rate

Let us consider an asset whose return $X_t^\epsilon := \log S_t^\epsilon$, under the so-called *risk neutral probability measure* $\mathbb{Q}$, evolves according to the following SDE

$$\begin{cases} dX_t^\epsilon = \left(r_t^\epsilon - \frac{\sigma_0^2}{2}\right)dt - \epsilon\lambda t\left(e^{\gamma + \frac{\delta^2}{2}} - 1\right) + \sigma_0 dW_t^1 + \epsilon \sum_{k=1}^{N_t} J_k, \\ X_0^\epsilon = x_0, \end{cases} \quad (6)$$

with $\sigma_0 \in \mathbb{R}_+$, while N_t is a standard Poisson process with intensity $\lambda > 0$, and $(J_i)_{i=1,\ldots,N_t}$ are i.i.d. Gaussian variables, i.e. $J_i \sim \mathcal{N}(\gamma, \delta^2)$, for some $\gamma \in \mathbb{R}$ and $\delta > 0$. Therefore, the Lévy measure $\nu(dz)$ of Z reads as

$$\nu(dz) = \frac{\lambda}{\sqrt{2\pi}\delta} e^{-\frac{(z-\gamma)^2}{2\delta^2}} dz, \quad z \in \mathbb{R}; \text{ with cumulant function } \kappa(\zeta) = \lambda \left(e^{\gamma\zeta + \frac{\delta^2\zeta^2}{2}} - 1\right).$$

Moreover, in Eq. (6), the r^ϵ-term represents a *stochastic interest rate* evolving according to

$$\begin{cases} dr_t^\epsilon = b(t, r_t^\epsilon)dt + \epsilon\nu(t, r^\epsilon)dW_t^2, \\ r_0 = r_0 \in \mathbb{R}, \end{cases} \quad (7)$$

for constant $\epsilon > 0$, b and ν being real functions satisfying suitable regularity and growth assumptions which guarantee both existence and uniqueness of the solution for (7). Further, W^1 and W^2 are assumed to be two standard Brownian motions with correlation $d\langle W^1 W^2 \rangle_t = \bar{\rho} dt$, for some constant $\bar{\rho} > 0$. In what follows we will assume that $b(t, r) = \kappa(\theta - r)$, for some real constants κ and θ, so that to preserve the *mean reverting* property usually required by standard interest rates model. Furthermore, we choose $v(t,r) = \bar{v}$ or $v(t,r) = \bar{v}\sqrt{r}$, for some constants $\bar{v}$, $r \in \mathbb{R}_+$.

Remark 2 Let us note that, although $v(t, r_t^\epsilon) = \bar{v}\sqrt{r_t^\epsilon}$ does not satisfy the smoothness assumptions we assumed up to now, our expansion works, at least as a *formal power series expansion*, because of the existence and uniqueness result, see [23], for this particular diffusion term component.

Assuming the above stated conditions, we have the following result.

Proposition 3.1 *Let us consider the return process X_t^ϵ evolving according to Eq. (6), where the stochastic interest rate r_t^ϵ evolves according to the SDE (7), with $v(t, r)$ satisfying standard conditions. Assume existence and uniqueness of solutions to (7), with $b(t, r) = \kappa(\theta - r)$. Assume that the jump process is independent of W_t^i, $i = 1, 2$, with*
$$d \langle W^1 W^2 \rangle_t = \bar{\rho} dt \,, \text{ for some constant } \bar{\rho} > 0.$$
Then the first order heuristic expansion for the stochastic interest rate r_t^ϵ reads as follow

$$r_t^0 = r_0 + \int_0^t \kappa \left(\theta - r_s^0\right) ds \,,$$
$$r_t^1 = -\int_0^t e^{-\kappa(t-s)} v(s, r_s^0) dW_s^2 \text{ with law } \mathcal{N}\left(0, Q_t^{r^1}\right),$$
(8)

where $Q_t^{r^1} := \int_0^t e^{-2\kappa(t-s)} v^2(s, r_s^0) ds$.

Furthermore the heuristic expansion up to the first order for the return X_t^ϵ is given by

$$X_t^0 = x_0 + \int_0^t \left(r_s^0 - \frac{\sigma_0^2}{2}\right) ds \,;\, X_t^1 = \int_0^t r_s^1 ds - \lambda t \left(e^{\gamma + \frac{\delta^2}{2}} - 1\right) + \sum_{k=1}^{N_t} J_k \,. \quad (9)$$

Proof Applying Theorem 2.2, we have that r_t^1 evolves according to

$$r_t^1 = -\int_0^t \kappa r_s^1 ds + \int_0^t v(s, r_s^0) dW_s^2 \,,$$

then, exploiting the Itô–Döblin lemma w.r.t. the function $re^{\kappa t}$, noticing that r_t^0 is a deterministic process, we find the desired solution. The distribution of r_t^1 directly follows by the fact that r_t^1 is obtained integrating w.r.t. a Brownian motion with deterministic integrand. □

To what regards the pricing problem of vanilla-type options, written on the underlying $e^{X_t^\epsilon}$ and with final payoff given by $\Phi(X_T)$, under the stochastic interest rate r^ϵ, we aim at finding the value $Pr(0; T)$ given by

$$Pr(0; T) = \mathbb{E}\left[e^{-\int_0^T r_s^\epsilon ds} \Phi\left(X_T^\epsilon\right)\right], \quad (10)$$

where the expectation is taken according to the joint measure generated by the two correlated Brownian motions W^1 and W^2. Therefore, to give an analytic expression for (10), we consider suitable transformations allowing to replace W^2 within the expression of r_t^1, by a random variable independent of W_T^1. Since r^1 is a Gaussian random variable, then X^1 is Gaussian, too. Therefore, defining

$$X_t^1 = \int_0^t r_s^1 ds - \lambda t \left(e^{\gamma + \frac{\delta^2}{2}} - 1 \right) + \sum_{k=1}^{N_t} J_k = \bar{X}_t^1 - \lambda t \left(e^{\gamma + \frac{\delta^2}{2}} - 1 \right) + \sum_{k=1}^{N_t} J_k, \tag{11}$$

we have $\bar{X}_T^1 \sim \mathcal{N}\left(0, Q_T^{X^1}\right)$, where $Q_T^{X^1} := \int_0^T \left(\int_u^T e^{\kappa s} ds \right)^2 e^{-2\kappa r} v(u, r_u^0)^2 du$.

Moreover $\Theta_T := Cov\left(\sigma_0 W_T^1, \bar{X}_T^1\right) = \Theta_T = \sigma_0 \bar{\rho} \int_0^T \int_u^T e^{\kappa s} ds e^{-\kappa r} v(u, r_u^0) du$, so that, exploiting the Gaussian distribution properties and following [36], we obtain

$$\int_0^T r_s^1 ds = \frac{\Theta_T}{\sigma_0 T} W_T^1 + \sqrt{\Lambda_T} Z, \tag{12}$$

with $\Lambda_T := Q_T^{X^1} - \frac{\Theta_T^2}{\sigma_0^2 T}$, for $\frac{\Theta_T^2}{\sigma_0^2 T} < Q_T^{X^1}$, being $Z \sim \mathcal{N}(0, 1)$ and independent from W_T^1, which allows us to state the following *pricing proposition*:

Proposition 3.2 *Let us consider a European call option with payoff* $\Phi(X_T^\epsilon) = \left(e^{X_T^\epsilon} - K\right)_+$, *then the associated first order (heuristic) correction to its fair price, is given by*

$$Pr_v^1(0; T) = P_{BS} + \epsilon e^{-\int_0^T r_s^0 ds} K \frac{\Theta_T}{\sigma_0 \sqrt{T}} \phi(-d_1) + \\ + \epsilon T e^{-\int_0^T r_s^0 ds} N(d_1) \lambda \left(e^{\gamma + \frac{\delta^2}{2}} - 1 \right) + \epsilon T e^{-\int_0^T r_s^0 ds} N(d_1) \delta \lambda, \tag{13}$$

ϕ, *resp.* N, *being the density, resp. the cumulative, function of the standard Gaussian law, while* d_1 *is defined as*

$$d_1 := \frac{1}{\sigma_0 \sqrt{T}} \left(\log \frac{s_0}{K} + \left(r - \frac{\sigma_0^2}{2} \right) T \right).$$

Proof For the sake of simplicity, let us first consider the case of a pure diffusive process. Then, applying Proposition 2.1 to the function $x \mapsto G(x) := e^{-x}$, with $X = \int_0^T r_s^\epsilon ds$, we get

$$e^{-\int_0^T r_s^\epsilon ds} = e^{-\int_0^T r_s^0 ds} - \epsilon e^{-\int_0^T r_s^0 ds} \int_0^T r_s^1 ds + R_1^G(\epsilon, T) = e^{-\int_0^T r_s^0 ds} \left(1 - \epsilon \int_0^T r_s^1 ds \right) + R_1^G(\epsilon, T),$$

with $|R_1^G(\epsilon, T)| \leq C_T \epsilon^2$, for some constant C_T, depending on the chosen G. Then, by applying Proposition 2.1 to the function Φ, we have

$$Pr^1(0;T) = \mathbb{E}\left[e^{-\int_0^T r_s^0 ds}\left(1 - \epsilon \int_0^T r_s^1 ds\right)\Phi(X_T^0)\right] + \epsilon \mathbb{E}\left[e^{-\int_0^T r_s^0 ds}\left(1 - \epsilon \int_0^T r_s^1 ds\right)\Phi'(X_T^0)X_T^1\right] =$$

$$= \mathbb{E}\left[e^{-\int_0^T r_s^0 ds}\Phi(X_T^0)\right] - \epsilon \mathbb{E}\left[e^{-\int_0^T r_s^0 ds}\int_0^T r_s^1 ds \Phi(X_T^0)\right] +$$

$$+ \epsilon \mathbb{E}\left[e^{-\int_0^T r_s^0 ds}\Phi'(X_T^0)X_T^1\right] - \epsilon^2 \mathbb{E}\left[e^{-\int_0^T r_s^0 ds}\int_0^T r_s^1 ds \Phi'(X_T^0)X_T^1\right] + R_1(\epsilon, T) =$$

$$= \mathbb{E}\left[e^{-\int_0^T r_s^0 ds}\Phi(X_T^0)\right] - \epsilon \mathbb{E}\left[e^{-\int_0^T r_s^0 ds}\int_0^T r_s^1 ds \Phi(X_T^0)\right] + \epsilon \mathbb{E}\left[e^{-\int_0^T r_s^0 ds}\Phi'(X_T^0)X_T^1\right],$$
(14)

where the last equality follows from incorporating the term with ϵ^2 into $R_1(\epsilon, T)$. Since $\Phi = (e^{X_T^0} - K)\mathbb{1}_{[W_T^1 > -\sqrt{T}d(1)]}$, and also considering that $\Phi' = e^{X_T^0}\mathbb{1}_{[W_T^1 > -\sqrt{T}d(1)]}$, we eventually obtain

$$Pr^1(0;T) = \mathbb{E}\left[e^{-\int_0^T r_s^0 ds}\Phi(X_T^0)\right] + \epsilon \mathbb{E}\left[e^{-\int_0^T r_s^0 ds}KX_T^1 \mathbb{1}_{[W_T^1 > -\sqrt{T}d(1)]}\right].$$

Therefore, by (12), exploiting the independence of W_T^1 and Z, and since Z has zero mean, we complete the proof, in the pure diffusive setting, by explicitly computing the expectation. While, also considering the jump terms, the result follows by Eq. (11) in Eq. (14), and by the very definition of the Gaussian cumulative density function. $\square$

Remark 3 We would like to underline that the study of the expansion related to the validation of interest rate contingent claim has been already discussed in [38, 39], where *only* Gaussian noise-driven models have been considered, and the expansion itself has been performed around an unperturbed and deterministic model, while, more recently, different asymptotic approaches have been provided to consider the specific CEV-model as well as some type of two-factor (stochastic) volatility interest models, see, e.g., [24, 34, 53].

4 Conclusions

In the present paper we have focused on the analysis of the small noise asymptotic expansions for a generalization of the standard Black–Scholes (BS) model, namely considering it under stochastic interest rates. This has been motivated by one of the major BS-model drawback, namely the unrealistic assumption that there exists a unique constant interest rate, which is shown to be an unrealistic assumption within current financial scenarios. Dropping latter assumption causes the loose of a closed formula for European-type options. Therefore, we have provided rigorous explicit expressions for a corrections around the classical BS-formula, when the interest rate is perturbed by a small random noise, also allowing it to have jumps.

Acknowledgements The authors would like to thank the *Gruppo Nazionale per l'Analisi Matematica, la Probabilità e le loro Applicazioni* (GNAMPA) for the financial support that has funded the present research within the project called *Set-valued and optimal transportation theory methods to model financial markets with transaction costs both in deterministic and stochastic frameworks*.

References

1. Albeverio, S., Cordoni, F., Di Persio, L., Pellegrini, G.: Asymptotic expansion for some local volatility models arising in finance. Decis. Econ. Fin. **42**(2), 527–573 (2019)
2. Albeverio, S., Di Persio, L., Mastrogiacomo, E.: Small noise asymptotic expansion for stochastic PDE's, the case of a dissipative polynomially bounded non linearity I. Tôhôku Math. J. **63**, 877–898 (2011)
3. Albeverio, S., Di Persio, L., Mastrogiacomo, E., Smii, B.: A class of Lévy driven SDEs and their explicit invariant measures. Potential Anal. **45**(2), 229–259 (2016)
4. Albeverio, S., Di Persio, L., Mastrogiacomo, E., Smii, B.: Invariant measures for SDEs driven by Lévy noise. A case study for dissipative nonlinear drift in infinite dimension. Commun. Math. Sci. **15**(4), 957–983 (2016)
5. Albeverio, S., Hilbert, A., Kolokoltsov, V.: Uniform asymptotic bounds for the heat kernel and the trace of a stochastic geodesic flow. Stoch. Int. J. Probab. Stoch. Proc. **84**, 315–333 (2012)
6. Albeverio, S., Smii, B.: Asymptotic expansions for SDE's with small multiplicative noise. Stoch. Proc. Appl. **125**(3), 1009–1031 (2013)
7. Albeverio, S., Steblovskaya, V.: Asymptotics of Gaussian integrals in infinite dimensions. Infin. Dimens. Anal., Quantum Probab. Relat Top. **22**(01), 1950004 (2019)
8. Albeverio, S., Schmitz, M., Steblovskaya, V., Wallbaum, K.: A model with interacting assets driven by Poisson processes. Stoch. Anal. Appl. **24**(1), 241–261 (2006)
9. Andersen, L., Lipton, A.: Asymptotics for exponential Lévy processes and their volatility smile: survey and new results. Int. J. Theor. Appl. Finance **16**(135), 0001 (2012)
10. Applebaum, D.: Lévy Processes and Stochastic Calculus. Cambridge Studies in Advanced Mathematics, vol. 116. Cambridge University Press, Cambridge (2009)
11. Arnold, L.: Stochastic Differential Equations: Theory and Applications. Wiley, New York (1974)
12. Bayraktar, E., Cayé, T., Ekren, I.: Asymptotics for small nonlinear price impact: a PDE approach to the multidimensional case. Math. Finance (2020). To appear (https://onlinelibrary.wiley.com/doi/abs/10.1111/mafi.12283)
13. Barletta, A., Nicolato, E., Pagliarani, S.: The short-time behavior of VIX-implied volatilities in a multifactor stochastic volatility framework. Math. Finance **29**(3), 928–966 (2019)
14. Bayer, C., Laurence, P.: Asymptotics beats Monte Carlo: the case of correlated local vol baskets. Commun. Pure Appl. Math. **67**(10), 1618–1657 (2014)
15. Benarous, A., Laurence, P.: Second Order Expansion for Implied Volatility in Two Factor Local Stochastic Volatility Models and Applications to the Dynamic λ–Sabr Model. Large Deviations and Asymptotic Methods in Finance, pp. 89–136. Springer International Publishing, Berlin (2013)
16. Benhamou, E., Gobet, E., Miri, M.: Smart expansion and fast calibration for jump diffusions. Finance Stoch. **13**, 563–589 (2009)
17. Black, F., Scholes, M.: The pricing of options and corporate liabilities. J. Polit. Econ. **81**, 637–654 (1973)
18. Bonollo, M., Di Persio, L., Pellegrini, G.: Polynomial Chaos Expansion approach to interest rate models. J. Probab. Stat. (2015)
19. Bonollo, M., Di Persio, L., Pellegrini, G.: A computational spectral approach to interest rate models (2015). arXiv:1508.06236
20. Breitung, K.: Asymptotic Approximations for Probability Integrals. Springer, Berlin (1994)

21. Brigo, D., Mercurio, F.: Interest Rate Models: Theory and Practice. Springer Finance. Springer, Berlin (2006)
22. Cordoni, F., Di Persio, L.: Small noise expansion for the Lévy perturbed Vasicek model. Int. J. Pure Appl. Math. **98**, 2 (2015)
23. Cox, J.C., Ingersoll, J.E., Ross, S.A.: A theory of the term structure of interest rates. Econometrica **53**, 385–407 (1985)
24. Di Francesco, M., Diop, S., Pascucci, A.: CDS calibration under an extended JDCEV model. Int. J. Comput. Math. **96**(9), 1735–1751 (2019)
25. Fouque, J.P., Papanicolau, G., Sircar, R.: Derivatives in Financial Markets with Stochastic Volatility. Cambridge University Press, Cambridge (2000)
26. Friz, P.K., Gatheral, J., Guliashvili, A., Jacquier, A., Teichman, J.: Large Deviations and Asymptotic Methods in Finance. Springer Proceedings in Mathematics & Statistics, vol. 110 (2015)
27. Fuji, M., Akihiko, T.: Perturbative expansion of FBSDE in an incomplete market with stochastic volatility. Q. J. Finance **2**, 03 (2012)
28. Gardiner, C.W.: Handbook of Stochastic Methods for Physics, Chemistry and Natural Sciences. Springer Series in Synergetics. Springer, Berlin (2004)
29. Gatheral, J., Hsu, E.P., Laurence, P., Ouyang, C., Wang, T.H.: Asymptotics of implied volatility in local volatility models. Math. Finance **4**, 591–620 (2012)
30. Gihman, I.I., Skorokhod, A.V.: Stochastic Differential Equations. Springer, New York (1972)
31. Grishchenko, O., Han, X., Nistor, V.: A volatility-of-volatility expansion of the option prices in the SABR stochastic volatility model. Int. J. Theor. Appl. Finance **23**(3), art. no. 2050018 (2010)
32. Grunspan, C., Van Der Hoeven, J.: Effective Asymptotics Analysis for Finance. Int. J. Theor. Appl. Finance **23**(2), art. no. 2050013 (2020)
33. Gulisashvili, A.: Analytically Tractable Stochastic Stock Price Models. Springer, Berlin (2012)
34. He, Y., Chen, P.: Optimal investment strategy under the CEV model with stochastic interest rate. Math. Problems Eng. art. no. 7489174 (2020)
35. Imkeller, P., Pavlyukevich, I., Wetzel, T.: First exit times for Lévy-driven diffusions with exponentially light jumps. Ann. Probab. **37**(2), 530–564 (2009)
36. Kim, Y.J., Kunitomo, N.: Pricing options under stochastic interest rates: a new approach. Asia-Pacific Financ. Markets **6**(1), 49–70 (1999)
37. Kumar, R., Nasralah, H.: Asymptotic approximation of optimal portfolio for small time horizons. SIAM J. Financ. Math. **9**(2), 755–774 (2018)
38. Kunitomo, N., Takahashi, A.: On validity of the asymptotic expansion approach in contingent claim analysis. Ann. Appl. Probab. **13**(3), 914–952 (2003)
39. Kunitomo, N., Takahashi, A.: The asymptotic expansion approach to the valuation of interest rate contingent claims. Math. Finance **11**(1), 117–151 (2001)
40. Kusuoka, S., Yoshida, N.: Malliavin calculus, geometric mixing, and expansion of diffusion functionals. Probab. Theory Related Fields **116**(4), 457–484 (2000)
41. Lorig, M.: Local Lévy models and their volatility smile (2012). arXiv:1207.1630v1
42. Lütkebohmert, E.: An asymptotic expansion for a Black-Scholes type model. Bulletin des sciences mathématiques **128**(8), 661–685 (2004)
43. Mandrekar, V., Rüdiger, B.: Stochastic integration in banach spaces theory and applications. Springer, Berlin (2015)
44. Pagliarani, S., Pascucci, A., Riga, C.: Adjoint expansions in local Lévy models. SIAM J. Financ. Math. **4**(1), 265–296 (2013)
45. Park, S.-H., Lee, K.: Hedging with liquidity risk under CEV diffusion. Risks **8**(2), art. no. 62, 1–12 (2020)
46. Shiraya, K., Takahashi, A.: An asymptotic expansion for local-stochastic volatility with jump models. Stochastics **89**(1), 65–88 (2017)
47. Shreve, S.E.: Stochastic calculus for finance II. Springer Finance. Springer, New York (2004)
48. Takahashi, A.: An asymptotic expansion approach to pricing financial contingent claims. Asia-Pacific Financ. Markets **6**(2), 115–151 (1999)

49. Takahashi, A., Tsuruki, Y.: A new improvement solution for approximation methods of probability density functions. No. CIRJE-F-916. CIRJE, Faculty of Economics, University of Tokyo (2014)
50. Uchida, M., Yosida, N.: Asymptotic expansion for small diffusions applied to option pricing. Stat. Inference Stoch. Process **3**, 189–223 (2004)
51. Yoshida, N.: Conditional expansions and their applications. Stoch. Proc. Appl. **107**(1), 53–81 (2003)
52. Zhang, S.M., Feng, Y.: American option pricing under the double Heston model based on asymptotic expansion. Quant. Finance **19**(2), 211–226 (2019)
53. Zhang, S., Zhang, J.: Asymptotic expansion method for pricing and hedging American options with two-factor stochastic volatilities and stochastic interest rate. Int. J. Computer Math. **97**(3), 546–563 (2020)

Stochastic Geodesics

Ana Bela Cruzeiro and Jean-Claude Zambrini

Abstract We describe, in an intrinsic way and using the global chart provided by Itô's parallel transport, a generalisation of the notion of geodesic (as critical path of an energy functional) to diffusion processes on Riemannian manifolds. These stochastic processes are no longer smooth paths but they are still critical points of a regularised stochastic energy functional. We consider stochastic geodesics on compact Riemannian manifolds and also on (possibly infinite dimensional) Lie groups. Finally the question of existence of such stochastic geodesics is discussed: we show how it can be approached via forward-backward stochastic differential equations.

Keywords Geodesics · Diffusions on manifolds · Stochastic variational principles

1 Introduction

The notion of geodesic in Riemannian manifolds appeared first in a lecture of Riemann, in 1854. Originally, it was referring to the shortest path between two points on Earth's surface. Nowadays, given an affine connection like the one of Levi-Civita, it can also be defined as a curve whose tangent vectors remain parallel when transported along the curve. In Theoretical Physics it is in General Relativity that this notion played a key rôle.

In a stochastic framework, a generalisation of geodesic curve is described. It corresponds to a critical path for some generalised action functional. The concept is reminiscent of Feynman path integral approach to Quantum Mechanics [1] but for

A. B. Cruzeiro (✉)
Grupo de Física-Matemática and Departamento de Matemática I.S.T., Universidade de Lisboa, Av. Rovisco Pais, 1049-001 Lisboa, Portugal
e-mail: ana.cruzeiro@tecnico.ulisboa.pt

J.-C. Zambrini
Grupo de Física-Matemática, Faculdade Ciências Universidade de Lisboa, Campo Grande Ed. C6, 1749-016 Lisboa, Portugal
e-mail: jczambrini@fc.ul.pt

© Springer Nature Switzerland AG 2021
S. Ugolini et al. (eds.), *Geometry and Invariance in Stochastic Dynamics*, Springer Proceedings in Mathematics & Statistics 378,
https://doi.org/10.1007/978-3-030-87432-2_4

well defined probability measures on path spaces. It involves, in particular, regularisation of the second order in time classical dynamical equations, which is not traditional in Stochastic Analysis.

The derived equations of motion are of Burgers type. When considering flows which keep the volume measure invariant one obtains Navier–Stokes equations. This point of view was developed in [2, 3, 9] in particular. It is currently being investigated (c.f. [8] as well as [10] for a review on this subject).

After a short survey of classical geodesics on Riemannian manifolds, Cartan's frame bundle approach and its relation with the horizontal and Laplace–Beltrami operators are recalled.

Stochastic Analysis of diffusions on manifolds along the line of Itô–Ikeda–Watanabe is given, together with Itô's associated notion of parallel transport. Then one comes back to one of the historic definitions of geodesics, namely as critical points of an Action functional. The regularisations associated with the critical diffusion provide the appropriate generalised energy functional. The same strategy applies to geodesics on Lie groups.

It is also shown how, if needed, stochastic geodesics can be characterised via stochastic forward-backward SDEs.

It is a special pleasure to dedicate this paper to Sergio Albeverio as a modest sign of recognition for his faithful friendship along the years.

2 Geodesics on Riemannian Manifolds

We shall denote by M a d-dimensional compact Riemannian manifold and g its metric tensor. Given $m \in M$, if u, v are vectors in the tangent space $T_m(M)$ the Riemannian inner product is given in local chart by

$$g_m(u, v) = (g_{i,j} u^i v^j)(m)$$

Here and in the rest of the paper we adopt Einstein summation convention.

The Levi-Civita covariant derivative of a vector field z has the expression

$$[\nabla_k z]^j = \frac{\partial}{\partial m^k} z^j + \Gamma_{k,l}^j z^l,$$

where Γ denotes the corresponding Christoffel symbols in the local chart; explicitly,

$$\Gamma_{k,l}^j = \frac{1}{2} \left(\frac{\partial}{\partial m^k} g_{i,l} + \frac{\partial}{\partial m^l} g_{k,i} - \frac{\partial}{\partial m^i} g_{k,l} \right) g^{j,i} \qquad (1)$$

Given a smooth curve $t \to \varphi(t) \in M$, the parallel transport of a vector field z along this curve is defined by the condition of zero covariant derivative of z in the $\dot{\varphi}$ direction,

$$\nabla_{\dot\varphi(t)} z(t) = 0 \text{ or } \dot z^j = -\Gamma^j_{k,l}\dot\varphi^k z^l. \tag{2}$$

Its solution, $z(t) = t^\varphi_{t\leftarrow 0}(z(0))$, the parallel transport of z along the curve, provides an Euclidean isomorphism between tangent spaces:

$$t^\varphi_{t\leftarrow 0} : T_{\varphi(0)}(M) \to T_{\varphi(t)}(M).$$

Consider the curves minimising the length

$$\mathscr{I}(\gamma) = \int l(\gamma,\dot\gamma)dt, \quad l(\gamma,\dot\gamma) = \sqrt{g_{i,j}\dot\gamma^i\dot\gamma^j}$$

and therefore satisfying the Euler Lagrange variational equation

$$\frac{d}{dt}\left(\frac{g_{i,j}\dot\gamma^j}{l}\right) = \frac{1}{2l}\partial_i(g_{j,k})\dot\gamma^j\dot\gamma^k.$$

Replacing dt by ds (where s is the arc length) we obtain

$$\frac{d}{dt}\left(g_{i,j}\dot\gamma^j\frac{1}{ds}\right) - \frac{1}{2}\frac{1}{ds}\partial_i(g_{j,k})\dot\gamma^j\dot\gamma^k = 0$$

and also

$$g_{i,j}\frac{d^2\gamma^j}{ds^2} + \partial_k(g_{i,j})\frac{d\gamma^k}{ds}\frac{d\gamma^j}{ds} - \frac{1}{2}\partial_i(g_{j,k})\frac{d\gamma^k}{ds}\frac{d\gamma^j}{ds} = 0$$

Multiplying both members by $g^{\alpha,i}$ we obtain the following classical form of the geodesic equation:

$$\frac{d^2\gamma^\alpha}{ds^2} + \Gamma^\alpha_{j,k}\frac{d\gamma^j}{ds}\frac{d\gamma^k}{ds} = 0 \tag{3}$$

or $\nabla_{\dot\gamma}\dot\gamma = 0$.

A curve satisfying the last equation is called a *geodesic* for the corresponding Riemannian metric. It is also well known that geodesics (defined in a time interval $[0, T]$) are characterised as being critical paths of the (kinetic) energy functional

$$\mathscr{E}(\gamma) = \int_0^T \|\dot\gamma(t)\|^2 dt = \int_0^T g_{i,j}(\gamma(t))\dot\gamma^i(t)\dot\gamma^j(t)dt \tag{4}$$

By critical it is meant that, for every family of smooth curves (variations of γ) γ_ϵ starting (at time 0) and ending (at time T) at $\gamma(0)$ and $\gamma(T)$ resp., we have $\frac{d}{d\epsilon}|_{\epsilon=0}\mathcal{E}(\gamma_\epsilon) = 0$.

3 The Frame Bundle and the Laplacians

The bundle of orthonormal frames over M is defined by

$$O(M) = \{(m,r) : m \in M, r : R^d \to T_m(M) \text{ is an Euclidean isometry}\}$$

The map $\pi : O(M) \to M$, $\pi(m,r) = m$ is the canonical projection.

Let e_i, $i = 1, \ldots, d$ denote the vectors of the canonical basis of R^d and γ_i denote the (unique) geodesic on M such that $\gamma_i(0) = m$, $\frac{d}{dt}|_{t=0} \gamma_i(t) = r(e_i)$. Let $(\gamma_i(t), r_i(t))$ represent the parallel transport of r along γ_i, $\nabla_{\dot\gamma_i} r_i = 0$, $r_i(0) = \text{Id}$. Then

$$A_i(m,r) = \frac{d}{dt}\bigg|_{t=0} r_i(t)$$

are called the horizontal vector fields on M.

Denote by Θ the one-form defined on $O(M)$ with values in $R^d \times so(d)$ such that $<\Theta, A_i> = (e_i, 0)$; $\Theta = (\theta, \omega)$, with $\omega(m,r) = r^{-1}dr$ the Maurer–Cartan form on the orthogonal group $O(d)$. Its structure equations are given by

$$\begin{cases} d\theta = \omega \wedge \theta \\ d\omega = \omega \wedge \omega + \Omega(\theta \wedge \theta), \end{cases}$$

where Ω denotes the curvature tensor:

$$\Omega(X, Y, Z) = (\nabla_X \nabla_Y - \nabla_Y \nabla_X - \nabla_{[X,Y]})Z,$$

and where $[X, Y]$ denotes the bracket of two vector fields. Recall also that the Ricci tensor (Ricci_{kl}) is the trace of the curvature, taken in the second and third entries.

In particular $\theta(A_k) = e_k$ and $\omega(A_k) = 0$. The horizontal Laplacian on $O(M)$ is the second order differential operator

$$\Delta_{O(M)} = \sum_{k=1}^{d} \mathcal{L}_{A_k}^2 \qquad (5)$$

where $\mathcal{L}_{A_k}$ denotes the Lie derivative along the vector field A_k. For every smooth function f defined on M we have

$$\Delta_{O(M)}(f \circ \pi) = (\Delta_M f) \circ \pi$$

where Δ_M is the Laplace–Beltrami operator on M. This operator is expressed in local coordinates by

$$\Delta_M f = g^{i,j} \left[\frac{\partial^2 f}{\partial m^i \partial m^j} - \Gamma_{i,j}^k \frac{\partial f}{\partial m^k} \right]. \tag{6}$$

4 Stochastic Analysis on Manifolds

We are going to consider stochastic diffusions associated to elliptic operators on M of the form

$$L_u f := \frac{1}{2} \Delta_M f + \partial_u f \tag{7}$$

in the sense of Itô stochastic calculus. Here u is a possibly time-dependent, smooth (at least C^2) vector field on M. In local coordinates the diffusion with generator L_u can be written as

$$dm^j(t) = \sigma_k^j dx^k(t) - \left(\frac{1}{2} g^{m,n} \Gamma_{m,n}^j - u^j \right) dt \tag{8}$$

where $\sigma = \sqrt{g}$ and x_k are independent real-valued Brownian motions.

We consider the horizontal lift of these M-valued diffusion processes. Denote by u_k the functions defined on $O(M)$ by

$$U_k(r) = r(e_k).u_{\pi(r)}.$$

Then $\tilde{u} = \sum_k U_k A_k$ satisfies $\pi'(\tilde{u}) = u$ (π' being the derivative of the canonical projection π).

Denoting by x a sample path of the standard Brownian motion on R^d, $x(t), t \in [0, T]$, $x(0) = 0$, we consider the following Stratonovich stochastic differential equation on $O(M)$:

$$dr_x(t) = \sum_{k=1}^{d} A_k(\circ dx^k(t) + U_k dt), \quad r_x(0) = 0 \tag{9}$$

with $\pi(r_0) = m_0$. In local coordinates (m^i, e_α^i) on $O(M)$ and if $r(t) = (m(t), e(t))$ we have,

$$\begin{cases} dm^i(t) = e^i_\alpha \circ (dx^\alpha(t) + u^\alpha dt) \\ de^i_\alpha(t) = -\Gamma^i_{j,k}(m(t))e^k_\alpha(t) \circ dm^j(t), \end{cases}$$

If $a \in M$, we denote the path space of the manifold-valued paths starting from a by

$$P_a(M) = \{p : [0, T] \to M, p(0) = a, p \text{ continuous}\}.$$

The diffusion $m(t)$ has for generator the operator L_u. We refer to [16] for a detailed exposition of diffusions on Riemannian manifolds constructed on the frame bundle.

For each vector field u the operator L_u and the operator $\mathcal{L}_U = \frac{1}{2}\Delta_{O(M)} + \partial_U$ induce on the path spaces $P_{m_0}(M)$ and $P_{m_0}(O(M))$, respectively, two probability measures, namely the laws of the corresponding diffusion processes. The projection map π realizes an isomorphism between these two probability spaces.

Let the path space $P_0(R^d)$ be endowed with the law of the process $dy(t) = (\circ dx(t) + U)(y(t))$, $t \in [0, T]$ and $P_{m_0}(M)$ with the law of the diffusion p with generator L_u). Consider the Itô map $I : P_0(R^d) \to P_{m_0}(M)$ defined by

$$I(x)(t) = \pi(r_x(t))$$

This map is a.s. bijective and provides an isomorphism between the corresponding probability measures [18].

Even though p is not differentiable in time, Itô has shown that one can still define a parallel transport along p, which is the isomorphism from $T_{p(s)}(M) \to T_{p(t)}(M)$ given by

$$t^p_{t \leftarrow s} := r_x(t)r_x(s)^{-1}.$$

The differentiability of $r_x(t)$ with respect to variations of the Brownian motion x was studied in [12, 17] within the framework of Malliavin Calculus [5, 19] (c.f. also [13] for the case of the Brownian motion with drift).

Denote $\mathcal{D}^\beta_\alpha = \mathcal{L}_{A_\alpha} u^\beta$. The following result holds:

Proposition 1 *Given a process of bounded variation in time $h : P_0(R^d) \times [0, T] \to R^d$, we have, using the notations of Sect. 3,*

$$< \theta, \frac{d}{d\epsilon}\Big|_{\epsilon=0} r_{x+\epsilon h} >= \zeta, \quad < \omega, \frac{d}{d\epsilon}\Big|_{\epsilon=0} r_{x+\epsilon h} >= \rho \tag{10}$$

where ζ and ρ are determined by the Itô (and Stratonovich) stochastic differential equations

$$d\zeta(t) = \dot{h}(t)dt - [\frac{1}{2}Ricci + \mathcal{D}](h(t))dt - \rho(t)dx(t) \tag{11}$$

$$d\rho = \Omega(\circ dx + u dt, h)$$

with initial conditions $\zeta(0) = 0$, $\rho(0) = 0$.

The result above is still valid for pinned Brownian motion, namely when $p(T)$ is fixed. Then the variations are equal to zero not only at the initial but also at this final time. The sigma-algebra and filtration on the corresponding path space are the usual ones, generated by the coordinate maps and generated by the coordinate maps up to time T, respectively. We refer to [15] for more details.

5 Stochastic Geodesics

We shall consider stochastic geodesics as processes which are critical points of some energy functional generalising the classical deterministic one. Since the stochastic processes, diffusions on the manifold, are no longer differentiable in time, some notion of generalised velocity has to replace the usual time derivative.

If $\xi(\cdot)$ is a semimartingale with respect to an increasing filtration $\mathscr{P}_t$, $t \in [0, T]$ and with values in a manifold M, we consider the process η defined by the Stratonovich integral

$$\eta(t) := \int_0^t t_{0 \leftarrow s}^{\xi} \, od\xi(s)$$

This is a semimartingale taking values in $T_{\xi(0)}(M)$. We consider its (generalised) right-hand time derivative (or drift) by taking conditional expectations:

$$D_t \eta(t) = \lim_{\epsilon \to 0} E^{\mathscr{P}_t} \left[\frac{\eta(t + \epsilon) - \eta(t)}{\epsilon} \right]$$

Notice that if ξ is a differentiable deterministic path, this notion of derivative reduces to the usual one.

Then we define the *generalised (forward) derivative*

$$D_t^\nabla \xi(t) := t_{t \leftarrow 0} D_t \eta(t) \tag{12}$$

We use the symbol ∇ to stress that the derivative depends on the choice of covariant derivative used to define the parallel transport, although in this work we are only consider the Levi-Civita covariant derivative.

For a (possibly time dependent) vector field Z computed along a semimartingale ξ, the generalised derivative is defined as

$$D_t^\nabla Z(t) = \lim_{\epsilon \to 0} \frac{1}{\epsilon} E^{\mathscr{P}_t}[t_{t \leftarrow t+\epsilon} Z(t + \epsilon, \xi(t + \epsilon)) - Z(t, \xi(t))]$$

Let us consider our base manifold M and, for a M-valued semimartingale ξ, define the corresponding kinetic energy by

$$\mathcal{E}(\xi) = E \int_0^T ||D^\nabla \xi(t)||^2 dt \tag{13}$$

Next Theorem characterises the critical paths of $\mathcal{E}$. Allowed variations are processes of bounded variation h satisfying $h(0) = h(T) = 0$. We have the following result:

Theorem 1 *A diffusion process $m(\cdot)$ with generator L_u, $u \in C^2(M)$, is a critical path for the energy functional $\mathcal{E}$ if and only if $D_t^\nabla u(t, m(t)) = 0$ almost everywhere or, equivalently,*

$$\frac{\partial}{\partial t} u + (\nabla_u u) + \frac{1}{2}[(\Delta u) + Ricci(u)] = 0 \tag{14}$$

Notice that, in particular, we obtain the expression derived in [20] using local coordinates.

It is shown in [4] (c.f., more generally, [20]) that the symmetries of the critical diffusion coincide with the regularisation of its classical counterpart. In other words, if the diffusion coefficient in (8), regarded now as variable, tends to zero, $D^\nabla \xi$ in (12) reduces to an ordinary (strong) derivative, Eq. (14) and the symmetries of the critical diffusion reduce to those of the classical functional (4).

Proof We first write the energy functional via the lift of the process to the frame bundle, as explained in the last paragraph:

$$\mathcal{E} = E \int_0^T ||D_t \pi(r_x(t))||^2 dt$$

where D_t refers to the generalised derivative for processes defined on the flat space (of the Brownian motion x). Then we perform variations of the Brownian motion x along directions $h(\cdot)$, processes of bounded variation with $h(0) = h(T) = 0$. Using Proposition 1, these variations will give rise to variations on the path space of the manifold M along semimartingales $\zeta(\cdot)$, where ζ is given by (11). More precisely we have,

$$\frac{d}{d\epsilon}\Big|_{\epsilon=0} E \int_0^T ||D_t \pi(r_{x+\epsilon h}(t))||^2 dt = 2E \int_0^T < D_t \pi(r_x(t)), D_t \pi' \left(\frac{d}{d\epsilon}\Big|_{\epsilon=0} r_{x+\epsilon h}(t)\right) > dt$$

$$= 2E \int_0^T < D_t \pi(r_x(t)), D_t(\zeta)(t)) > dt$$

$$= 2E \int_0^T < D_t \pi(r_x(t)), \dot{h} - \frac{1}{2} Ricci(h) - \mathcal{D}(h)(t) > dt$$

Using integration by parts in time, the assumption $h(0) = h(T) = 0$ and the fact that there is no Itô's extra term in the integration since h is of bounded variation, the first term is equal to $-2E \int_0^T < D_t D_t \pi(r_x), h(t) >$.

We arrive to the conclusion that a process r_x of the form (9) is critical for the action functional $\mathscr{E}$ if and only if $D_t^\nabla u = 0$ almost everywhere, which proves the Theorem.

6 Stochastic Geodesics on Lie Groups

Let G denote a Lie group endowed with a left invariant metric $< \ >$ and a left invariant connection ∇, that we assume here to be the Levi-Civita connection. The corresponding Lie algebra $\mathscr{G}$ can be identified with the tangent space $T_e G$, where e is the identity element of the group. Taking a sequence of vectors $H_k \in \mathscr{G}$, consider the following Stratonovich stochastic differential equation on the group:

$$dg(t) = T_e L_{g(t)} \left(\sum_k H_k \circ dx^k(t) - \frac{1}{2} \nabla_{H_k} H_k dt + u(t) dt \right) \tag{15}$$

with $g(0) = e$, where $T_a L_{g(t)} : T_a G \to T_{g(t)a} G$ is the differential of the left translation $L_{g(t)}(x) = g(t)x$, $x \in G$ and where $x^k(t)$ are independent real valued Brownian motions. The vector $u(\cdot)$ is assumed to be non random, $u(\cdot) \in C^2([0, T]; \mathscr{G})$.

The stochastic energy functional for a general G-valued semimartingale $\xi(t)$, $t \in [0, T]$, reads:

$$\mathscr{E}(\xi) = E \int_0^T ||T_{\xi(t)} L_{\xi(t)^{-1}} D_t^\nabla \xi(t)||^2 dt \tag{16}$$

Assume furthermore that $\nabla_{H_k} H_k = 0$ for all k (in particular the Stratanovich integral in (15) coincides with the Itô one). Then the following result holds:

Theorem 2 ([2]) *A G-valued semimartingale of the form (15) is critical for the energy functional (16) if and only if the vector field $u(\cdot)$ satisfies the equation*

$$\frac{d}{dt} u(t) = ad_{u(t)} u(t) - \frac{1}{2} \left(\sum_k \nabla_{H_k} \nabla_{H_k} u(t) + Ricci(u(t)) \right)$$

When $H_k = 0$ for all k the equation reduces to the well known Euler–Poincaré equation for (deterministic) geodesics in Lie groups $\frac{d}{dt} u(t) = ad_{u(t)} u(t)$.

Up to some sign changes, the right invariant case is analogous.

The theorem also holds for infinite-dimensional Lie groups and allows, as a particular case, to derive the Navier–Stokes equation, when the problem is formulated on the diffeomorphisms group (c.f. [2]).

7 Relation with Stochastic Forward-Backward Differential Equations

Deterministic geodesics solve second order differential equations and as such can be obtained using standard methods for such equations, with given initial position and velocity as well as with initial and final given positions. The meaning of "second order" stochastic differential equations is not so clear. A possible method is its characterisation via stochastic forward-backward differential equations. In local coordinates (c.f. notations defined in (8), a stochastic geodesic in the time interval $[0, T]$ reads

$$m^j(t) = m^j(0) + \int_0^t \sigma_k^j(m(s))dx^k(s) - \int_0^t \left(\frac{1}{2}g^{m,n}\Gamma_{m,n}^j(m(s)) - y^j(s)\right)ds$$

$$y^j(t) = y^j(T) - \int_t^T Z_k^j(s)dx^k(s) - \frac{1}{2}\int_t^T \text{Ricci}^j(m(s))ds$$

Given $m^j(0)$ and $y(T) = u(T, m(T))$ these kind of systems may provide solutions of the form $(m(t), y(t))$ with $y(t) = u(t, m(t))$ corresponding to our stochastic geodesics (c.f., for example, [14]). The term Z is an a priori unknown of the equation, but is in fact determined a posteriori by the solution (m, y).

In the case of stochastic geodesics on Lie groups, the characterisation via forward-backward equations was described in [6]. An extension to infinite dimensional Lie groups and, in particular, to the Navier–Stokes equation framework, is also possible [7, 11].

Acknowledgements The authors acknowledge the support of the FCT Portuguese grant PTDC/MAT-STA/28812/2017.

References

1. Albeverio, S., Hoegh-Krohn, R., Mazzucchi, S., Mathematical Theory of Feynman Path Integrals: An Introduction, 2nd edn. Lecture Notes in Mathematics, vol. 523. Springer, Berlin (2008)
2. Arnaudon, M., Chen, X., Cruzeiro, A.B.: Stochastic Euler Poincaré reduction. J. Math. Phys. **55**, 081507 (2014)
3. Arnaudon, M., Cruzeiro, A.B.: Lagrangian Navier-Stokes diffusions on manifolds: variational principle and stability. Bull. des Sc. Mathématiques **136**(8), 857–881 (2012)
4. Arnaudon, M., Zambrini, J.-C.: A stochastic look at geodesics on the sphere. In: Nielsen, F., Barbaresco, F. (eds.) GSI 2017, Springer Lecture Notes in Computer Sciences 10589, pp. 470–476. Springer, Berlin (2017)
5. Bismut, J.-M.: Large deviations and the Malliavin calculus. Progress in Mathematics, vol. 45. Birkhäuser Boston Inc, Boston (1984)
6. Chen, X., Cruzeiro, A.B.: Stochastic geodesics and forward-backward stochastic differential equations on Lie groups. Disc. Cont. Dyn. Syst. 115–121 (2013)

7. Chen, X., Cruzeiro, A.B., Qian, Z.: Navier-Stokes equation and forward-backward stochastic differential system in the Besov spaces. arXiv:abs/1305.0647
8. Chen, X., Cruzeiro, A.B., Ratiu, T.: Stochastic variational principles for dissipative equations with advected quantities. arXiv:pdf/1506.05024.pdf
9. Cipriano, F., Cruzeiro, A.B.: Navier-Stokes equation and diffusions on the group of homeomorphisms of the torus. Comm. Math. Phys. **275**, 255–269 (2007)
10. Cruzeiro, A.B.: Stochastic approaches to deterministic fluid dynamics: a selective review. Water **12**(3), 864 (2020)
11. Cruzeiro, A.B., Shamarova, E.: Navier-Stokes equations and forward-backward SDEs on the group of diffeomorphisms of a torus. Stoch. Proc. Appl. **119**, 4034–4060 (2009)
12. Cruzeiro, A.B., Malliavin, P.: Renormalized differential geometry on path spaces: structural equation and curvature. J. Funct. Anal. **139**, 119–181 (1996)
13. Cruzeiro, A.B., Malliavin, P.: Nonperturbative construction of invariant measure through confinement by curvature. J. Math. Pures Appl. **139**, 119–181 (1998)
14. Delarue, F.: On the existence and uniqueness of solutions to FBSDEs in a non-degenerate case. Stoch. Proc. Appl. **99**, 209–286 (2002)
15. Driver, B.K.: A Cameron-Martin type quasi-invariance theorem for pinned Brownian motion on a compact Riemannian manifold. Trans. Am. Math. Soc. **342**(1) (1994)
16. Ikeda, N., Watanabe, S.: Stochastic differential equations and diffusion processes. vol. 24. North-Holland Math. Library (1989)
17. Fang, S., Malliavin, P.: Stochastic calculus on the path space of a Riemannian manifold. J. Funct. Anal. **118**, 249–274 (1993)
18. Malliavin, P.: Formule de la moyenne, calcul de perturbations et théorème d'annulation pour les formes harmoniques. J. Funct. Anal. **169**, 321–354 (1995)
19. Malliavin, P.: Stochastic Analysis. Grund. der Mathem. Wissen. vol. 313. Springer, New York (1997)
20. Zambrini, J.-C.: Probability and quantum symmetries in a Riemannian manifold. Progress Probab. **45**, 283–300 (1999)

A Note on Supersymmetry and Stochastic Differential Equations

Francesco C. De Vecchi and Massimiliano Gubinelli

Abstract We obtain a dimensional reduction result for the law of a class of stochastic differential equations using a supersymmetric representation first introduced by Parisi and Sourlas.

Keywords Invariant measures of sdes · Dimensional reduction · Supersymmetric field theories

AMS: 60H10 · 81T60 · 46L53

1 Introduction

In this paper we want to exploit a supersymmetric representation of scalar stochastic differential equations (SDEs) with additive noise and nonlinear drift V' in order to prove the well known relation between the invariant law of these SDEs and the Gibbs measure $e^{-2V(x)}\mathrm{d}x$.

The supersymmetric representation of SDEs or more generally SPDEs was first noted by Parisi and Sourlas [15, 16] and it is well known and used in the physics literature (see, e.g. [21]) where the relation between supersymmetry, SDEs and Gibbs measures (called dimensional reduction) was formally established [7, 10]. In the case of elliptic SPDEs these formal arguments have been rigorously exploited and proved [3, 14] and applied to the stochastic quantization program of quantum field theory [2, 11]. In the present paper we want to propose a similar rigorous version of dimensional reduction for one dimensional SDEs. The proof proposed here follows more closely the methods used for dimensional reduction of elliptic equations

F. C. De Vecchi (✉) · M. Gubinelli
Institute for Applied Mathematics & Hausdorff Center for Mathematics,
University of Bonn, Bonn, Germany
e-mail: fdevecch@uni-bonn.de

M. Gubinelli
e-mail: mgubinel@uni-bonn.de

used in [3] (see also [14]) rather than the formal proofs of the physics literature (see, e.g. [7, 10]). The dimensional reduction of parabolic and elliptic stochastic differential equations with additive noise and gradient type non-linearity is an example of a more general phenomenon involving a supersymmetric representation of generic stochastic differential equations with Gaussian white noise. Although there are some formal arguments for proving this (conjectured) relation between dimensional reduction, supersymmetry and generic stochastic differential equations (see, e.g., [21] Chap. 15), outside the elliptic case with additive noise cited above and the standard stochastic differential equation with additive noise treated here, to the best of our knowledge, there is no proof in the general setting.

We describe in more details the result proved in this paper. Here we consider the following SDE

$$\partial_t \phi(t) + m^2 \phi(t) + f(t) V'(\phi(t)) = \xi(t), \qquad t \in \mathbb{R}, \tag{1}$$

where $m > 0$, $f : \mathbb{R} \to \mathbb{R}_+$ is a compactly supported positive even smooth function such that $f(0) = 1$, $V : \mathbb{R} \to \mathbb{R}$ is a smooth bounded function with all derivatives bounded and ξ is a Gaussian white noise on $\mathbb{R}$. Equation (1) has a unique solution $\phi : \mathbb{R} \to \mathbb{R}$ which coincides for sufficiently negative times with the Ornstein–Uhlenbeck process $\varphi = \mathcal{G} * \xi$ where

$$\mathcal{G}(t) = e^{-m^2 t} \mathbb{I}_{t>0}.$$

This solution satisfies the integral equation

$$\phi(t) + \mathcal{G} * (f V'(\phi))(t) = \varphi(t), \qquad t \in \mathbb{R}, \tag{2}$$

and moreover its law is invariant under the inversion $t \mapsto -t$ of the time variable.

The aim of this note is to prove the following theorem.

Theorem 1 *For any bounded measurable function $F : \mathbb{R} \to \mathbb{R}$ we have*

$$\mathbb{E}\left[F(\phi(0)) e^{-2 \int_{-\infty}^{0} f'(t) V(\phi(t)) dt} \right] = \frac{1}{\mathcal{Z}} \int_{\mathbb{R}} F(x) e^{-m^2 x^2 - 2 V(x)} dx$$

where

$$\mathcal{Z} = \frac{\mathbb{E}\left[e^{-2 \int_{-\infty}^{0} f'(t) V(\phi(t)) dt} \right]}{\int_{\mathbb{R}} e^{-m^2 x^2 - 2 V(x)} dx}.$$

Proof Let μ_φ be the law of the Gaussian field $\varphi = \mathcal{G} * \xi$ on the space $C(\mathbb{R}; \mathbb{R})$ endowed with the topology of uniform convergence on bounded intervals. Girsanov theorem implies that for any measurable bounded function $F : \mathbb{R} \to \mathbb{R}$

$$\mathbb{E}\left[F(\phi(0)) e^{-2 \int_{-\infty}^{0} f'(t) V(\phi(t)) dt} \right] = \int F(\varphi(0)) \exp(S(\varphi)) \mu_\varphi(d\varphi). \tag{3}$$

with

$$S(\varphi) = \int_{-\infty}^{0}\left[\frac{f(t)}{2}V''(\varphi(t)) - \frac{1}{2}(f(t)V'(\varphi(t)))^2 - 2f'(t)V(\phi(t))\right] dt +$$
$$- \int_{-\infty}^{0} f(t)V'(\varphi(t)) \circ dB(t).$$

Here $(B(t))_{t\in\mathbb{R}}$ is the double sided Brownian motion (adapted with respect to φ) such that $\partial_t B = \xi = (\partial_t + m^2)\varphi$ and $\circ dB$ denotes the corresponding Stratonovich integral.

Parisi and Sourlas [16] observed long ago that the r.h.s. of Eq. (3) admits a representation using a Gaussian super-field Φ defined on the superspace $(t, \theta, \bar{\theta})$, where $t \in \mathbb{R}$ is the usual time variable and $\theta, \bar{\theta}$ are two Grassmann variables playing the role of additional "fermionic" spatial coordinates (see Sect. 2 for the necessary notions and notations). For the moment let us simply remark that Φ can be rigorously constructed as a random field on a non-commutative probability space with expectation denoted by $\langle \cdot \rangle$ in such a way that the expectation of polynomials in Φ can be reduced, via an analog of Wick's theorem, to linear combinations of products of covariances. If the covariance of the super-field has the form

$$\langle \Phi(t, \theta, \bar{\theta})\Phi(s, \theta', \bar{\theta}')\rangle = \frac{1}{2m^2}\mathcal{G}(|t-s|) + \mathcal{G}(t-s)(\theta'-\theta)\bar{\theta}' - \mathcal{G}(s-t)(\theta'-\theta)\bar{\theta}, \quad (4)$$

then we will prove in Theorem 8 below that the following representation formula holds

$$\int F(\varphi(0)) \exp(S(\varphi))\mu_\varphi(d\varphi) = \left\langle F(\Phi(0)) \exp\left(\int_{-\infty}^{0} f(t + 2\theta\bar{\theta})V(\Phi(t, \theta, \bar{\theta}))dt d\theta d\bar{\theta}\right)\right\rangle. \quad (5)$$

Note that on the l.h.s. we have usual (commutative) probabilistic objects while the r.h.s. is expressed in the language of non-commutative probability.

The interest of this reformulation lies in the fact that on the superspace $(t, \theta, \bar{\theta})$ one can define supersymmetric transformations which preserve the quantity $t + 2\theta\bar{\theta}$. Integrals of supersymmetric quantities satisfy well known localization (also called dimensional reduction) formulas [5, 6, 12, 14, 17] which express integrals over the superspace as evaluations in zero, more precisely if $F = f(t + 2\theta\bar{\theta}) \in \mathcal{S}(\mathfrak{S})$ is a supersymmetric function and $T \in \mathcal{S}'(\mathfrak{S})$ is a supersymmetric distribution we have that

$$\int_{-\infty}^{K} T(t, \theta, \bar{\theta}) \cdot F(t, \theta, \bar{\theta})dt d\theta d\bar{\theta} = -2T_\emptyset(K)F_\emptyset(K)$$

for any $K \in \mathbb{R}$ (see Theorem 9 for a precise statement).

We cannot apply Theorem 9 directly to expression (5) since the superfield Φ is not supersymmetric. On the other hand the correlation function (4) is supersymmetric with respect to $(s, \theta', \bar{\theta}')$ when $t \geq s$ and with respect to $(t, \theta, \bar{\theta})$ when $t \leq s$. This property and the Markovianity of the kernel $\mathcal{G}$, namely

$$\mathcal{G}(t-s)\mathcal{G}(s-t) = 0$$

when $s \neq t$, allows us to prove a localization property for the *expectation of super-symmetric linear functionals of* Φ (see Theorem 14), namely we prove that

$$\left\langle F(\Phi(0)) \exp\left(\int_{-\infty}^{0} f(t+2\theta\bar{\theta}) V(\Phi(t,\theta,\bar{\theta})) \mathrm{d}t \mathrm{d}\theta \mathrm{d}\bar{\theta}\right)\right\rangle = \langle F(\Phi(0)) \exp[-2V(\Phi(0))]\rangle.$$

Since $\Phi(0)$ is distributed as a Gaussian with mean 0 and variance $2m^{-2}$ this implies the claim. □

The rest of the paper contains details on the definition of the super-fields and the proofs of the intermediate results.

2 Super-Geometry and Gaussian Super-Fields

2.1 Some Notions of Super-Geometry

We denote by $\mathfrak{S}$ an infinite dimensional Grassmannian algebra generated by an enumerable number of free generators $\{1, \theta_1, \theta_2, \ldots, \theta_n, \ldots\}$. By this we mean that any element of $\Theta \in \mathfrak{S}$ can be written in a unique way using a finite number of sum and products between the generators θ_i. The product between the θ_i is anti-commuting which means that $\theta_i \theta_j = -\theta_j \theta_i$, for all $i, j \in \mathbb{N}$, and they commute with 1. We call $\mathfrak{S}_0 = \text{span}\{1\}$, $\mathfrak{S}_1 = \text{span}\{\theta_1, \theta_2, \ldots, \theta_n, \ldots\}$ and with $\mathfrak{S}_k = \text{span}\{\theta_{i_1} \cdots \theta_{i_k} | \theta_{i_j} \in \mathfrak{S}_1\}$.

If $\theta_1, \ldots, \theta_h \in \mathfrak{S}_1$ we denote by $\mathfrak{S}(\theta_1, \ldots, \theta_h)$ the finite dimensional Grassmannian sub-algebra of $\mathfrak{S}$ generated by $\{1, \theta_1, \ldots, \theta_h\}$, and we denote by $\mathfrak{D}_h$ the universal Grassmannian algebra generated by h elements. We suppose that there is an order between the generators of $\mathfrak{D}_h$. Once we fixed an order between $\theta_1, \ldots, \theta_h$ there is a natural isomorphism between $\mathfrak{D}_h$ and $\mathfrak{S}(\theta_1, \ldots, \theta_h)$.

We can define a notion of smooth function $F : \mathbb{R}^n \times \mathfrak{S}_1^h \to \mathfrak{S}$. Let $\tilde{F}$ be a smooth function from $\mathbb{R}^n$ taking values in $\mathfrak{D}_h$ which means an object of the form

$$\tilde{F}(x) = \tilde{F}_\emptyset(x) 1 + \sum_{i=1}^{h} \tilde{F}_i(x) t_i + \sum_{1 \leq i < j \leq h} \tilde{F}_{i,j}(x) t_i t_j + \cdots + \tilde{F}_{1,2,\ldots,h}(x) t_1 \cdots t_h,$$

where $t_1, \ldots, t_h$ are a set of generators of $\mathfrak{D}_h$ and $x \in \mathbb{R}^n$. We define F associated with $\tilde{F}$ in the following way: F associates to $(x, \theta_1, \ldots, \theta_h) \in \mathbb{R}^n \times \mathfrak{S}_1^h$ the element $\tilde{F}(x) \in \mathfrak{S}(\theta_1, \ldots, \theta_h)$ where we make the identification of $\mathfrak{S}(\theta_1, \ldots, \theta_h)$ with $\mathfrak{D}_h$, i.e.

$$F(x,\theta_1,\ldots,\theta_h) = \widetilde{F}_\emptyset(x)1 + \sum_{i=1}^{h}\tilde{F}_i(x)\theta_i + \sum_{1\leqslant i<j\leqslant h}\tilde{F}_{i,j}(x)\theta_i\theta_j + \cdots + \tilde{F}_{1,2,\ldots,h}(x)\theta_1\cdots\theta_h. \tag{6}$$

Hereafter we use the notation $F_\emptyset, F_{\theta_1}, \ldots$ for denoting $F_\emptyset = \widetilde{F}_\emptyset, F_{\theta_1} = \tilde{F}_1, \ldots$. We say that F is a Schwartz function if $F_\emptyset, F_i, \ldots$ are Schwartz functions. We denote by $\mathcal{S}(\mathfrak{S}^h)$ the set of Schwartz functions with h anti-commuting variables.

If $H : \mathbb{R} \to \mathbb{R}$ is a smooth function we can define the composition $H \circ F$ in the following way

$$\begin{aligned}H \circ F(x,\theta_1,\ldots,\theta_h) &= H(F_\emptyset(x))1 + H'(F_\emptyset(x))(F(x,\theta_1,\ldots,\theta_h) - F_\emptyset(x)1)\\ &\quad + \tfrac{1}{2}H''(F_\emptyset(x))(F(x,\theta_1,\ldots,\theta_h) - F_\emptyset(x)1)^2\\ &\quad \cdots + \tfrac{1}{h!}H^{(h)}(F_\emptyset(x))(F(x,\theta_1,\ldots,\theta_h) - F_\emptyset(x)1)^h.\end{aligned}$$

On $\mathfrak{S}$ it is possible to define a notion of integral called Berezin integral, in the following way $\int \theta d\theta = 1$, $\int \bar{\theta} d\theta = 0$ if $\bar{\theta} \in \mathfrak{S}_1$ and $\bar{\theta} \neq \theta$, $\int \Theta d\theta = \Theta$ where $\Theta = \theta_1 \cdots \theta_h \in \mathfrak{S}_h$ and $\theta_i \neq \theta$ and $\int \cdot d\theta$ is linear in its argument. The integral $\int \Theta d\theta_1 d\theta_2 \cdots d\theta_h$ is defined as $\int \left(\cdots \left(\int \left(\int \Theta d\theta_1\right) d\theta_2\right) \cdots\right) d\theta_h$.

If F is a smooth function, defined on $\mathbb{R}^n \times \mathfrak{S}_1^h$, we can define the integral of F with respect to $dx d\theta_1 \cdots d\theta_h$ in the following way $\int F(x,\theta_1,\ldots,\theta_h) dx d\theta_1 \cdots d\theta_h$ first applying the integral $\int \cdot dx$ to the set of functions $F_\emptyset, F_i, \ldots$ (which are the component of the function F by Eq. (6)) obtaining an element of $\mathfrak{S}(\theta_1,\ldots,\theta_h)$ and then applying the Berezin integral to this result. Using this notion of integral and the induced duality between smooth functions, it is possible to define the notion of tempered distribution $T \in \mathcal{S}'(\mathbb{R}^n \times \mathfrak{S}^h)$. The distribution T is an object of the form

$$T(x,\theta_1,\ldots,\theta_k) = T_\emptyset(x)1 + \sum_{i=1}^{h} T_{\theta_i}(x)\theta_i + \cdots + T_{\theta_1\ldots\theta_h}(x)\theta_1\cdots\theta_h$$

where $T_\emptyset(x), T_{\theta_i}(x), \ldots, T_{\theta_1\ldots\theta_h}(x)$ are Schwartz distributions on $\mathbb{R}^n$.

2.2 Construction of the Super-Field

Following the analogous construction in [3, 14] the super-field Φ is defined as

$$\Phi(t,\theta,\bar{\theta}) = \varphi(t) + \bar{\psi}(t)\theta + \psi(t)\bar{\theta} + \omega(t)\theta\bar{\theta},$$

where $t \in \mathbb{R}$, and $\varphi, \psi, \bar{\psi}, \omega$ are complex Gaussian fields realized as functional from $\mathcal{S}(\mathbb{R})$ into the set of operators $\mathcal{O}(\mathfrak{H})$ on a complex vector space $\mathfrak{H}$ with a fixed state Ω, and $\theta, \bar{\theta}$ are any pair of anti-commuting variables $\theta, \bar{\theta} \in \mathfrak{S}$ commuting with the operators ω, φ and anti-commuting with the operators $\psi, \bar{\psi}$. Hereafter we shall use the notation denote by $\langle a \rangle_\Omega = \langle \Omega, a(\Omega) \rangle_\mathfrak{H}$ for any $a \in \mathcal{O}(\mathfrak{H})$, where $\langle \cdot, \cdot \rangle_\mathcal{H}$ is the scalar product in $\mathcal{H}$. For a background on superfield, supermanifolds and Berezin integral see, e.g., [4, 8] and furthere referefnces in [2, 3].

The Gaussian fields $\varphi, \psi, \bar{\psi}, \omega$ mush be realized as operators defined from $\mathcal{S}(\mathbb{R})$ taking values in $\mathcal{O}(\mathfrak{H})$, for a suitable Hilbert space $\mathfrak{H}$ with a state $\Omega \in \mathfrak{H}$ such that the condition (4) holds. Making a formal computation we obtain that

$$\langle \Phi(t,\theta,\bar{\theta})\Phi(s,\theta',\bar{\theta}')\rangle_\Omega = \langle \varphi(t)\varphi(s)\rangle_\Omega - \langle \bar{\psi}(t)\psi(s)\rangle_\Omega \theta\bar{\theta}' - \langle \psi(t)\bar{\psi}(s)\rangle_\Omega \bar{\theta}\theta' + $$
$$+ \langle \varphi(t)\omega(s)\rangle_\Omega \theta'\bar{\theta}' + \langle \omega(t)\varphi(s)\rangle_\Omega \theta\bar{\theta} + \langle \omega(t)\omega(s)\rangle_\Omega \theta\bar{\theta}\theta'\bar{\theta}'$$

from which we get

$$\langle \varphi(t)\varphi(s)\rangle_\Omega = \tfrac{1}{2m^2}\mathcal{G}(|t-s|) \quad \langle \bar{\psi}(t)\psi(s)\rangle_\Omega = \mathcal{G}(t-s) \quad \langle \varphi(t)\omega(s)\rangle_\Omega = \mathcal{G}(t-s)$$
$$\langle \omega(t)\omega(s)\rangle_\Omega = 0. \tag{7}$$

Using the commutation relations

$$\{\varphi(t),\varphi(s)\}_+ = 0 \quad \{\varphi(t),\omega(s)\}_+ = 0 \quad \{\omega(t),\omega(s)\}_+ = 0 \tag{8}$$

$$\{\bar{\psi}(t),\psi(s)\}_- = \{\psi(t),\psi(s)\}_- = \{\bar{\psi}(t),\bar{\psi}(s)\}_- = 0$$
$$\{\varphi(t)\psi(s)\}_+ = \{\varphi(t)\bar{\psi}(s)\}_+ = \{\omega(t),\psi(s)\}_+ = \{\omega(t),\bar{\psi}(s)\}_+ = 0 \tag{9}$$

where $\{K_1,K_2\}_+ = K_1K_2 - K_2K_1$ and $\{K_1,K_2\}_- = K_1K_2 + K_2K_1$ (where $K_1, K_2 \in \mathcal{B}(\mathfrak{H})$) are the commutator and the anti-commutator of closed operators having a non void common core. By Wick's theorem (see, e.g. [9] Chap. 3 Sect. 8) the expectation of arbitrary polynomials in $\varphi, \psi, \bar{\psi}, \omega$ is completely determined.

The bosonic field φ is a standard (real and commutative) Gaussian field with covariance $\mathcal{G}(|t-s|)$. Also ω is a standard (complex and commutative) Gaussian field of the form

$$\omega(t) = \xi(t) + i\eta(t),$$

where $\xi = (\partial_t + m^2)\varphi$ and η is a Gaussian white noise with Cameron-Martin space $L^2(\mathbb{R})$ independent of φ. We can realize the Gaussian field φ, ω as (unbounded) operators defined on a Hilbert space $\mathfrak{H}_{\varphi,\omega}$ and with a state $\Omega_{\varphi,\omega}$. We can take $\mathfrak{H}_{\varphi,\omega} = L^2(\mu_{\varphi,\omega})$ where $\mu_{\varphi,\omega}$ is the law of (φ,ω) on $C(\mathbb{R}) \times \mathcal{S}'_C(\mathbb{R})$ and $\Omega_{\varphi,\omega} = 1$.

The fermionic fields $\psi, \bar{\psi}$ are build as follows (for a different construction of fermionic fields see also [1]). Let a, b and a^*, b^* be two creation and annihilation operators defined as bounded functionals on $\mathcal{S}(\mathbb{R})$ taking values in $\mathcal{B}(\mathfrak{H}_{\psi,\bar{\psi}})$ (where $\mathfrak{H}_{\psi,\bar{\psi}}$ is a suitable Hilbert space with a fixed state $\Omega_{\psi,\bar{\psi}}$) such that

$$\{a(f),a(g)\}_- = \{b(f),b(g)\}_- = 0$$
$$\{a(f),b(g)\}_- = \{a^*(f),b(g)\}_- = 0$$
$$\{a^*(g),a(f)\}_- = \{b^*(g),b(f)\}_- = \left(\int_\mathbb{R} f(t)g(t)dt\right) I_{\mathfrak{H}_{\psi,\bar{\psi}}},$$

for any $f, g \in \mathcal{S}(\mathbb{R})$, and such that

$$\langle a(f)K\rangle_{\Omega_{\psi,\bar{\psi}}} = \langle Ka^*(f)\rangle_{\Omega_{\psi,\bar{\psi}}} = \langle b(f)K\rangle_{\Omega_{\psi,\bar{\psi}}} = \langle Kb^*(f)\rangle_{\Omega_{\psi,\bar{\psi}}} = 0,$$

where K is any bounded operator $K \in \mathcal{B}(\mathfrak{H}_{\psi,\bar{\psi}})$. We define $\mathcal{U} : \mathcal{S}(\mathbb{R}) \to \mathcal{S}(\mathbb{R})$ by

$$\mathcal{U}(f)(t) = \frac{1}{2\pi} \int_\mathbb{R} \frac{e^{-i\xi t}}{i\xi + m^2} \hat{f}(\xi) d\xi.$$

We then write

$$\psi(f) = a^*(\mathcal{U}^*(f)) + b(f), \quad \bar{\psi}(f) = b^*(\mathcal{U}(f)) - a(f),$$

where $\mathcal{U}^*$ is the adjoint of $\mathcal{U}$ in $L^2(\mathbb{R})$ with respect to the Lebesgue measure on $\mathbb{R}$. In this way, we have

$$\{\bar{\psi}(t), \psi(s)\}_- = \{\psi(t), \psi(s)\}_- = \{\bar{\psi}(t), \bar{\psi}(s)\}_- = 0,$$

and also

$$\langle \bar{\psi}(f)\psi(g) \rangle_{\Omega_{\psi,\bar{\psi}}} = \langle b^*(f)a^*(g) \rangle_{\Omega_{\psi,\bar{\psi}}} + \langle b^*(f)b(g) \rangle_{\Omega_{\psi,\bar{\psi}}} - \langle a(f)a^*(g) \rangle_{\Omega_{\psi,\bar{\psi}}} +$$
$$-\langle a(f)b(g) \rangle_{\Omega_{\psi,\bar{\psi}}} = \int_\mathbb{R} \mathcal{U}(f)(t) g(t) dt$$
$$= \int_\mathbb{R} g(t) \int_{-\infty}^t e^{-m^2(t-s)} f(s) ds dt = \int_{\mathbb{R}^2} g(t) \mathcal{G}(t-s) f(s) ds dt.$$

In other words we have $\langle \bar{\psi}(t)\psi(s) \rangle_{\Omega_{\psi,\bar{\psi}}} = \mathcal{G}(t-s)$ as required (for a more detailed proof see, e.g., [1, 15, 16, 21]). We can define the operators $\varphi, \psi, \bar{\psi}, \omega$ as acting on a unique (quantum) probability space, by taking

$$\mathfrak{H} = \mathfrak{H}_{\varphi,\omega} \otimes \mathfrak{H}_{\psi,\bar{\psi}} \quad \Omega = \Omega_{\varphi,\omega} \otimes \Omega_{\psi,\bar{\psi}}.$$

In order to realize the field Φ in a rigorous way we consider a complex sub-algebra $\mathfrak{A} \subset \mathcal{O}(\mathfrak{H})$ such that $\varphi, \psi, \bar{\psi}, \omega$ take values in $\mathfrak{A}$ and for any smooth function $V : \mathbb{R} \to \mathbb{R}$ we have $V(\varphi(g)) \in \mathfrak{A}$, where g is any function in $\mathcal{S}(\mathbb{R})$. This sub-algebra $\mathfrak{A}$ is generated (from an algebraic point of view) by operators of the form $V(\varphi(g))$, $\omega(g), \psi(g), \bar{\psi}(g)$ and $I_\mathfrak{H}$. We consider the vector space $\mathcal{A} = \mathfrak{A} \times \mathfrak{S}$. There are two preferred hyperplanes $\mathcal{A}_\mathfrak{A}$ and $\mathcal{A}_\mathfrak{S}$ defined by

$$\mathcal{A}_\mathfrak{A} = \{(a, 1_\mathfrak{S}) | a \in \mathfrak{A}\}, \quad \mathcal{A}_\mathfrak{S} = \{(I_\mathfrak{H}, \theta) | \theta \in \mathfrak{S}\},$$

with the natural immersions $i_\mathfrak{A} : \mathfrak{A} \to \mathcal{A}$ and $i_\mathfrak{S} : \mathfrak{S} \to \mathcal{A}$ defined by $i_\mathfrak{A}(a) = (a, 1_\mathfrak{S})$ and $i_\mathfrak{S}(\theta) = (I_\mathfrak{H}, \theta)$ (we note that $\mathcal{A}_\mathfrak{A} := i_\mathfrak{A}(\mathfrak{A})$ and $\mathcal{A}_\mathfrak{S} := i_\mathfrak{S}(\mathfrak{S})$). It is clear that $\mathcal{A}_\mathfrak{A}, \mathcal{A}_\mathfrak{S}$ generates the whole $\mathcal{A}$. On $\mathcal{A}$ we define the following product $\cdot$, in such a way that the maps $i_\mathfrak{A}$ and $i_\mathfrak{S}$ respect the product (i.e. $i_\mathfrak{A}(ab) = i_\mathfrak{A}(a) \cdot i_\mathfrak{A}(b)$ and $i_\mathfrak{S}(\theta_1 \theta_2) = i_\mathfrak{S}(\theta_1) \cdot i_\mathfrak{S}(\theta_2)$) and such that

$$(V(\varphi(g)), 1_\mathfrak{S}) \cdot (I_\mathfrak{H}, \theta) = (I_\mathfrak{H}, \theta) \cdot (V(\varphi(g)), 1_\mathfrak{S}) = (V(\varphi(g)), \theta)$$

$$(\omega(g), 1_\mathfrak{S}) \cdot (I_\mathfrak{H}, \theta) = (I_\mathfrak{H}, \theta) \cdot (\omega(g), 1_\mathfrak{S}) = (\omega(g), \theta)$$

$$(\psi(g), 1_{\mathfrak{S}}) \cdot (I_{\mathfrak{H}}, \theta) = -(I_{\mathfrak{H}}, \theta) \cdot (\psi(g), 1_{\mathfrak{S}}) = (\psi(g), \theta)$$

$$(\bar{\psi}(g), 1_{\mathfrak{S}}) \cdot (I_{\mathfrak{H}}, \theta) = -(I_{\mathfrak{H}}, \theta) \cdot (\bar{\psi}(g), 1_{\mathfrak{S}}) = (\bar{\psi}(g), \theta)$$

where $g \in \mathcal{S}(\mathbb{R})$ and $\theta \in \mathfrak{S}_1$ (not in $\mathfrak{S}$). The product $\cdot$ can be uniquely extended (in a associative way) on $\mathcal{A}$ since $\mathcal{A}_\mathfrak{A}$, $\mathcal{A}_\mathfrak{S}$ generates the whole $\mathcal{A}$, operators of the form $V(\varphi(g))$, $\omega(g)$, $\psi(g)$, $\bar{\psi}(g)$ generates the whole $\mathfrak{A}$ and $\mathfrak{S}_1$ generates the whole $\mathfrak{S}$. Hereafter we will omit to explicitly write the product $\cdot$ if this omission does not cause any confusion.

On $\mathcal{A}$ we can define a linear operator $\langle \cdot \rangle : \mathcal{A} \to \mathcal{A}_\mathfrak{S} \simeq \mathfrak{S}$ by

$$\langle (a, \theta) \rangle = \langle a \rangle_\Omega (I_\mathfrak{H}, \theta).$$

Furthermore for any $\theta_1, \ldots, \theta_n \in \mathfrak{S}_1$ we define the linear operator $\int \cdot d\theta_1 \ldots d\theta_n : \mathcal{A} \to \mathcal{A}_\mathfrak{A}$ such $\int \cdot d\theta_1 \ldots d\theta_n|_{\mathcal{A}_\mathfrak{S}}$ is the usual Berezin integral induced by the identification $\mathcal{A}_\mathfrak{S} \simeq \mathfrak{S}$

$$\int (a, \theta) d\theta_1 \ldots d\theta_n = \left(\int \theta d\theta_1 \ldots d\theta_n \right) (a, 1_\mathfrak{S}).$$

Hereafter we identify the space $\mathfrak{A}$ and $\mathfrak{S}$ with $\mathcal{A}_\mathfrak{A}$ and $\mathcal{A}_\mathfrak{S}$ respectively, and we write instead of $(a, 1_\mathfrak{S})$, $(I_\mathfrak{H}, \theta)$, $(I_\mathfrak{H}, 1_\mathfrak{S})$ simply a, θ and 1 respectively (in this way we take also the tacit identification of $\text{span}\{1_\mathfrak{S}\} = \mathfrak{S}_0$ with $\mathbb{R}$). Furthermore we identify $\varphi, \psi, \bar{\psi}, \omega$ with $i_\mathfrak{A} \circ \varphi$, $i_\mathfrak{A} \circ \psi$, $i_\mathfrak{A} \circ \bar{\psi}$, $i_\mathfrak{A} \circ \omega$.

Remark 2 Since $\psi, \bar{\psi}$ are "independent" with respect to φ and ω (since they can be realized on a space of the form $\mathfrak{H} = \mathfrak{H}_{\varphi, \omega} \otimes \mathfrak{H}_{\psi, \bar{\psi}}$) the expectation only with respect to the fields $\psi, \bar{\psi}$ is well defined, namely there exixts an operator $\langle \cdot \rangle_{\psi, \bar{\psi}} : \mathcal{O}_\mathfrak{H} \to \mathcal{O}_{,!}$ such that

$$\langle V(\varphi(t_1), \ldots, \varphi(t_k)) \psi(t'_1) \bar{\psi}(t''_1) \cdots \psi(t'_{k'}) \bar{\psi}(t''_{k'}) \rangle_{\psi, \bar{\psi}}$$
$$= V(\varphi(t_1), \ldots, \varphi(t_k)) \langle \psi(t'_1) \bar{\psi}(t''_1) \cdots \psi(t'_{k'}) \bar{\psi}(t''_{k'}) \rangle.$$

This operator $\langle \cdot \rangle_{\psi, \bar{\psi}}$ extends to $\mathcal{A}$ in the same way in which the operator $\langle \cdot \rangle$ is extended on the whole $\mathcal{A}$.

2.3 Relation with SDEs

In this section we want to use the super-field Φ for representing the solution to the SDE (1) through the integral (3).

First of all we have to define the notion of composition of the super-field Φ with smooth functions. Consider the smooth function $H : \mathbb{R} \to \mathbb{R}$ growing at most

A Note on Supersymmetry and Stochastic Differential Equations

exponentially at infinity. We can formally expand H in Taylor series and using the properties of $\theta, \bar{\theta}$ we obtain

$$H(\Phi(t, \theta, \bar{\theta})) = H(\varphi(t)) + H'(\varphi(t))\bar{\psi}(t)\theta + H'(\varphi(t))\psi(t)\bar{\theta} + \\ + (H'(\varphi(t))w(t) + H''(\varphi(t))\psi(t)\bar{\psi}(t))\theta\bar{\theta}.$$

Unfortunately the products $H'(\varphi(t))w(t)$ and $H''(\varphi(t))\psi(t)\bar{\psi}(t)$ are ill defined since the factors are not regular enough. For this reason we consider a symmetric mollifier $\rho : \mathbb{R} \to \mathbb{R}_+$ (with $\rho(t) = \rho(-t)$) and the field $\Phi_\epsilon = \rho_\epsilon * \Phi$, where $\rho_\epsilon(t) = \epsilon^{-1}\rho(t\epsilon^{-1})$. If G is a super-function, $F : \mathbb{R} \to \mathbb{R}$ is a smooth function and $\mathbb{K}$ is an entire function we define

$$\left\langle F(\varphi(0))\mathbb{K}\left(\int G(t, \theta, \bar{\theta})H(\Phi(t, \theta, \bar{\theta}))dt d\theta d\bar{\theta}\right)\right\rangle \\ := \lim_{\epsilon \to 0}\left\langle F(\varphi_\epsilon(0))\mathbb{K}\left(\int G(t, \theta, \bar{\theta})H(\Phi_\epsilon(t, \theta, \bar{\theta}))dt d\theta d\bar{\theta}\right)\right\rangle. \quad (10)$$

We want to prove that the previous expression is well defined and does not depend on ρ.

Remark 3 It is important to note that the expression (10) does not depend on ρ only if ρ is reflection symmetric (i.e. $\rho(t) = \rho(-t)$). If we choose a different ρ (such that for example $\int_{-\infty}^{0} \rho dt \neq \int_{0}^{+\infty} \rho dt$) we will obtain a different limit in (10). This is due to the fact that the products $H'(\varphi(t))w(t)$ and $H''(\varphi(t))\psi(t)\bar{\psi}(t)$ are ill defined and it is analogous to the possibility to obtain Itô or Stratonovich integral in stochastic calculus considering different approximations of the stochastic integral.

Lemma 4 *Let $F_1, \ldots, F_n : \mathbb{R} \times \mathbb{R} \to \mathbb{R}$ be smooth functions with compact support in the first variable and growing at most exponentially at infinity in the second variable then we have*

$$\lim_{\epsilon \to 0}\left\langle \prod_{i=1}^{n} \int F_i(t, \varphi_\epsilon(t))\bar{\psi}_\epsilon(t)\psi_\epsilon(t)dt\right\rangle_{\psi,\bar{\psi}} = \int \prod_{i=1}^{n} F_i(t_i, \varphi(t))\mathfrak{G}_n(t_1, \ldots, t_n)dt_1 \ldots dt_n. \quad (11)$$

in $L^p(\mu_\varphi)$ for all $1 \leq p < +\infty$. Here $\mathfrak{G}_n(t_1, \ldots, t_n) = \det((G_{i,j})_{i,j=1,\ldots,n})$ with $G_{i,j} = \mathcal{G}(t_j - t_i)$ if $i \neq j$ and $G_{i,i} = 1/2$.

Proof It is simple to see that $\lim_{\epsilon \to 0}\langle \bar{\psi}_\epsilon(t_1)\psi_\epsilon(t_2)\rangle_{\psi,\bar{\psi}} = \mathcal{G}(t_1 - t_2)$ when $t_1 \neq t_2$ and $\lim_{\epsilon \to 0}\langle \bar{\psi}_\epsilon(t)\psi_\epsilon(t)\rangle_{\psi,\bar{\psi}} = \frac{1}{2}$ (this is due to the fact that $\rho(t) = \rho(-t)$). Since $\langle \bar{\psi}_\epsilon(t_1)\psi_\epsilon(t_1)\cdots\bar{\psi}_\epsilon(t_2)\psi_\epsilon(t_n)\rangle_{\psi,\bar{\psi}}$ is uniformly bounded in t and ϵ and $F_i(t, \varphi_\epsilon(t))$ is uniformly bounded in $L^p(\mu_\varphi)$ in t and ϵ the claim follows. □

Remark 5 Since only one between $\mathcal{G}(t - s)$ and $\mathcal{G}(s - t)$ is non zero if $F_1 = F_2 = \cdots = F_n$ then we get

$$\lim_{\epsilon \to 0}\left\langle \left(\int F_1(t, \varphi_\epsilon(t))\bar{\psi}_\epsilon(t)\psi_\epsilon(t)dt\right)^n\right\rangle_{\psi,\bar{\psi}} = \left(\frac{1}{2}\int F_1(t, \varphi(t))dt\right)^n.$$

Lemma 6 *Let $F : \mathbb{R} \times \mathbb{R} \to \mathbb{R}$ be smooth functions with compact support in the first variable and growing at most exponentially at infinity in the second variable then we have in $L^p(\mu_{\varphi,\omega})$, for all $1 \le p < +\infty$,*

$$\lim_{\epsilon \to 0} \int F(t, \varphi_\epsilon(t))\omega_\epsilon(t) = \int F(t, \varphi(t)) \circ \mathrm{d}B(t) + i \int F(t, \varphi(t))\mathrm{d}W(t)$$

where the first one is Stratonovich integral and the second one is Itô integral with respect to (double sided) Brownian motions $(B(t), W(t))_{t \in \mathbb{R}}$ such that $\partial_t B(t) = \xi(t)$ and $\partial_t W(t) = \eta(t)$ with $B_0 = W_0 = 0$.

Proof This is the Wong–Zakai theorem [13, 18–20]. □

Theorem 7 *When $\mathbb{K}$ is a polynomial and H grows at most exponentially at infinity, or $\mathbb{K}$ is entire and H is bounded with first and second derivative bounded, the limit (10) is well defined and does not depend on the symmetric mollifier ρ.*

Proof When $\mathbb{K}$ is a polynomial the thesis follows directly from Lemmas 4 and 6. If $\mathbb{K}$ is an entire function and H is a bounded function with first and second derivatives bounded it is possible to exchange the limit in ϵ with the power series, since

$$\left\langle \left| \left(\int G(t, \theta, \bar{\theta}) H(\Phi_\epsilon(t, \theta, \bar{\theta})) \right)^k \mathrm{d}t \mathrm{d}\theta \mathrm{d}\bar{\theta} \right|^p \right\rangle_{\psi, \bar{\psi}}$$

is uniformly bounded in ϵ for any $p \ge 1$. □

Theorem 8 *Suppose that $G(t, \theta, \bar{\theta}) = G_\emptyset(t) + G_{\theta\bar{\theta}}(t)\theta\bar{\theta}$ and that H is bounded with the first and second derivatives bounded then*

$$\left\langle F(\varphi(0)) \exp\left(\int G(t, \theta, \bar{\theta}) H(\Phi(t, \theta, \bar{\theta}))\mathrm{d}t\mathrm{d}\theta\mathrm{d}\bar{\theta} \right) \right\rangle =$$
$$= \int F(\varphi(0)) \exp\left(\frac{1}{2} \int G_\emptyset(t) H''(\varphi(t))\mathrm{d}t - \int G_\emptyset(t) H'(\varphi(t)) \circ \mathrm{d}\xi(t) + \right.$$
$$\left. - \frac{1}{2} \int (G_\emptyset(t) H'(\varphi(t)))^2 \mathrm{d}t - \int G_{\theta\bar{\theta}}(t) H(\varphi(t)) \mathrm{d}t \right) \mu_\varphi(\mathrm{d}\varphi).$$
(12)

Proof The proof follows from Theorem 7, the multiplicative property of exponentials, Remark 5, and the fact that the Fourier transform of a process integrated with respect to an independent Gaussian white noise can be computed explicitly and in this case gives the factor $\exp(-\frac{1}{2} \int (G_\emptyset(t) H'(\varphi(t)))^2 \mathrm{d}t)$. □

3 Supersymmetry and the Supersymmetric Field

3.1 The Supersymmetry

On $C^\infty(\mathbb{R} \times \mathfrak{S}_1^2)$ one can introduce the (graded) derivations

$$Q := 2\theta\partial_t + \partial_{\bar\theta}, \qquad \bar Q := 2\bar\theta\partial_t - \partial_\theta,$$

which are such that

$$Q(t + 2\theta\bar\theta) = 0 = \bar Q(t + 2\theta\bar\theta),$$

namely they annihilate the function $t + 2\theta\bar\theta$ defined on $\mathbb{R} \times \mathfrak{S}_1^2$. Moreover if $QF = \bar Q F = 0$, for F in $C^\infty(\mathbb{R} \times \mathfrak{S}_1^2)$, then we must have

$$0 = QF(x, \theta, \bar\theta) = 2\partial_t f_\emptyset(t)\theta + f_{\bar\theta}(t) + \partial_t f_{\bar\theta}(x)\theta\bar\theta - f_{\theta\bar\theta}(t)\theta$$

$$0 = \bar Q F(x, \theta, \bar\theta) = 2\partial_t f_\emptyset(t)\bar\theta + f_\theta(t) - \partial_t f_\theta(x)\theta\bar\theta - f_{\theta\bar\theta}(t)\bar\theta$$

and therefore

$$\partial_t f_\emptyset(t) = \frac{1}{2} f_{\theta\bar\theta}(t) \qquad \text{and} \qquad f_\theta(t) = f_{\bar\theta}(t) = 0.$$

This means that there exists an $f \in C^\infty(\mathbb{R}, \mathbb{R})$ such that

$$f(t + 2\theta\bar\theta) = f(t) + 2f'(t)\theta\bar\theta = f_\emptyset(t) + f_{\theta\bar\theta}(t)\theta\bar\theta = F(t, \theta, \bar\theta).$$

Namely any function satisfying these two equations can be written in the form

$$F(t, \theta, \bar\theta) = f(t + 2\theta\bar\theta).$$

Suppose that $t > 0$, if we introduce the linear transformations

$$\tau(b, \bar b) \begin{pmatrix} t \\ \theta \\ \bar\theta \end{pmatrix} = \begin{pmatrix} t + 2\bar b\theta\rho + 2b\bar\theta\rho \\ \theta - b\rho \\ \bar\theta + \bar b\rho \end{pmatrix} \in \mathfrak{S}(\theta, \bar\theta, \rho)$$

for $b, \bar b \in \mathbb{R}$ and where $\rho \in \mathfrak{S}_1$ is a new odd variable different from $\theta, \bar\theta$, then we have

$$\frac{d}{da}\bigg|_{a=0} \tau(ab, a\bar b) F(t, \theta, \bar\theta) = \frac{d}{da}\bigg|_{a=0} F(\tau(ab, a\bar b)(t, \theta, \bar\theta)) = (b \cdot \bar Q + \bar b \cdot Q) F(t, \theta, \bar\theta)$$

so $\tau(b, \bar b) = \exp(b \cdot \bar Q + \bar b \cdot Q)$ and $\tau(ab, a\bar b)\tau(cb, c\bar b) = \tau((a+c)b, (a+c)\bar b)$.

In particular $F \in C^\infty(\mathbb{R} \times \mathfrak{S}^2)$ is supersymmetric if and only if for any $b, \bar{b} \in \mathbb{R}$ we have $\tau(b, \bar{b})F = F$.

By duality the operators Q, $\bar{Q}$ and $\tau(b, \bar{b})$ also act on the space $\mathcal{S}'(\mathfrak{S})$ and we say that the distribution $T \in \mathcal{S}'(\mathfrak{S})$ is supersymmetric if it is invariant with respect to rotations in space and $QT = \bar{Q}T = 0$. For supersymmetric functions and distributions the following fundamental theorem holds.

Theorem 9 *Let $F \in \mathcal{S}(\mathfrak{S})$ and $T \in \mathcal{S}'(\mathfrak{S})$ such that $T_\emptyset$ is a continuous function. If both F and T are supersymmetric. Then for any $K \in \mathbb{R}$ we have the reduction formula*

$$\int_{-\infty}^{K} T(t, \theta, \bar{\theta}) \cdot F(t, \theta, \bar{\theta}) dt d\theta d\bar{\theta} = -2T_\emptyset(K) F_\emptyset(K). \tag{13}$$

Proof The proof can be found in [14], Lemma 4.5 for $\mathbb{R}^2$ and in [17] for the case of a general super-manifold. Here we give the proof only for the case where T is a super-function. In this case we have that $T(t, \theta, \bar{\theta}) = T_\emptyset(t) + 2T'_\emptyset(t)\theta\bar{\theta}$ and $F(t, \theta, \bar{\theta}) = F_\emptyset(t) + 2F'_\emptyset(t)\theta\bar{\theta}$, from which we have

$$T(t, \theta, \bar{\theta}) \cdot F(t, \theta, \bar{\theta}) = T_\emptyset(t) F_\emptyset(t) + 2(T'_\emptyset(t) F_\emptyset(t) + T_\emptyset(t) F'_\emptyset(t))\theta\bar{\theta}$$
$$= T_\emptyset(t) F_\emptyset(t) + 2\partial_t(T_\emptyset F_\emptyset)(t)\theta\bar{\theta}.$$

By definition of Berezin integral we have

$$\int_{-\infty}^{K} T(t, \theta, \bar{\theta}) \cdot F(t, \theta, \bar{\theta}) dx d\theta d\bar{\theta} = -2 \int_{-\infty}^{K} \partial_t(T_\emptyset F_\emptyset)(t) dt$$
$$= -2T_\emptyset(K) F_\emptyset(K).$$

$\square$

Remark 10 In Theorem 9 we can assume that $F = F_\emptyset(t) + F_{\theta\bar{\theta}}(t)\theta\bar{\theta}$ and $T(t, \theta, \bar{\theta}) = T_\emptyset(t) + T_{\theta\bar{\theta}}(t)\theta\bar{\theta}$ where $F_{\theta\bar{\theta}}(t) = 2F'_\emptyset(t)$ and $T_{\theta\bar{\theta}}(t) = 2T'_\emptyset(t)$ only for $t \leqslant K$. In this way we can consider supersymmetric functions only on the set $(-\infty, K]$.

3.2 Localization of Supersymmetric Averages

Remark 11 We note that the correlation function

$$C^\Phi(t, s, \theta, \bar{\theta}) := \langle \varphi(t)\Phi(s, \theta, \bar{\theta}) \rangle = \frac{1}{2m^2} \mathcal{G}(t-s) + \mathcal{G}(t-s)\theta\bar{\theta}$$

is a supersymmetric function when $t \geqslant s$.

Lemma 12 *Let $g(t)$ be smooth function with compact support, $t \in \mathbb{R}$, let P be a polynomial. Then for $t_1 > t_2 > \cdots > t_k$ and $M = (m_1, \ldots, m_k) \in \mathbb{N}^k$ we have*

$$\mathcal{H}_{\ell,P}^{M,G}(t_1, \ldots, t_k) =$$

$$= \left\langle \prod_{j=1}^{k} \varphi(t_j)^{m_j} \int_{-\infty}^{t_k} \int_{-\infty}^{\tau_1} \cdots \int_{-\infty}^{\tau_\ell} \prod_{i=1}^{\ell} g\left(\tau_i + 2\theta_i \bar{\theta}_i\right) P(\Phi(\tau_i, \theta_i, \bar{\theta}_i)) \mathrm{d}\tau_i \mathrm{d}\theta_i \mathrm{d}\bar{\theta}_i \right\rangle =$$

$$= \frac{(-2\,g(t_k))^\ell}{\ell!} \left\langle \prod_{j=1}^{k} \varphi(t_j)^{m_j} P(\varphi(t_k))^\ell \right\rangle.$$

Proof We prove the lemma by induction on ℓ and for simplicity we assume that $P(x) = x^n$, the general case being a straightforward generalization. Since the proof is essentially of combinatorial nature in the following we consider some ill defined objects like the products $\varphi(t)\omega(s)$ or $\psi(t)\bar{\psi}(s)$. This fact does not change the main idea of the proof since all the expectations with respect to the previous products are defined using the symmetric regularization proposed in Lemma 4 and Lemma 6, i.e. all the following computations can be made rigorous replacing φ, ω, ψ and $\bar{\psi}$ by the regularized Gaussian fields $\varphi_\epsilon, \omega_\epsilon, \psi_\epsilon$ and $\bar{\psi}_\epsilon$ (as defined in Lemma 4 and Lemma 6) and then taking the limit as $\epsilon \to 0$. The main difference between the proof below and the one involving the regularized fields is that in the regularized case we have also to consider the contractions of the form $\omega_\varepsilon(t)\varphi_\varepsilon(s)$ and $\psi_\varepsilon(t)\bar{\psi}_\varepsilon(s)$ when $s < t$ and $|s - t| < \varepsilon$. Since the contributions of this kind of terms are proportional to the support of the mollifier ρ_ε, they go to zero as $\varepsilon \to 0$. Let

$$Y^M(t_1, \ldots, t_k) := \prod_{j=1}^{k} \varphi(t_j)^{m_j}$$

We have

$$\mathcal{H}_{1,x^n}^{M,G}(t_1, \ldots, t_k) = \left\langle Y^M(t_1, \ldots, t_k) \int_{-\infty}^{t_k} g(\tau + 2\theta\bar{\theta})(\Phi(\tau, \theta, \bar{\theta}))^n \mathrm{d}\tau \mathrm{d}\theta \mathrm{d}\bar{\theta} \right\rangle =$$

$$= \int_{-\infty}^{t_k} g(\tau + 2\theta\bar{\theta}) \langle Y^M(t_1, \ldots, t_k)(\Phi(\tau, \theta, \bar{\theta}))^n \rangle \mathrm{d}\tau \mathrm{d}\theta \mathrm{d}\bar{\theta}.$$

Since Φ and φ are Gaussian fields, by Wick theorem and by Remark 11, we have that $\langle Y^M(t_1, \ldots, t_k)(\Phi(\tau, \theta, \bar{\theta}))^n \rangle$ is supersymmetric in $(\tau, \theta, \bar{\theta})$ when $\tau \leqslant t_k$. Moreover, given that $G = g(t + 2\theta\bar{\theta})$ is a supersymmetric function by Remark 10, we have the thesis.

Suppose now that the lemma holds for $\ell - 1 \in \mathbb{N}$, then letting

$$H(\tau_1) := \int_{-\infty}^{\tau_1} \cdots \int_{-\infty}^{\tau_\ell} \prod_{i=2}^{\ell} g(\tau_i + 2\theta_i \bar{\theta}_i)(\Phi(\tau_i, \theta_i, \bar{\theta}_i))^n \mathrm{d}\tau_i \mathrm{d}\theta_i \mathrm{d}\bar{\theta}_i,$$

we have

$$\mathcal{H}^{M,g}_{\ell,x^n}(t_1,\ldots,t_k) =$$

$$= \left\langle Y^M(t_1,\ldots,t_k) \int_{-\infty}^{t_k} \int_{-\infty}^{\tau_1} \cdots \int_{-\infty}^{\tau_\ell} \prod_{i=1}^{\ell} g(\tau_i + 2\theta_i\bar{\theta}_i)(\Phi(\tau_i,\theta_i,\bar{\theta}_i))^n \, \mathrm{d}\tau_i \mathrm{d}\theta_i \mathrm{d}\bar{\theta}_i \right\rangle$$

$$= \int_{-\infty}^{t_k} g(\tau_1 + 2\theta_1\bar{\theta}_1) \langle Y^M(t_1,\ldots,t_k) \Phi(\tau_1,\theta_1,\bar{\theta}_1)^n H(\tau_1)\rangle \mathrm{d}\tau_1 \mathrm{d}\theta_1 \mathrm{d}\bar{\theta}_1$$

$$= \int_{-\infty}^{t_k} g'(\tau_1) \mathcal{H}^{(M,n),g}_{\ell-1,x^n}(t_1,\ldots,t_k,\tau_1) \mathrm{d}\tau_1 - n \int_{-\infty}^{t_k} \langle Y^M(t_1,\ldots,t_k) \varphi(\tau_1)^{n-1} \omega(\tau_1) H(\tau_1)\rangle \cdot$$

$$\cdot g(\tau_1) \mathrm{d}\tau_1 - n(n-1) \int_{-\infty}^{t_k} \langle Y^M(t_1,\ldots,t_k) \varphi(\tau_1)^{n-2} \psi(\tau_1) \bar{\psi}(\tau_1) H(\tau_1)\rangle g(\tau_1) \mathrm{d}\tau_1.$$

Here $(M,n) = (m_1,\ldots,m_k,n)$. By the induction hypothesis the first term in the sum is exactly

$$\int_{-\infty}^{t_k} g'(\tau_1) \mathcal{H}^{(M,n),g}_{\ell-1,x^n}(t_1,\ldots,t_k,\tau_1) \mathrm{d}\tau_1 = \int_{-\infty}^{t_k} g'(\tau_1) \frac{(2g(\tau_1))^{\ell-1}}{(\ell-1)!} \langle \varphi(\tau_1)^{\ell n} Y^M(t_1,\ldots,t_k)\rangle \mathrm{d}\tau_1.$$

For the second term we note that

$$\left\langle \varphi(\tau_1)^{n-1}\omega(\tau_1) Y^M(t_1,\ldots,t_k) \int_{-\infty}^{\tau_1} \cdots \int_{-\infty}^{\tau_\ell} \prod_{i=2}^{\ell} g(\tau_i + 2\theta_i\bar{\theta}_i)(\Phi(\tau_i,\theta_i,\bar{\theta}_i))^n \, \mathrm{d}\tau_i \mathrm{d}\theta_i \mathrm{d}\bar{\theta}_i \right\rangle =$$

$$= \sum_{j=1}^{k} m_j \langle \omega(\tau_1)\varphi(t_j)\rangle \mathcal{H}^{(M-1_j,n-1),g}_{\ell-1,x^n}(t_1,\ldots,t_k,\tau_1) + (n-1)\langle \varphi(\tau_1)\omega(\tau_1)\rangle \mathcal{H}^{(M,n-2),g}_{\ell-1,x^n}(t_1,\ldots,t_k,\tau_1)$$

where $1_j = (0,\ldots,1,0,\ldots,0) \in \mathbb{N}^k$ with 1 in the j-th position and where we used Wick's theorem and the fact that

$$\langle \varphi(\tau_1)\omega(\tau_1)\rangle = \frac{1}{2} \quad \text{and} \quad \left\langle \omega(\tau_1) \int_{-\infty}^{\tau_1} \cdots \int_{-\infty}^{\tau_\ell} \prod_{i=2}^{\ell} g(\tau_i + 2\theta_i\bar{\theta}_i)(\Phi(\tau_i,\theta_i,\bar{\theta}_i))^n \, \mathrm{d}\tau_i \mathrm{d}\theta_i \mathrm{d}\bar{\theta}_i \right\rangle = 0.$$

Furthermore for the third term we have

$$\left\langle \varphi(\tau_1)^{n-2}\psi(\tau_1)\bar{\psi}(\tau_1) \prod_{j=1}^{k} \varphi(t_j)^{m_j} \int_{-\infty}^{\tau_1} \cdots \int_{-\infty}^{\tau_\ell} \prod_{i=2}^{\ell} g(\tau_i + 2\theta_i\bar{\theta}_i)(\Phi(\tau_i,\theta_i,\bar{\theta}_i))^n \, \mathrm{d}\tau_i \mathrm{d}\theta_i \mathrm{d}\bar{\theta}_i \right\rangle =$$

$$= \langle \psi(\tau_1)\bar{\psi}(\tau_1)\rangle \mathcal{H}^{(M,n-2),g}_{\ell-1,x^n}(t_1,\ldots,t_k,\tau_1).$$

In this way we obtain that

$$\mathcal{H}^{M,g}_{\ell,x^n}(t_1,\ldots,t_k) = (-1)^{\ell-1} 2^{\ell-1} \int_{-\infty}^{t_k} g'(\tau_1) \frac{(g(\tau_1))^{\ell-1}}{(\ell-1)!} \langle \varphi(\tau_1)^{\ell n} Y^M(t_1,\ldots,t_k)\rangle \mathrm{d}\tau_1 +$$

$$- \sum_{j=1}^{k} m_j \langle \omega(\tau_1)\varphi(t_j)\rangle \cdot \mathcal{H}^{(M-1_j,n-1),g}_{\ell-1,x^n}(t_1,\ldots,t_k,\tau_1).$$

Here we use the fact that $\langle\varphi(\tau_1)w(\tau_1)\rangle = -\langle\psi(\tau_1)\bar\psi(\tau_1)\rangle = \frac{1}{2}$. Noting that

$$\langle\varphi^{\ell n-2}(\tau)\psi(\tau)\bar\psi(\tau)Y^M(t_1,\ldots,t_k)\rangle + \langle\varphi(\tau)w(\tau)\rangle\langle\varphi^{\ell n-2}(\tau)Y^M(t_1,\ldots,t_k)\rangle = 0$$

we obtain

$$\mathcal{H}^{M,g}_{\ell,x^n}(t_1,\ldots,t_k) = (-2)^{\ell-1}\left\langle Y^M(t_1,\ldots,t_k)\int_{-\infty}^{t_k}\frac{(g(\tau+2\theta\bar\theta))^\ell}{\ell!}\Phi^{n\ell}(\tau,\theta,\bar\theta)d\tau d\theta d\bar\theta\right\rangle =$$

$$= \frac{(-2)^{\ell-1}}{\ell!}\mathcal{H}^{M,g^\ell}_{1,x^{n\ell}}(t_1,\ldots,t_k)$$

Finally, the thesis follows from the induction hypothesis for $\mathcal{H}^{M,g^\ell}_{1,x^{n\ell}}(t_1,\ldots,t_k)$. □

Corollary 13 *Let G be a supersymmetric function with compact support, then we have*

$$\left\langle\varphi(0)^m\left(\int_{-\infty}^0 G(t,\theta,\bar\theta)P(\Phi(t,\theta,\bar\theta))dt d\theta d\bar\theta\right)^k\right\rangle = (-2G_\emptyset(0))^k\langle\varphi(0)^m P(\varphi(0))^k\rangle. \tag{14}$$

Proof Using the symmetry of the l.h.s. of (14) with respect to the exchanges $(\tau_i,\theta_i,\bar\theta_i) \longleftrightarrow (\tau_j,\theta_j,\bar\theta_j)$ we have that

$$\left\langle\varphi(0)^m\left(\int_{-\infty}^0 G(t,\theta,\bar\theta)P(\Phi(\tau,\theta,\bar\theta))dt d\theta d\bar\theta\right)^k\right\rangle =$$

$$= k!\left\langle\varphi(0)^m\int_{-\infty}^0\int_{-\infty}^{\tau_1}\cdots\int_{-\infty}^{\tau_{k-1}}\prod_{i=1}^k G(\tau_i,\theta_i,\bar\theta_i)P(\Phi(\tau_i,\theta_i,\bar\theta_i))d\tau_i d\theta_i d\bar\theta_i\right\rangle.$$

Then the claim follows directly from Lemma 12 taking $g = G_\emptyset$. □

Theorem 14 *Let F be a smooth bounded function, let G be a supersymmetric function with compact support, let H be a bounded function with all the derivatives bounded and let $\mathbb{K}$ be an entire function, then we have*

$$\left\langle F(\varphi(0))\mathbb{K}\left(\int_{-\infty}^0 G(t,\theta,\bar\theta)H(\Phi(t,\theta,\bar\theta))dt d\theta d\bar\theta\right)\right\rangle = \langle F(\varphi(0))\cdot\mathbb{K}(-2G_\emptyset(0)\cdot\varphi(0))\rangle.$$

Proof Using the density of polynomials in the set of smooth functions with respect to the topology given by the one of the Sobolev space with respect to the Gaussian law of $\varphi(t)$, Corollary 13 implies that for any $k\in\mathbb{N}$ and F,G,H satisfying the hypothesis of the theorem

$$\left\langle F(\varphi(0))\left(\int_{-\infty}^0 G(t,\theta,\bar\theta)H(\Phi(t,\theta,\bar\theta))dt d\theta d\bar\theta\right)^k\right\rangle = \langle F(\varphi(0))[-2G_\emptyset(0)H(\varphi(0))]^k\rangle.$$

Expanding $\mathbb{K}$ in power series, exploiting the fact that

$$\left\langle \left| \left(\left(\int G(t,\theta,\bar{\theta}) H(\Phi(t,\theta,\bar{\theta})) \mathrm{d}t \mathrm{d}\theta \mathrm{d}\bar{\theta} \right)^k \right\rangle_{\psi,\bar{\psi}} \right|^p \right\rangle$$

is uniformly bounded when H is bounded, for any $p \geq 1$, we can exchange the series with the expectation $\langle \cdot \rangle$, and obtain in this way the thesis. $\square$

Acknowledgements The authors are funded by the DFG under Germany's Excellence Strategy - GZ 2047/1, Project-ID 390685813. The second author is supported by DFG via CRC 1060.

References

1. Albeverio, S., Borasi, L., De Vecchi, F.C., Gubinelli, M.: Grassmannian stochastic analysis and the stochastic quantization of Euclidean Fermions (2020). arXiv:2004.09637
2. Albeverio, De Vecchi, F.C., Gubinelli, M.: The elliptic stochastic quantization of some two dimensional Euclidean QFTs. Ann. Inst. H. Poincaré Probab. Statist. **57**(4), 2372–2414 (2021)
3. Albeverio, S., De Vecchi, F.C., Gubinelli, M.: Elliptic stochastic quantization. Ann. Probab. **48**(4), 1693–1741 (2020)
4. Berezin, F.A.: Introduction to superanalysis, vol. 9 of *Mathematical Physics and Applied Mathematics*. D. Reidel Publishing Co., Dordrecht, 1987. Edited and with a foreword by A. A. Kirillov, With an appendix by V. I. Ogievetsky, Translated from the Russian by J. Niederle and R. Kotecký, Translation edited by Dimitri Leĭtes
5. Brydges, D., Imbrie, J.: Branched polymers and dimensional reduction. Ann. Math. **158**(3), 1019–1039 (2003)
6. Brydges, D.C., Imbrie, J.Z.: Dimensional reduction formulas for branched polymer correlation functions. J. Stat. Phys. **110**(3), 503–518 (2003)
7. Damgaard, P.H., Hüffel, H.: Stochastic Quantization. World Scientific (1988)
8. De Angelis, G.F., Jona-Lasinio, G., Sidoravicius, V.: Berezin integrals and Poisson processes. J. Phys. A **31**(1), 289–308 (1998)
9. Fetter, A.L., Walecka, J.D.: Quantum Theory of Many-Particle Systems. Dover Books on Physics, Dover Publications (2012)
10. Gozzi, E.: Dimensional reduction in parabolic stochastic equations. Phys. Lett. B **143**(1–3), 183–187 (1984)
11. Gubinelli, M., Hofmanová, M.: Global solutions to elliptic and parabolic Φ^4 models in Euclidean space. Comm. Math. Phys. **368**(3), 1201–1266 (2019)
12. Helmuth, T.: Dimensional reduction for generalized continuum polymers. J. Stat. Phys. **165**(1), 24–43 (2016)
13. Ikeda, N., Watanabe, S.: Stochastic differential equations and diffusion processes, vol. 24 of North-Holland Mathematical Library. North-Holland Publishing Co., Amsterdam; Kodansha, Ltd., Tokyo, 2nd edn (1989)
14. Klein, A., Landau, L.J., Perez, J.F.: Supersymmetry and the Parisi-Sourlas dimensional reduction: a rigorous proof. Comm. Math. Phys. **94**(4), 459–482 (1984)
15. Parisi, G., Sourlas, N.: Random magnetic fields, supersymmetry, and negative dimensions. Phys. Rev. Lett. **43**(11), 744–745 (1979)
16. Parisi, G., Sourlas, N.: Supersymmetric field theories and stochastic differential equations. Nucl. Phys. B **206**(2), 321–332 (1982)
17. Schwarz, A., Zaboronsky, O.: Supersymmetry and localization. Comm. Math. Phys. **183**(2), 463–476 (1997)

18. Stroock, D.W., Varadhan, S.R.: On the support of diffusion processes with applications to the strong maximum principle. In: Proceedings of the Sixth Berkeley Symposium on Mathematical Statistics and Probability (Univ. California, Berkeley, Calif., 1970/1971), Vol. III: Probability Theory, pp. 333–359 (1972)
19. Wong, E., Zakai, M.: On the convergence of ordinary integrals to stochastic integrals. Ann. Math. Stat. **36**, 1560–1564 (1965)
20. Wong, E., Zakai, M.: On the relation between ordinary and stochastic differential equations. Int. J. Eng. Sci. **3**, 213–229 (1965)
21. Zinn-Justin, J.: Quantum field theory and critical phenomena, vol. 85 of International Series of Monographs on Physics. The Clarendon Press, Oxford University Press, New York, 2nd edn. Oxford Science Publications (1993)

Quasi-shuffle Algebras in Non-commutative Stochastic Calculus

Kurusch Ebrahimi-Fard and Frédéric Patras

Abstract This chapter is divided into two parts. The first is largely expository and builds on Karandikar's axiomatisation of Itô calculus for matrix-valued semimartingales. Its aim is to unfold in detail the algebraic structures implied for iterated Itô and Stratonovich integrals. These constructions generalise the classical rules of Chen calculus for deterministic scalar-valued iterated integrals. The second part develops the stochastic analog of what is commonly called chronological calculus in control theory. We obtain in particular a pre-Lie Magnus formula for the logarithm of the Itô stochastic exponential of matrix-valued semimartingales.

Keywords Quasi-shuffle · Itô integral · Stochastic exponential · Magnus formula · Stratonovich integral

1 Introduction

Algebra, renormalisation theory as well as numerical analysis are among a range of disparate fields that have seen a surge in interest for the study of various algebraic and combinatorial structures originating from the integration by parts formula, foundational to integral calculus. In particular, the theories of Rota–Baxter algebras, shuffle and quasi-shuffle products, pre- and post-Lie algebras as well as combinatorial bialgebras on rooted trees and words have undergone expansive phases in the last two decades. The following rather incomplete list of references provides some examples

---

K. Ebrahimi-Fard
Department of Mathematical Sciences, Norwegian University of Science and Technology – NTNU, 7491 Trondheim, Norway
e-mail: kurusch.ebrahimi-fard@ntnu.no
URL: https://folk.ntnu.no/kurusche/

F. Patras (✉)
Labo. J.-A. Dieudonné, UMR 7351, CNRS, Université Côte d'Azur,
06108 Nice Cedex 02, Parc Valrose, France
e-mail: patras@unice.fr
URL: https://www-math.unice.fr/~patras

© Springer Nature Switzerland AG 2021
S. Ugolini et al. (eds.), *Geometry and Invariance in Stochastic Dynamics*,
Springer Proceedings in Mathematics & Statistics 378,
https://doi.org/10.1007/978-3-030-87432-2_6

of these developments [11, 15, 17, 18, 21, 36, 38, 57, 63, 64, 66, 67, 70]. In the particular context of stochastic integration, interest in these structures concentrated largely in Lyons' seminal theory of rough paths [58, 59], which is based on Chen's iterated path integrals and shuffle algebra on words [22, 23, 72]. Gubinelli expanded Lyons' theory by generalising it to a certain combinatorial Hopf algebra of rooted trees. The resulting notion of branched rough paths [41, 42, 46] draws inspiration from Butcher's theory of B-series in numerical integration of differential equations [43, 60] as well as Connes and Kreimer's Hopf algebraic approach to renormalisation in perturbative quantum field theory [24]. The latter, moreover emphasised the pre-Lie algebraic perspective on rooted trees [20, 62]. Ideas from rough paths gave rise to various new developments, culminating in Hairer's celebrated theory of regularity structures [44, 45] and its algebraic renormalisation theory [12] used in the construction of solutions of very irregular S(P)DEs.

The relevance of such algebraic structures for classical—non-commutative—stochastic integration (in the sense of Itô–Stratonovich) [6, 71] has attracted less attention. Foundational papers in this field are Gaines's 1994 work on the algebra of iterated stochastic integrals [40], introducing what is now called quasi-shuffle product, as well as the equivalent sticky shuffle product formula for iterated quantum Itô integrals introduced in 1995 in the context of quantum stochastic calculus [8]. We refer to Hudson's review papers on Hopf-algebraic aspects of iterated stochastic integrals [49, 50]. The authors together with Charles Curry, Alexander Lundervold, Simon Malham, Hans Munthe-Kaas and Anke Wiese further developed the use of (quasi-)shuffle algebra in stochastic integration theory and numerical methods for SDEs in several joint works [25–27, 30, 33, 34].

The present article is divided into two main parts. The first part starts by recalling Karandikar's axiomatisation of Itô calculus for matrix-valued continuous semimartingales [53]. We discuss in detail the algebraic structures implied for iterated Itô and Stratonovich integrals. To the best of our knowledge, such an account does not seem to exist in the literature. In fact, Karandikar's ideas do not seem to be widely known, as the algebraic notions and techniques involved are not of common use in non-commutative stochastic calculus. Our presentation is written with a view toward operationality and therefore, as far as possible, in the language of theoretical probability theory. In modern algebraic terminology, Karandikar's axioms define the notion of non-commutative quasi-shuffle algebra. We proceed by building consistently on this structure aiming at unraveling properties of integration techniques, ranging from general matrix-valued semimartingales to more specific situations (continuous paths). We remark that introducing the Stratonovich integral in full generality requires extra axioms corresponding to the splitting of the quadratic covariation bracket into a continuous and a jump part. We hope that these ideas might be useful also in other settings.

The second part develops the stochastic analog of what is commonly called chronological calculus in control theory [1, 4, 5]. We feature in particular Agrachev and Gamkrelidze's [2, 3] study and systematical use of chronological algebra in control problems. In mathematics, chronological algebras are known as pre-Lie or

Vinberg algebras [13, 17]. This part of our work is a continuation of joint work with Charles Curry [26], where a pre-Lie Magnus formula for the logarithm of the Stratonovich stochastic exponential for continuous matrix-valued semimartingales was introduced.

A more detailed outline of the paper follows. The first section is divided into three subsections. We begin by focusing on general Itô calculus for matrix-valued semimartingales, introduce Karandikar's axioms, and point out connections with other theories, especially Rota–Baxter algebras. The second subsection introduces formally the splitting of the covariation bracket, leading to a tentative axiomatisation of Stratonovich calculus. The last subsection studies stochastic calculus for continuous semimartingales, a situation where the axioms simplify dramatically, allowing to replace quasi-shuffles by shuffles, that is, permitting the use the standard rules of calculus.

In the second section we study pre-Lie algebraic aspects in stochastic integral calculus. We consider the pre-Lie Magnus formula in the general context of enveloping algebras of pre-Lie algebras—the corresponding section can be understood also as an introduction to pre-Lie structures since we survey some of their most relevant properties for general integral calculus, following the chapter [35] written in the context of classical integration. The second and last subsection of this section changes focus by developing instead a pre-Lie point of view on Itô integral calculus, without restriction to continuous matrix-valued semimartingales. We obtain in particular a pre-Lie Magnus formula for the logarithm of the Itô stochastic exponential of matrix-valued semimartingales.

Conventions (i) With the aim of simplifying the presentation we shall always assume that the value of semimartingales is zero at $t = 0$. (ii) All structures are defined over a ground field k of characteristic zero.

2 Karandikar's Axioms and Quasi-shuffle Algebras

2.1 Itô Calculus for Semimartingales

This section discusses the formal properties of the integral calculus for semimartingales. Protter's textbook [71] will serve as the standard reference on stochastic integration. The central aim is to feature Karandikar's axioms for matrix-valued Itô integrals and the corresponding notion of non-commutative quasi-shuffle algebra.[1]

Recall first Itô's integration by parts formula for scalar semimartingales X, Y [71, chapter II.6]

$$X_t Y_t = \int_0^t X_{s^-} dY_s + \int_0^t Y_{s^-} dX_s + [X, Y]_t. \tag{1}$$

---

[1] Karandikar's axioms appeared in a 1982 paper [53]. They were (re-)discovered almost two decades later in a completely different context –the one of Stasheff polytopes– as axioms for dendriform trialgebras [56].

Here as well as in the rest of the paper, our conventions are in place, that is, we assume that $X_0 = Y_0 = 0$. Equation (1) defines the so-called quadratic covariation bracket, $[X, Y]_t$, the extra term that distinguishes Itô's integration by parts formula from the classical one.

To deal with matrix-valued semimartingales, one has to take into account the non-commutativity of matrix multiplication. The product in the second term on the righthand side of (1) is then in the wrong order. Following Protter (and Karandikar), we introduce left and right stochastic integrals

$$(X \succ Y)_t := \int_0^t X_{s^-} dY_s \qquad (X \prec Y)_t := \int_0^t dX_s Y_{s^-}. \tag{2}$$

Itô's integration by parts formula for matrix-valued semimartingales X, Y writes then [71, Chap. V.8, Theorem 47]

$$X_t Y_t = (X \succ Y)_t + (X \prec Y)_t + [X, Y]_t, \tag{3}$$

Notice that in the case of scalar-valued semimartingales we have that $X \prec Y = Y \succ X$ such that (3) can be changed back to (1).

Remark 1 Protter and Karandikar use a different notation: $(X \succ Y)_t = (X \cdot Y)_t$ and $(X \prec Y)_t = (X : Y)_t$. Our notation is in line with the one used in algebra. It is also convenient to identify the time-ordering of operations, see [35].

Hereafter, we account for various properties of (left and right) stochastic integrals of scalar-valued semimartingales. Even though some of the ternary formulas considered may seem redundant in the scalar case, we emphasise that they do not involve permutations of the variables. They thus hold immediately for $(n \times n)$ matrix-valued semimartingales.

Recall first [71, Chap. II.6, Theorem 29] that for H, K adapted processes with caglad (left continuous with right limits) paths and X, Y two semimartingales we have $[\int_0^t H_s dX_s, \int_0^t K_s dY_s]_t = \int_0^t H_s K_s d[X, Y]_s$.

Assuming now that L and M are semimartingales, we get $[L \succ X, M \succ Y] = (LM) \succ [X, Y]$ which simplifies to the ternary relation

$$[L \succ X, Y] = L \succ [X, Y].$$

Similarly, $[X \prec Y, Z] = [X, Y \succ Z]$ and $[X, Y] \prec Z = [X, Y \prec Z]$.

Other ternary relations satisfied by semimartingales follow directly from standard properties and from the definitions:

$$(X \succ (Y \succ Z))_t = \int_0^t X_{s^-} d(\int_0^s Y_{s^-} dZ_s) = \int_0^t X_{s^-} Y_{s^-} dZ_s = ((XY) \succ Z)_t.$$

Similarly, for G, H caglad and Y a semimartingale, $\int_0^t G_s d(\int_0^s H_s dY_s) = \int_0^t G_s H_s dY_s$ [71, Thm. II.19]. Dually, $(X \prec Y) \prec Z = X \prec (YZ)$. Finally,

$$((X \succ Y) \prec Z)_t = \int_0^t (d\int_0^s X_u \text{-} dY_u) Z_{s^-} = \int_0^t X_{s^-} dY_s Z_{s^-} = (X \succ (Y \prec Z))_t.$$
(4)

Notice that the associativity of the quadratic covariation bracket, $[X, [Y, Z]] = [[X, Y], Z]$, can be deduced from the associativity of the usual product of semimartingales together with the previous identities:

$$[[X, Y], Z] = [XY - X \prec Y - X \succ Y, Z]$$
$$= (XY)Z - (XY) \prec Z - (XY) \succ Z - [X \prec Y + X \succ Y, Z]$$
$$= X(YZ) - X \prec (YZ) - X \succ (Y \prec Z) - X \succ (Y \succ Z)$$
$$\quad - X \succ [Y, Z] - [X, Y \prec Z] - [X, Y \succ Z]$$
$$= [X, YZ] - [X, Y \prec Z] - [X, Y \succ Z] = [X, [Y, Z]].$$

Note that, as mentioned before, X, Y and Z appear always in the same order in the above formulas. This implies that they hold in the matrix-valued case. In that case, $[X, Y]$ is defined in terms of the component-wise quadratic covariation brackets, $[X, Y]_{ik} = \sum_{j=1}^n [X_{ij}, Y_{jk}]$, and similarly for the other products. Putting this together, yields the axiomatisation of Itô calculus for semimartingales, due to Karandikar.

Theorem 1 (Karandikar [53]) *The left- and right stochastic Itô integrals satisfy Karandikar's identities for matrix-valued semimartingales X, Y, Z*

$$(X \prec Y) \prec Z = X \prec (YZ) \tag{5}$$
$$X \succ (Y \succ Z) = (XY) \succ Z \tag{6}$$
$$(X \succ Y) \prec Z = X \succ (Y \prec Z) \tag{7}$$
$$(XY)Z = X(YZ) \tag{8}$$
$$[X \succ Y, Z] = X \succ [Y, Z] \tag{9}$$
$$[X \prec Y, Z] = [X, Y \succ Z] \tag{10}$$
$$[X, Y] \prec Z = [X, Y \prec Z]. \tag{11}$$

Remark 2 Karandikar considered in [53] the continuous case. Axiom 10 is stated in a slightly different way—namely in the case where Y is non-singular [53, Eq. (9) p. 1089]. However, these restrictions (to the continuous case and to non-singular Y) are not necessary, see Karandikar [55].

Definition 1 An associative algebra A equipped with three products $\prec$, $\succ$ and $[\,,\,]$, called respectively the left half-shuffle, the right half-shuffle and the bracket, such that $XY = X \prec Y + X \succ Y + [X, Y]$ and satisfying Karandikar's identities in Theorem 1 is called a quasi-shuffle algebra.

See Remark 4 below for an explanation of the terminology. Karandikar's axiomatisation of Itô calculus therefore says that the algebra of semimartingales is a non-commutative quasi-shuffle algebra. There are many examples of quasi-shuffle algebras besides the one coming from stochastic calculus and Karandikar's results apply immediately to them. Conversely, general results from abstract quasi-shuffle algebra apply to stochastic calculus. Examples of application of this strategy can be found in our joint works with Simon Malham and Anke Wiese [33, 34].

Remark 3 We have seen that the associativity of the bracket operation [,]

$$[X, [Y, Z]] = [[X, Y], Z] \tag{12}$$

follows from the associativity of product in the algebra A. One can define equivalently a quasi-shuffle algebra to be a vector space A equipped with three products $\prec, \succ$ and [,] satisfying Eqs. (5)–(11) and (8) replaced by (12), where one sets $XY = X \prec Y + X \succ Y + [X, Y]$. The associativity of the product XY results then automatically from these axioms (the proof parallels the proof showing that the quadratic covariation bracket is associative for semimartingales):

$$(XY)Z = (XY) \prec Z + (XY) \succ Z + [X \prec Y + X \succ Y + [X, Y], Z]$$
$$= X \prec (YZ) + X \succ (Y \prec Z) + X \succ (Y \succ Z)$$
$$\quad + [X, Y \prec Z] + [X, Y \succ Z] + X \succ [Y, Z] + [X, [Y, Z]]$$
$$= X \prec (YZ) + X \succ (Y \prec Z) + X \succ (Y \succ Z) + X \succ [Y, Z] + [X, YZ] = X(YZ).$$

Remark 4 The terminology "quasi-shuffle" algebra is used in algebra and combinatorics. It reflects the close similarity with the classical shuffle algebra [73]. The work [38] explores the relation between the two families of algebras from a deformation theoretical viewpoint. Hoffman [47], independently of Karandikar's seminal work, largely initiated the development of the Hopf algebraic theory of commutative quasi-shuffle algebra. However, as we mentioned in the introduction, Gaines [40] as well as Hudson et al. [8, 49, 50] introduced quasi-shuffle products to study properties of products of iterated Itô integrals. We note that Cartier, back in 1972 [17], used a quasi-shuffle product in the construction of free (Rota-)Baxter algebra.

When dealing with scalar-valued semimartingales, we already noticed that $X \prec Y = Y \succ X$. Correspondingly, Karandikar's axioms simplify, leading to the notion of commutative quasi-shuffle algebra studied in detail by Hoffman, see [37] for a recent account. Some extra properties are then available such as Hoffman's isomorphism linking shuffle and quasi-shuffle products. Features of the commutative theory have been exploited recently in stochastic calculus, for example in [25, 34, 39].

From a purely algebraic viewpoint, the fundamental example of a quasi-shuffle algebra is the linear span of words $X = x_1 \cdots x_n$ where the x_i belong to a monoid M with (not necessarily commutative) product denoted $\times$. The axioms for the products, $\prec, \succ$, and [,], used to define inductively the associative product $X * Y := X \prec Y + X \succ Y + [X, Y]$ of words X, Y, are given by:

$$x_1 \cdots x_n \prec y_1 \cdots y_m := x_1(x_2 \cdots x_n * y_1 \cdots y_m)$$

$$x_1 \cdots x_n \succ y_1 \cdots y_m := y_1(x_1 \cdots x_n \ast y_2 \cdots y_m)$$
$$[x_1 \cdots x_n, y_1 \cdots y_m] := (x_1 \times y_1)(x_2 \cdots x_n \ast y_2 \cdots y_m).$$

For example, taking M to be the monoid of the integers, we have

$$2\,3 \prec 1 = 2(3 \ast 1) = 2(3 \prec 1 + 3 \succ 1 + [3, 1]) = 2\,3\,1 + 2\,1\,3 + 2\,4.$$

Remark 5 In the following, quasi-shuffle algebra shall mean non-commutative quasi-shuffle algebra. The latter are also called tridendriform algebras in the literature. However, the quasi-shuffle terminology, besides being close to other ones that have been used in stochastics (modified shuffle product, sticky shuffle product, ...) has the advantage of underlining the connection to the familiar shuffle calculus for Chen's iterated integrals and the related product of simplices, as well as many more topics (such as quasi-symmetric functions, multizeta values, etc.). We refer to [31, 37, 68, 69] for accounts on the combinatorial theory of quasi-shuffle algebras as well as further bibliographical references and various examples. See [25, 26, 33, 34, 40, 49, 50] and references therein for more details and references on quasi-shuffle calculus in probability.

We introduce now Rota–Baxter algebras, which provide a more general approach to the algebraic axiomatisation of integral calculus [77] and therefore an important class of examples for quasi-shuffle algebras. Indeed, Theorem 2 below shows that any Rota–Baxter algebra is a quasi-shuffle algebra [28]. We refer to the survey [35] for further details and references about Rota–Baxter algebras and their use in integral calculus, probability theory, renormalisation in perturbative quantum field theory and classical integrable systems.

Definition 2 A Rota–Baxter algebra of weight $\theta \in k$ consists of an associative k-algebra A equipped with a linear operator $R: A \to A$ satisfying the Rota–Baxter relation of weight θ:

$$R(x)R(y) = R\big(R(x)y + xR(y) + \theta xy\big) \quad \forall x, y \in A. \tag{13}$$

Note that if R is a Rota–Baxter map of weight θ, then the map $R' := \beta R$ for $\beta \in k$ different from zero is of weight $\beta\theta$. This permits to rescale the original weight $\theta \neq 0$ to the standard weight $\theta' = +1$ (or $\theta' = -1$). The argument of the map R on the righthand side of (13) consists of a sum of three terms; one can show that it defines a new associative product on A.

Definition 3 (*Rota–Baxter product*) The Rota–Baxter associative product is defined by

$$x \ast_\theta y := R(x)y + xR(y) + \theta xy. \tag{14}$$

The Rota–Baxter relation originated in the work of the mathematician Glen Baxter [7]. Rota [74, 76] followed by Cartier [16] made important contributions to the algebraic foundations of Baxter's work, among others, by providing different constructions of free commutative objects. The idea of quasi-shuffle product is actually often traced back to Cartier's 1972 article.

Theorem 2 ([28]) *Assume now that $\theta = 1$. Writing $a \cdot b := ab$ for the usual associative product on A and setting $a \prec b := aR(b)$ and $a \succ b := R(a)b$, so that $\ast := \prec + \succ + \cdot$, then the quasi-shuffle algebra identities hold:*

$$(a \prec b) \prec c = a \prec (b \ast c), \quad (a \succ b) \cdot c = a \succ (b \cdot c)$$
$$a \succ (b \succ c) = (a \ast b) \succ c, \quad (a \prec b) \cdot c = a \cdot (b \succ c) \quad (15)$$
$$(a \succ b) \prec c = a \succ (b \prec c), \quad (a \cdot b) \prec c = a \cdot (b \prec c).$$

Remark 6 Without the normalisation to the standard weight, one obtains an example of a θ-quasi-shuffle algebra, studied in greater detail in [14]. See also [48].

Example 1 (*Fluctuation theory*) Baxter's work was motivated by problems in the theory of fluctuations [75]. The latter deals with extrema of sequences of real valued random variables. Their distribution can be studied using operators on random variables such as $X \to X^+ := \max(0, X)$. This motivates to define the operator

$$R(F)(t) := \mathbf{E}[\exp(itX^+)]$$

on characteristic functions $F(t) := \mathbf{E}[\exp(itX)]$ of real valued random variables X, which is a Rota–Baxter map of weight $\theta = 1$.

Example 2 (*Finite summation operators*) On functions f defined on $\mathbb{N}$ and with values in an associative algebra A, the summation operator $R(f)(n) := \sum_{k=0}^{n-1} f(k)$ is a Rota–Baxter map of weight one. It is the right inverse of the finite difference operator $\Delta(f)(n) := f(n+1) - f(n)$.

Remark 7 (*Shuffle algebras in classical calculus*) Before concluding this section, we apply the previous ideas to the case of deterministic matrix-valued semimartingales. Even in that seemingly simple case the relations put forward by Karandikar prove to be interesting and useful.

We consider for example the algebra A of matrices whose entries are continuous functions of finite variation. Then, since the quadratic covariation bracket vanishes, Karandikar's identities reduce to an algebra equipped with an associative product $XY = X \prec Y + X \succ Y$ and

$$(X \prec Y) \prec Z = X \prec (YZ)$$
$$X \succ (Y \succ Z) = (XY) \succ Z \quad (16)$$
$$(X \succ Y) \prec Z = X \succ (Y \prec Z).$$

Our previous arguments show that the associativity of the product XY can be recovered formally from these identities. These relations have been used first by Eilenberg and MacLane to give an abstract proof of the associativity of the shuffle product of simplices in topology.

Definition 4 The three identities (16) define the structure of (non-commutative) shuffle algebra (aka dendriform algebra).

From Theorem 2 is clear that on a Rota–Baxter algebra of weight $\theta = 0$ one can define left and right half-shuffle products satisfying the three identities (16). We refer to [35] for a survey and more details and applications in classical integral calculus as well as general references on the subject. We will come back to these relations later as they encode the algebra structure underlying Stratonovich calculus for semimartingales.

2.2 Singular Quasi-shuffle Algebras and Stratonovich Calculus

In this section we extend Karandikar's axiomatisation beyond the setting of continuous semimartingales. Namely, we include in the algebraic description of Itô calculus the decomposition into continuous and jump parts of the quadratic covariation bracket [71, Chap. II.6].

We write $\Delta(X)$ for the process of jumps of a semimartingale X, i.e., $\Delta(X)_s = (X - X_-)_s$, and introduce the corresponding decomposition of the bracket into continuous and jump parts, $[X, Y] = [X, Y]^c + [X, Y]^j$. The definition extends from the scalar-valued to the continuous matrix-valued case components-wise. For scalar-valued processes, $[X, X]_t^j = \sum_{0 \leq s \leq t} (\Delta(X)_s)^2$, a term that appears frequently in stochastic calculus, for example, in the study of the stochastic or Doléans-Dade exponential [71, Chap. II.8, Theorem 37]. A semimartingale X is called quadratic pure jump if $[X, X] = [X, X]^j$.

Recall first a fundamental property of Δ acting on scalar-valued processes. Since the bracket of two semimartingales has paths of finite variation on compact sets [71, Chap. II, Cor.1], it is a quadratic pure jump semimartingale, that is, $[[X, Y], [X, Y]] = [[X, Y], [X, Y]]^j$ by [71, Chap. II.6, Theorem 26]. This implies by [71, chap. II.6. Theorem 28] that for arbitrary semimartingales X, Y, Z we have $[[X, Y], Z] = \sum_{0 \leq s \leq t} \Delta([X, Y])_s \Delta(Z)_s$. In particular,

$$[[X, Y]^c, Z] = [[X, Y]^c, Z]^c = [[X, Y]^c, Z]^j = 0,$$
$$[[X, Y], Z]^c = [[X, Y]^j, Z]^c = 0,$$
$$[[X, Y], Z] = [[X, Y]^j, Z] = [[X, Y]^j, Z]^j.$$

As a corollary, we notice for further use that for continuous semimartingales $[[X, Y], Z] = 0$. These identities hold for matrix-valued semimartingales (since the splitting of processes into a continuous and a pure jump part is linear—it commutes with taking linear combinations of brackets).

A full axiomatisation of Itô calculus taking into account such phenomena would require the introduction of the operator Δ, those identities, and most likely other aspects of standard stochastic calculus. We propose a lighter version that provides an axiomatic framework allowing to relate formally Itô and Stratonovich calculi.

Definition 5 A singular quasi-shuffle algebra is a quasi-shuffle algebra, $(A, \succ, \prec, [-,-])$, such that the associative bracket splits into $[-,-] = [-,-]^c + [-,-]^j$ and furthermore the following relations hold:

$$[[X, Y]^c, Z]^c = [X, [Y, Z]^c]^c = 0 \tag{17}$$

$$[[X, Y]^c, Z]^j = [X, [Y, Z]^c]^j = 0 \tag{18}$$

$$[[X, Y]^j, Z]^c = [X, [Y, Z]^j]^c = 0 \tag{19}$$

Notice that we also have then

$$[X \prec Y, Z]^c = [X, Y \succ Z]^c, \ [X \prec Y, Z]^j = [X, Y \succ Z]^j.$$

Recall that for matrix-valued semimartingales, the (left/right) Fisk–Stratonovich integrals are defined in terms of the Itô integral by

$$(X \succcurlyeq Y)_t := \int_0^t X_s \circ dY_s := \int_0^t X_s dY_s + \frac{1}{2}[X, Y]_t^c \tag{20}$$

$$(Y \preccurlyeq X)_t := \int_0^t \circ dY_s X_s := \int_0^t dY_s X_s + \frac{1}{2}[Y, X]_t^c. \tag{21}$$

Formally, in any singular quasi-shuffle algebra one can define the two productss

$$X \succcurlyeq Y := X \succ Y + \frac{1}{2}[X, Y]^c, \quad X \preccurlyeq Y := X \prec Y + \frac{1}{2}[X, Y]^c.$$

Then the integration by parts rule reads:

$$XY = X \succcurlyeq Y + X \preccurlyeq Y + [X, Y]^j. \tag{22}$$

Unfortunately, it seems to be difficult to find a simpler axiomatic framework than the one of singular quasi-shuffle algebras to account for Stratonovich calculus in the presence of jumps. Indeed, it is likely that a meaningful system of ternary relations involving only $\succcurlyeq$, $\preccurlyeq$ and $[\,,\,]^j$ is unavailable. Fortunately, these issues simplify considerably for continuous semimartingales.

2.3 Shuffle Algebra and Continuous Semimartingales

As we mentioned previously, stochastic integration simplifies dramatically when considering continuous semimartingales. The reason for this should be clear from our previous developments, that is, the jump part, $[-,-]^j$, of the bracket, $[-,-]$, vanishes, so that the latter reduces to its continuous part and becomes nilpotent of order 3: $[X, [Y, Z]] = [[X, Y], Z] = 0$. In that situation, the axioms of Itô calculus rewrite:

Lemma 1 *Continuous matrix-valued semimartingales equipped with the left and right half-shuffles and the covariation bracket obey the axioms (5)–(11) together with*

$$[X, [Y, Z]] = [[X, Y], Z] = 0 \tag{23}$$

Notice that the associativity of the product (axiom (8)) is then formally a consequence of the other axioms. This observation is of little interest when dealing with stochastic integrals for which the associativity of the product is somehow obvious. However, it is relevant with respect to the axiomatic point of view.

Definition 6 A regular quasi-shuffle algebra is a quasi-shuffle algebra such that the bracket satisfies the extra axiom (23).

The continuity hypothesis has more interesting consequences when dealing with Fisk–Stratonovich integrals. We follow closely the exposition in [26]. The Stratonovich formula is indeed then the usual integration by parts formula

$$X_t Y_t = (X \succcurlyeq Y)_t + (X \preccurlyeq Y)_t, \tag{24}$$

The classical statement that the Stratonovich integral for continuous semimartingales obeys the usual laws of calculus translates formally into the

Theorem 3 *For continuous semimartingales X, Y, Z, the left and right Fisk–Stratonovich integrals satisfy the half-shuffle identities*

$$(X \preccurlyeq Y) \preccurlyeq Z = X \preccurlyeq (YZ) \tag{25}$$
$$(X \succcurlyeq Y) \preccurlyeq Z = X \succcurlyeq (Y \preccurlyeq Z) \tag{26}$$
$$X \succcurlyeq (Y \succcurlyeq Z) = (XY) \succcurlyeq Z. \tag{27}$$

In particular, the algebra of continuous matrix-valued semimartingales is a non-commutative shuffle algebra.

Proof

$$(X \preccurlyeq Y) \preccurlyeq Z = \left(X \prec Y + \frac{1}{2}[X, Y]\right) \preccurlyeq Z$$
$$= (X \prec Y) \prec Z + \frac{1}{2}[X, Y] \prec Z + \frac{1}{2}[X \prec Y, Z] + \frac{1}{4}[[X, Y], Z]$$
$$= X \prec (YZ) + \frac{1}{2}[X, Y \prec Z] + \frac{1}{2}[X, Y \succ Z]$$
$$= X \preccurlyeq (YZ).$$

Identity (27) is proved similarly.

$$(X \succcurlyeq Y) \preccurlyeq Z = \left(X \succ Y + \frac{1}{2}[X, Y]\right) \preccurlyeq Z$$
$$= (X \succ Y) \prec Z + \frac{1}{2}[X, Y] \prec Z + \frac{1}{2}[X \succ Y, Z] + \frac{1}{4}[[X, Y], Z]$$
$$= X \succ (Y \prec Z) + \frac{1}{2}[X, Y \prec Z] + \frac{1}{2}X \succ [Y, Z]$$
$$= X \succcurlyeq (Y \preccurlyeq Z).$$

$\square$

In general, the same argument show

Theorem 4 *The map* $(A, \prec, \succ, [\,,\,]) \longmapsto (A, \preccurlyeq, \succcurlyeq)$ *is a functor from the category of regular quasi-shuffle algebras to the category of shuffle algebras.*

3 Chronological Calculus for Stochastic Integration

In this section, the second part of this work, we start by briefly reviewing the classical chronological calculus following Agrachev, Gamkrelidze and collaborators [1–4]. The aim is to show how chronological calculus can be applied in the context of stochastic calculus. The key idea is to use the notion of pre-Lie (or chronological) algebra instead of that of usual Lie algebra, to analyse group- and Lie-theoretical phenomena associated to evolution equations. We refer to [35] where this point of view is developed in more detail.

3.1 Chronological Calculus and Pre-Lie Algebra

Time- or path-ordered products are ubiquitous, especially in theoretical physics and control theory, and form the basis for Agrachev and Gamkrelidze's chronological calculus [1]. These authors understood that the combination of Lie algebra and integration by parts permits to define the useful notion of chronological algebra [2], better known as pre-Lie or Vinberg algebra in algebra and geometry [13, 17, 62]. Concepts from chronological calculus apply in the context of stochastic integration as far as iterated Stratonovich integrals for continuous semimartingales are concerned, because they obey the usual rules of calculus.

However, for Itô and Stratonovich integrals in the non-continuous case, the usual ideas of chronological calculus do not apply immediately, due to the terms arising from the jump component of the covariation bracket. It turns out that in this case, one must appeal to results originating in the study of non-commutative Rota–Baxter algebras. We refer to [35] for more details as well as to joint works [26, 34] for results in that direction related to stochastic exponentials in the context of Itô calculus.

In a nutshell, chronological calculus is based on the idea of time-ordering of operators. Consider for example two time-dependent operators, $M(t)$ and $N(t)$ (with $M(0) = N(0) = 0$), in a non-unital algebra A of operators—having suitable regularity properties allowing to compute derivatives, integrals, and so on. The classical integration by parts rule is satisfied

$$M(t)N(t) = \int_0^t ds \int_0^s du \dot{M}(s)\dot{N}(u) + \int_0^t ds \int_0^s du \dot{M}(u)\dot{N}(s)$$
$$=: (M \succ N)(t) + (M \prec N)(t).$$

We recognise in $\prec$ and $\succ$ the usual operations of left/right integration—restricted to the context of deterministic processes. In particular, they satisfy the shuffle algebra axioms (16). Agrachev and Gamkrelidze observed that the binary operation

$$(M \triangleright N)(t) := (M \succ N)(t) - (N \prec n)(t) = \int_0^t ds \int_0^s du [\dot{M}(s), \dot{N}(u)]$$

has particular properties defining a chronological algebra structure on A. The latter is known as pre-Lie or Vinberg algebra in the mathematical literature.

Consider a vector space A with a binary product $\triangleright: A \otimes A \to A$ and the associated bracket product $[a, b]_\triangleright := a \triangleright b - b \triangleright a$. Write L_x for the linear endomorphism of A defined by left multiplication, $L_x(y) := x \triangleright y$, and define the usual commutator bracket of linear endomorphisms of A, $[L_x, L_y]_\circ := L_x \circ L_y - L_y \circ L_x$.

Definition 7 ([2]) The pair $(A, \triangleright)$ is a pre-Lie algebra if and only if for any $x, y \in A$, the identity $[L_x, L_y]_\circ = L_{[x,y]_\triangleright}$ holds, which is equivalent to the (left) pre-Lie relation

$$x \triangleright (y \triangleright z) - (x \triangleright y) \triangleright z = y \triangleright (x \triangleright z) - (y \triangleright x) \triangleright z.$$

The notion of pre-Lie algebra is finer than that of Lie algebra (it contains more information). Indeed, pre-Lie algebras are Lie admissible, that is, if A is a pre-Lie algebra, then $(A, [-, -]_\triangleright)$ is a Lie algebra.

The link with classical chronological calculus is as follows. Consider an algebra $\mathscr{A}$ of matrix-valued continuous semimartingales equipped with the left/right Fisk–Stratonovich integrals, $\succcurlyeq$ and $\preccurlyeq$, defined in (20) respectively (21). We write $[\![-, -]\!]$ for its commutator bracket:

$$[\![X, Y]\!]_t := X_t Y_t - Y_t X_t.$$

According to our previous developments, computing in this algebra amounts to computing with time-dependent operators. The—Fisk–Stratonovich—integration by parts formula implies that

$$[\![X,Y]\!]_t = \int_0^t X_s \circ dY_s + \int_0^t \circ dX_t Y_t - \int_0^t Y_s \circ dX_t - \int_0^t \circ dY_s X_s, \quad (28)$$

which can be written as the difference of:

$$(X \triangleright Y)_t := (X \succ Y - Y \prec X)_t = \int_0^t X_s \circ dY_s - \int_0^t \circ dY_s X_s$$

and $(Y \triangleright X)_t$ so that $[\![X,Y]\!] = [X,Y]_\triangleright$. That the algebra $\mathscr{A}$ is indeed a pre-Lie algebra, that is,

$$([X,Y]_\triangleright \triangleright Z)_t = (X \triangleright (Y \triangleright Z))_t - (Y \triangleright (X \triangleright Z))_t. \quad (29)$$

follows from the Jacobi identity of the commutator bracket on the non-commutative algebra $\mathscr{A}$.

Remark 8 The same argument shows that, more generally, there is a forgetful functor from shuffle to pre-Lie algebras, that is, any shuffle algebra $(A, \prec, \succ)$ has the structure of a pre-Lie algebra with pre-Lie product: $x \triangleright y := x \succ y - y \prec x$.

Let us apply these ideas in the context of stochastic exponentials. Recall a fundamental object in the classical analysis of differential equations, known as the Magnus formula [61] and its pre-Lie interpretation [2, 29]. It follows from studying the formal properties of the flow associated to a matrix differential equation using a Lie theoretic approach, for theoretical and numerical reasons. Consider for instance the evolution operator solution of the linear differential equation $\dot{X}(t) = X(t)H(t)$ with initial value $X(0) = \mathbf{1}$, the identity matrix. Its logarithm is computed by a Lie series. Truncating the expansion of this logarithm, $\Omega(t) := \log(X(t))$, and taking its exponential is a classical and efficient way to approximate $X(t)$ numerically, while preserving group-theoretic properties [10, 51].

The logarithm can be computed using the Baker–Campbell–Hausdorff formula (see, e.g., [65]) or Magnus' non-linear differential equation [61]

$$\dot{\Omega}(t) = \frac{ad_\Omega}{e^{ad_\Omega} - 1} H(t) = H(t) + \sum_{n>0} \frac{B_n}{n!} ad^n_{\Omega(t)}(H(t)),$$

where ad stands for the usual Lie adjoint representation, $ad_N(M) := NM - MN$, $ad^0_N(M) = M$, and the B_n are the Bernoulli numbers.

Let us explain how the formula adapts to Stratonovich integrals using recently developed algebraic tools that are most likely not familiar in the context of stochastic integration. The following developments are based on [35]. We omit here the group-theoretical perspective that relies on two underlying Hopf algebra structures, existing on the enveloping algebra of any pre-Lie algebra [21].

Let $\mathscr{A}$ be our usual algebra of continuous matrix-valued semimartingales, now equipped with the pre-Lie product $\triangleright$. The algebra of polynomials over $\mathscr{A}$ is denoted $\mathbb{R}[\mathscr{A}]$ and we identify m-multilinear maps symmetric in the m entries with maps from the degree m component of this polynomial algebra. To avoid confusion between the product of two matrix-valued semimartingales in $\mathscr{A}$ and their (commutative) product in $\mathbb{R}[\mathscr{A}]$, we denote the latter $X \odot Y$. The brace map on $\mathscr{A}$ is the family of symmetric multilinear maps into $\mathscr{A}$

$$\mathbb{R}[\mathscr{A}] \otimes \mathscr{A} \longrightarrow \mathscr{A}, \quad P \otimes X \longmapsto \{P\}X,$$

defined inductively by

$$\{Y\}X := Y \triangleright X,$$

for $Y, X \in \mathscr{A}$, and for $Y_1, Y_2, \ldots, Y_n, X \in \mathscr{A}$ we have

$$\{Y_1, \ldots, Y_n\}X := \{Y_n\}(\{Y_1, \ldots, Y_{n-1}\}X) - \sum_{i=1}^{n-1} \{Y_1, \ldots, \{Y_n\}Y_i, \ldots, Y_{n-1}\}X.$$

Observe that for $n = 2$ the last equality encodes the pre-Lie identity (29) as $\{Y_1, Y_2\}X = \{Y_2, Y_1\}X$. Following Guin and Oudom [70], we introduce a product $*$ on $\mathbb{R}[\mathscr{A}]$ in terms of the brace map. For elements $X_1, \ldots, X_n$ and $Y_1, \ldots, Y_m$ in $\mathscr{A}$,

$$(Y_1 \odot \cdots \odot Y_m) * (X_1 \odot \cdots \odot X_n) := \sum_f W_0 \odot \{W_1\}X_1 \odot \cdots \odot \{W_n\}X_n, \quad (30)$$

where the sum is over all maps f from $\{1, \ldots, m\}$ to $\{0, \ldots, n\}$ and the $W_i := \prod_{j \in f^{-1}(i)} Y_j$. For example, $Y * X = YX + \{Y\}X$, for $X, Y \in \mathscr{A}$.

Recall now that the enveloping algebra, $U(L)$, of a Lie algebra L is an associative algebra (uniquely defined up to isomorphism) such that [73]:

- the Lie algebra L embeds in $U(L)$ (as a Lie algebra, where the Lie algebra structure on $U(L)$ is induced by the associative product, that is, in terms of the ususal commutator bracket,
- for any associative algebra A (which is a Lie algebra, L_A, when equipped with the commutator bracket), there is a natural bijection between Lie algebra maps from L to A and associative algebra maps from $U(L)$ to A.

The central result of the work of Oudom and Guin [70] is the next theorem.

Theorem 5 ([70]) $\mathbb{R}[\mathscr{A}]$ *with the product $*$ defined in (30) is a non-commutative, associative and unital algebra. The product makes $\mathbb{R}[\mathscr{A}]$ the enveloping algebra of the Lie algebra $L_{\mathscr{A}}$ associated to $\mathscr{A}$.*

Applying Theorem 5 to $\mathscr{A}$ we see that the commutator bracket in $\mathscr{A}$ identifies with the pre-Lie bracket: $[\![X, Y]\!] = [\ ,\ ]_{\triangleright}$. On the other hand, by the universal properties of enveloping algebras, there is a unique associative algebra map ι from $(\mathbb{R}[A], *)$ to $\mathscr{A}$ which is the identity on $\mathscr{A}$. In degree two we have:

$$\iota(Y \odot X) = \iota(Y * X) - \iota(\{Y\}X) = YX - Y \triangleright X$$
$$= Y \preccurlyeq X + Y \succcurlyeq X - (Y \succcurlyeq X - X \preccurlyeq Y)$$
$$= Y \preccurlyeq X + X \preccurlyeq Y =: \mathscr{T}\langle Y, X \rangle,$$

where, using now the language of theoretical physics, we call time-ordered product of two elements in $\mathscr{A}$ the product $\mathscr{T}\langle Y, X \rangle := X \preccurlyeq Y + Y \preccurlyeq X$. In general, for $X_1, \ldots, X_n \in \mathscr{A}$,

$$\mathscr{T}\langle X_1, X_2, \ldots, X_n \rangle := \sum_{\sigma \in S_n} X_{\sigma(1)} \preccurlyeq (X_{\sigma(2)} \preccurlyeq (\cdots \preccurlyeq (X_{\sigma(n-1)} \preccurlyeq X_{\sigma(n)}) \cdots)),$$

where S_n denotes the symmetric group of order n. The degree two calculation is a particular instance of a general phenomenon. The following Theorem relating pre-Lie products and time-ordered exponentials was obtained in [32, p. 1291]:

Theorem 6 *The image in $\mathscr{A}$ of a monomial $X_1 \odot \cdots \odot X_n \in \mathbb{R}[\mathscr{A}]$ by the canonical map ι is the time-ordered product of the X_is in $\mathscr{A}$:*

$$\iota(X_1 \odot \cdots \odot X_n) = \mathscr{T}\langle X_1, \ldots, X_n \rangle. \tag{31}$$

Notice that, in particular,

$$\frac{1}{n!}\iota(X^{\odot n}) = \frac{1}{n!}\mathscr{T}\langle X, \ldots, X \rangle = X \preccurlyeq (X \preccurlyeq (\cdots \preccurlyeq (X \preccurlyeq X) \cdots)).$$

Let us apply these ideas to the study of the stochastic exponential and its logarithm in the Stratonovich framework. We address these problems at a purely formal level. Regarding the existence of the stochastic exponential and the convergence issues of the related series we refer to [71] for the Itô case and to Ben Arous [9] and Castell [19] for the Stratonovich one.

Definition 8 For a continuous matrix-valued semimartingale, $X \in \mathscr{A}$, the (Stratonovich) right stochastic exponential is defined through

$$\mathscr{E}_{\preccurlyeq}(X) = 1 + \bigl(X \preccurlyeq \mathscr{E}_{\preccurlyeq}(X)\bigr),$$

or, by a Picard iteration, as a series

$$\mathscr{E}_{\preccurlyeq}(X) = 1 + X + X \preccurlyeq X + \cdots + X \preccurlyeq (X \preccurlyeq (\cdots \preccurlyeq (X \preccurlyeq X) \cdots)) + \cdots$$

We are interested in the stochastic analogue of the classical Baker–Campbell–Hausdorff problem of computing the logarithm $\Omega(X)$ of the solution

$$\Omega(X) = \log\bigl(\mathscr{E}_{\preccurlyeq}(X)\bigr). \tag{32}$$

Since Stratonovich calculus obeys the usual integration by parts rule, the well-known Strichartz formula holds [9, 65, 78]. We are interested here in the stochastic analog of the so-called Magnus solution.

By Theorem 6, the equation

$$\mathcal{E}_\preccurlyeq(X) = \exp(\Omega(X))$$

lifts in $\mathbb{R}[\mathscr{A}]$ to an equality of exponentials:

$$\exp^\odot(X) = \exp^*(\tilde{\Omega}(X)), \tag{33}$$

where $\exp^\odot(X)$ (resp. $\exp^*(\tilde{\Omega}(X))$) denotes the exponential of $X \in \mathscr{A}$ (resp. $\tilde{\Omega}(X)$) for the $\odot$ (resp. $*$) product. Theorem 6 together with the general properties of enveloping algebras insure that this identity maps to (32) by ι and that $\iota(\tilde{\Omega}(X)) = \Omega(X)$. The next proposition was shown in [21], using the fact that the maps $\exp^\odot$ and $\exp^*$ have a Lie theoretic interpretation.

Proposition 1 *The element $\tilde{\Omega}(X) = \log^* \circ \exp^\odot(X)$ belongs to $\mathscr{A}$ and satifies the fixed point equation:*

$$\tilde{\Omega}(X) = \left\{ \frac{\tilde{\Omega}(X)}{\exp^*(\tilde{\Omega}(X)) - 1} \right\} X, \tag{34}$$

where $\tilde{\Omega}(X)/(\exp^*(\tilde{\Omega}(X)) - 1)$ is computed in $\mathbb{R}[\mathscr{A}]$ using the $*$ product.

Note that the brace map is in place on the righthand side in (34). We set

$$\ell_{X\triangleright}^{(n)}(Y) := X \triangleright (\ell_{X\triangleright}^{(n-1)}(Y)), \quad \ell_{X\triangleright}^{(0)}(Y) := Y.$$

The $B_n/n!$ are the coefficients of the formal power series expansion of $x/(\exp(x) - 1)$ and, by formal properties of the enveloping algebra, we have $\iota(\{X^{*n}\}Y) = \ell_{X\triangleright}^{(n)}(Y)$. We refer, e.g., to [35] for an explanation of this general phenomenon in the context of enveloping algebras of pre-Lie algebras. We recover finally the pre-Lie Magnus expansion of the logarithm of the right Stratonovich exponential obtained in [26].

Theorem 7 *The continuous matrix-valued semimartingale $\Omega(X)$ satisfies the fixed point equation*

$$\Omega(X) = \sum_{n \geq 0} \frac{B_n}{n!} \ell_{\Omega(X)\triangleright}^{(n)}(X). \tag{35}$$

Remark 9 The left stochastic exponential is defined similarly through $\mathcal{E}_\succcurlyeq(X)_t = 1 + (\mathcal{E}_\succcurlyeq(X) \succcurlyeq X)_t$. It satisfies

$$\mathcal{E}_\succcurlyeq(X) = \exp(-\Omega(-X)). \tag{36}$$

3.2 Chronological Itô Calculus

In the present subsection we will apply the machinery developed in the previous subsection in the context of Itô calculus. As remarked earlier, our arguments are purely algebraic. We do neither address the question of existence of stochastic exponentials nor do we discuss convergence issues of the associated series. On these questions, the reader is referred to the standard reference [71].

The existence of a continuous Baker–Campbell–Hausdorff, or Strichartz formula, the presence of pre-Lie structures as well as a Magnus formula could be expected in Stratonovich calculus due to the fact that the latter satisfies the usual rules of calculus. In Itô calculus, however, things are not so simple due to the presence of the covariation bracket and the fact that the usual shuffle algebra structure must be replaced by Karandikar's axioms, i.e., a quasi-shuffle algebra. For iterated Itô integrals of matrix-valued semimartingales, a Strichartz-type formula was obtained in [33, 34]. The difference with the classical formula reflect the fact that one must take into account the covariation bracket. This is achieved by replacing bijections, that is, permutations and their descent statistics as they appear in the classical formula, by surjections and a suitable notion of descents in this new context.

Here, we focus again on pre-Lie structures and the Magnus formula in the context of Itô calculus. Our results are obtained by adapting ideas from the theory of Rota–Baxter algebras to quasi-shuffle algebras. Our presentation is almost self-contained. On Rota–Baxter algebras and integral calculus, we refer to [35] for a general survey combined with references.

In this subsection, $\mathscr{A}$ denotes an algebra of matrix-valued semimartingales (notice that we do not require continuity anymore).

Proposition 2 *For $X, Y \in \mathscr{A}$, set $X \succ\!\!\!\succ Y := X \succ Y + [X, Y]$ and $X \blacktriangleright Y := X \succ\!\!\!\succ Y - Y \prec X$, then the pair $(\mathscr{A}, \blacktriangleright)$ is a pre-Lie algebra. Moreover, $[\![X, Y]\!] = [X, Y]_{\blacktriangleright}$.*

Proof Indeed, the quasi-shuffle axioms imply that

$$XY = X \prec Y + X \succ Y + [X, Y] = X \prec Y + X \succ\!\!\!\succ Y.$$

This yields

$$\begin{aligned}[] [\![X, Y]\!] &= XY - YX = X \prec Y + X \succ\!\!\!\succ Y - Y \prec X - Y \succ\!\!\!\succ X \\ &= (X \succ\!\!\!\succ Y - Y \prec X) - (Y \succ\!\!\!\succ X - X \prec Y) \\ &= X \blacktriangleright Y - Y \blacktriangleright X = [X, Y]_{\blacktriangleright}. \end{aligned}$$

Using the Jacobi identity (to avoid any notational ambiguity, recall that [,] denotes in this article the covariation bracket, not to be confused with the Lie bracket $[\![\ ,\ ]\!]$):

$$([X,Y]_\blacktriangleright \blacktriangleright Z)_t = ([\![X,Y]\!] \blacktriangleright Z)_t = \int_0^t [\![[\![X_s, Y_s]\!], \circ dZ_s]\!] + [\![[X,Y], Z]\!]_t$$

$$= (\int_0^t [\![X_s, [\![Y_s, \circ dZ_s]\!]]\!] + [XY, Z]_t) - (\int_0^t [\![Y_s, [\![X_s, \circ dZ_s]\!]]\!] + [XY, Z]_t)$$

$$= \int_0^t [\![X_s, [\![Y_s, \circ dZ_s]\!]]\!] + [X \succ Y + X \prec Y + [X,Y], Z]_t$$

$$- (\int_0^t [\![Y_s, [\![X_s, \circ dZ_s]\!]]\!] + [Y \succ X + Y \prec X + [Y,X], Z]_t)$$

$$= \int_0^t [\![X_s, [\![Y_s, \circ dZ_s]\!]]\!] + [X, Y \succ Z]_t + (X \succ [Y,Z])_t + [[X,Y], Z]_t$$

$$- (\int_0^t [\![Y_s, [\![X_s, \circ dZ_s]\!]]\!] + [Y, X \succ Z]_t + (Y \succ [X,Z])_t + [[Y,X], Z]_t)$$

$$= (X \blacktriangleright (Y \blacktriangleright Z))_t - (Y \blacktriangleright (X \blacktriangleright Z))_t.$$

□

Remark 10 The triple $(\mathcal{A}, \prec, \succ)$ is a shuffle algebra. More generally, in [28] it was shown that any quasi-shuffle algebra gives rise to a shuffle algebra. Indeed, $(X \prec Y) \prec Z = X \prec (YZ)$ and

$$(X \succ Y) \prec Z = (X \succ Y) \prec Z + [X,Y] \prec Z$$
$$= X \succ (Y \prec Z) + [X, Y \prec Z] = X \succ (Y \prec Z)$$

$$(XY) \succ Z = (XY) \succ Z + [XY, Z]$$
$$= (XY) \succ Z + [X \succ Y, Z] + [X \prec Y, Z] + [[X,Y], Z]$$
$$= X \succ (Y \succ Z) + X \succ [Y,Z] + [X, Y \succ Z] + [X, [Y,Z]]$$
$$= X \succ (Y \succ Z) + [X, Y \succ Z] = X \succ (Y \succ Z).$$

We note that this property is also common in Rota–Baxter algebras, where a shuffle algebra structure is defined similarly starting from the operations $R(X)Y$, $XR(Y)$ and XY instead of $\prec, \succ, [-,-]$.

Definition 9 For a matrix-valued semimartingale $X \in \mathcal{A}$ the (Itô) right stochastic exponential is defined through

$$\mathcal{E}_\prec(X) = \mathbf{1} + (X \prec \mathcal{E}_\prec(X)),$$

or, by a Picard iteration, as a series

$$\mathcal{E}_\prec(X) = \mathbf{1} + X + X \prec X + \cdots + X \prec (X \prec (\cdots \prec (X \prec X) \cdots)) + \ldots$$

We are interested again in the stochastic analogue of the Baker–Campbell–Hausdorff problem of computing the logarithm $\Gamma(X)$ of the solution:

$$\mathcal{E}_\prec(X) = \exp(\Gamma(X)). \tag{37}$$

Let us denote now by $\mathbb{R}_{\blacktriangleright}[\mathscr{A}]$ the enveloping algebra of $\mathscr{A}$ constructed exactly as in the previous section but using the new pre-Lie product $\blacktriangleright$ instead of $\triangleright$. In particular, as a vector space, $\mathbb{R}_{\blacktriangleright}[\mathscr{A}] = \mathbb{R}[\mathscr{A}]$, the algebra of polynomials over $\mathscr{A}$.

To avoid notational ambiguities, we write $*_{\blacktriangleright}$ for the associative product making $\mathbb{R}_{\blacktriangleright}[\mathscr{A}]$ the enveloping algebra of $L_{\mathscr{A}}$ and $\{P\}_{\blacktriangleright} X$ the brace map for $P \in \mathbb{R}_{\blacktriangleright}[\mathscr{A}]$. We also write $\mathscr{T}_{\blacktriangleright}$ for the corresponding time-ordered product, associated to the left-half shuffle $\prec$, e.g., $\mathscr{T}_{\blacktriangleright}[X, Y] := X \prec Y$, and so on. Lastly, we denote $\iota_{\blacktriangleright}$ the canonical algebra map from $(\mathbb{R}_{\blacktriangleright}[\mathscr{A}], *_{\blacktriangleright})$ to $\mathscr{A}$ obtained from the universal properties of the enveloping algebra. Theorem 6 holds *mutatis mutandis* in the new context and equation (37) lifts in $\mathbb{R}_{\blacktriangleright}[\mathscr{A}]$ to:

$$\exp^{\circ}(X) = \exp^{*_{\blacktriangleright}}(\tilde{\Gamma}(X)). \tag{38}$$

Theorem 6 and the general properties of enveloping algebras insure that this identity maps to (37) by $\iota_{\blacktriangleright}$ and that $\iota_{\blacktriangleright}(\tilde{\Gamma}(X)) = \Gamma(X)$.

We obtain finally the analogous of Proposition 1:

Proposition 3 *The element $\tilde{\Gamma}(X) = \log^*_{\blacktriangleright} \circ \exp^{\circ}(X)$ belongs to $\mathscr{A}$ and satifies the fixed point equation:*

$$\tilde{\Gamma}(X) = \left\{\frac{\tilde{\Gamma}(X)}{\exp^*_{\blacktriangleright}(\tilde{\Gamma}(X)) - \mathbf{1}}\right\}_{\blacktriangleright} X, \tag{39}$$

where $\tilde{\Gamma}(X)/(\exp^*_{\blacktriangleright}(\tilde{\Gamma}(X)) - \mathbf{1})$ is computed in $\mathbb{R}_{\blacktriangleright}[\mathscr{A}]$ using the $*_{\blacktriangleright}$ product.

Setting

$$\ell^{(n)}_{X\blacktriangleright}(Y) := X \blacktriangleright (\ell^{(n-1)}_{X\blacktriangleright}(Y)),$$

$\ell^{(0)}_{X\blacktriangleright}(Y) := Y$, we get finally a pre-Lie Magnus expansion of the logarithm of the right Itô stochastic exponential:

Theorem 8 *The matrix-valued semi-martingale $\Gamma(X)$, which is the logarithm of the (Itô) stochastic exponential, satisfies the fixed point equation*

$$\Gamma(X) = \sum_{n \geq 0} \frac{B_n}{n!} \ell^{(n)}_{\Gamma(X)\blacktriangleright}(X). \tag{40}$$

Acknowledgements The second author would like to express his gratitude for the warm hospitality he experienced during the Verona meeting, with special thoughts for S. Albeverio and S. Ugolini. This work was supported by the French government, managed by the ANR under the UCA JEDI Investments for the Future project, reference number ANR-15-IDEX-01.

References

1. Agrachev, A., Gamkrelidze, R.: The exponential representation of flows and chronological calculus, Math. sbornik **107**(149), 467–532 (1978); English transl. in Math. USSR Sbornik **35**, 727–785 (1979)
2. Agrachev, A., Gamkrelidze, R.: Chronological algebras and nonstationary vector fields. J. Sov. Math. **17**, 1650–1675 (1981)
3. Agrachev, A., Gamkrelidze, R., Sarychev, V.: Local Invariants of Smooth Control Systems. Acta Applicandae Mathematicae **14**, 191–237 (1989)
4. Agrachev, A., Sachkov, Y.: Control Theory from the Geometric Viewpoint, Encyclopaedia of Mathematical Sciences, vol. 84. Springer, Berlin, Heidelberg (2004)
5. Agrachev, A., Barilari, D., Boscain, U.: A Comprehensive Introduction to Sub-Riemannian Geometry, Cambridge Studies in Advanced Mathematics, vol. 181. Cambridge University Press (2019)
6. Baldeaux, J., Platen, E.: Functionals of Multidimensional Diffusions with Applications to Finance, in Bocconi & Springer Series vol. 5. Springer (2013)
7. Baxter, G.: An analytic problem whose solution follows from a simple algebraic identity. Pac. J. Math. **10**, 731–742 (1960)
8. Beasley Cohen, P., Eyre, T.W.M., Hudson, R.L.: Higher order Itô product formula and generators of evolutions and flows. Int. J. Theor. Phys. **34**, 1–6 (1995)
9. Ben Arous, G.: Flots et séries de Taylor stochastiques. Probab. Th. Rel. Fields **81**, 29–77 (1989)
10. Blanes, S., Casas, F., Oteo, J.A., Ros, J.: Magnus expansion: mathematical study and physical applications. Phys. Rep. **470**, 151–238 (2009)
11. Brouder, Ch.: Runge-Kutta methods and renormalization. Europ. Phys. J. C **12**, 512–534 (2000)
12. Bruned, Y., Hairer, M., Zambotti, L.: Algebraic renormalisation of regularity structures. Invent. Math. **215**(3), 1039–1156 (2019)
13. Burde, D.: Left-symmetric algebras, or pre-Lie algebras in geometry and physics. Central Eur. J. Math. **4**(3), 323–357 (2006)
14. Burgunder, E., Ronco, M.: Tridendriform structure on combinatorial Hopf algebras. J. Algebra **324**(10), 2860–2883 (2010)
15. Calaque, D., Ebrahimi-Fard, K., Manchon, D.: Two interacting Hopf algebras of trees: A Hopf-algebraic approach to composition and substitution of B-series. Advances in Applied Mathematics **47**, 282–308 (2011)
16. Cartier, P.: On the structure of free Baxter algebras. Adv. Math. **9**(2), 253–265 (1972)
17. Cartier, P.: Vinberg algebras. Lie groups and combinatorics, Clay Math. Proc. **11**, 107–126 (2011)
18. Cartier, P., Patras, F.: Classical hopf algebras and their applications. Springer (2021)
19. Castell, F.: Asymptotic expansion of stochastic flows. Probab. Th. Rel. Fields **96**, 225–239 (1993)
20. Chapoton, F., Livernet, M.: Pre-Lie algebras and the rooted trees operad. Int. Math. Res. Notices **2001**, 395–408 (2001)
21. Chapoton, F., Patras, F.: Enveloping algebras of preLie algebras, Solomon idempotents and the Magnus formula. Int. J. Algebra Comput. **23**(4), 853–861 (2013)
22. Chen, K.T.: Integration of paths, geometric invariants and a generalized Baker-Hausdorff formula. Ann. Math. **65**, 163–178 (1957)
23. Chen, K.T.: Algebras of iterated path integrals and fundamental groups. Trans. Am. Math. Soc. **156**, 359–379 (1971)
24. Connes, A., Kreimer, D.: Hopf algebras, Renormalization and Noncommutative Geometry. Commun. Math. Phys. **199**, 203–242 (1998)
25. Curry, C., Ebrahimi-Fard, K., Malham, S.J.A., Wiese, A.: Lévy processes and quasi-shuffle algebras. Stochastics **86**(4), 632–642 (2014)
26. C. Curry, K. Ebrahimi-Fard, F. Patras, On non-commutative stochastic exponentials, in proceedings volume ENUMATH2017 conference, Springer's Lecture Notes in Computational Science and Engineering, vol. 126 (2018)

27. Curry, C., Ebrahimi-Fard, K., Malham, S.J.A., Wiese, A.: Algebraic Structures and Stochastic Differential Equations driven by Lévy processes. Proc. R. Soc. A **475**, 20180567 (2019)
28. Ebrahimi-Fard, K.: Loday-type algebras and the Rota-Baxter relation. Lett. Math. Phys. **61**(2), 139–147 (2002)
29. Ebrahimi-Fard, K., Manchon, D.: A Magnus- and Fer-type formula in dendriform algebras. Found. Comput. Math. **9**, 295–316 (2009)
30. K. Ebrahimi-Fard, A. Lundervold, S.J.A. Malham, H. Munthe-Kaas, A. Wiese, Algebraic structure of stochastic expansions and efficient simulation. Proc. R. Soc. A (2012)
31. Ebrahimi-Fard, K., Manchon, D.: The tridendriform structure of a discrete Magnus expansion. Discret. Contin. Dyn. Syst.-A **34**(3), 1021–1040 (2014)
32. Ebrahimi-Fard, K., Patras, F.: The Pre-Lie Structure of the Time-Ordered Exponential. Lett. Math. Phys. **104**, 1281–1302 (2014)
33. Ebrahimi-Fard, K., Malham, S.J.A., Patras, F., Wiese, A.: The exponential Lie series for continuous semimartingales. Proc. R. Soc. A **471**, 20150429 (2015)
34. Ebrahimi-Fard, K., Malham, S.J.A., Patras, F., Wiese, A.: Flows and stochastic Taylor series in Ito calculus. J. Phys. A: Math. Theor. **48**, 495202 (2015)
35. Ebrahimi-Fard, K., Patras, F.: From iterated integrals and chronological calculus to Hopf and Rota–Baxter algebras, Encyclopedia in Algebra and Applications (to appear) arXiv:1911.08766
36. Fauvet, F., Menous, F.: Ecalle's arborification-coarborification transforms and Connes-Kreimer Hopf algebra. Ann. Sci. Éc. Norm. Supér. **50**, 39–83 (2017)
37. Foissy, L., Patras, F.: Lie theory for quasi-shuffle bialgebras. in Periods in Quantum Field Theory and Arithmetic, (Burgos Gil. et al., eds) Springer Proceedings in Mathematics and Statistics, vol. 314 (2020)
38. Foissy, L., Patras, F., Thibon, J.-Y.: Deformations of shuffles and quasi-shuffles. Ann. Inst. Fourier **66**(1), 209–237 (2016)
39. Friedrich, R.: Operads in Itô calculus, arXiv:1604.08547
40. Gaines, J.: The algebra of iterated stochastic integrals. Stochast. Stochast. Rep. **49**, 169–179 (1994)
41. Gubinelli, M.: Ramification of rough paths. J. Differ. Equ. **248**, 693–721 (2010)
42. M. Gubinelli, Abstract integration, combinatorics of trees and differential equations. In: Proceedings of the Workshop Combinatorics and Physics, 2007. MPI Bonn. Combinatorics and physics, Contemporary Mathematics, vol. 539, pp. 135–151. American Mathematical Society, Providence, RI (2011)
43. Hairer, E., Lubich, C., Wanner, G.: Geometric numerical integration Structure-preserving algorithms for ordinary differential equations, vol. 31. Springer Series in Computational Mathematics. Springer, Berlin (2002)
44. Hairer, M.: Solving the KPZ equation. Ann. Math. **178**, 559–664 (2013)
45. Hairer, M.: A theory of regularity structures. Invent. Math. **198**(2), 269–504 (2014)
46. M. Hairer, D. Kelly, Geometric versus non-geometric rough paths. Ann. de l'I.H.P. Probabilités et Statistiques **51**(1), 207–251 (2015)
47. Hoffman, M.E.: Quasi-Shuffle Products. J. Algebr. Combinator. **11**(1), 49–68 (2000)
48. Hoffman, M.E., Ihara, K.: Quasi-shuffle products revisited. J. Algebra **481**(1), 293–326 (2017)
49. Hudson, R.L.: Hopf-algebraic aspects of iterated stochastic integrals. Infinite Dimens. Anal. Quantum Probab. Relat. Top. **12**, 479–496 (2009)
50. Hudson, R.L.: Sticky shuffle product Hopf algebras and their stochastic representations. In: New Trends in Stochastic Analysis and Related Topics. Interdiscipilanary Mathematics Science, vol. 12, pp. 165–181. World Scientific Publishing, Hackensack, NJ (2012)
51. Iserles, A., Munthe-Kaas, H.Z., Nørsett, S.P., Zanna, A.: Lie-group methods. Acta Numer. **9**, 215–365 (2000)
52. R. L. Karandikar, A.s. approximation results for multiplicative stochastic integration, Séminaire de Probabilitiés XVI. Lecture Notes in Mathematics, vol. 920, pp. 384–391, Springer (1981)
53. Karandikar, R.L.: Multiplicative decomposition of non-singular matrix valued continuous semimartingales. Ann. Probab. **10**, 1088–1091 (1982)

54. Karandikar, R.L.: Girsanov type formula for a lie group valued brownian motion, Séminaire de Probabilitiés XVII. Lecture Notes in Mathematics vol. 986, pp. 198–204. Springer, Berlin (1982)
55. Karandikar, R.L.: Multiplicative decomposition of nonsingular matrix valued semimartingales. In: Azéma, J., Yor, M., Meyer, P. (eds.) Séminaire de Probabilités XXV. Lecture Notes in Mathematics, vol. 1485, pp. 262–269. Springer, Berlin (1991)
56. Loday, J.-L., Ronco, M.: Une dualité entre simplexes standards et polytopes de Stasheff. C. R. Acad. Sci. Paris Série I(333), 81–86 (2001)
57. Lundervold, A., Munthe-Kaas, H.Z.: Hopf algebras of formal diffeomorphisms and numerical integration on manifolds. Contemporary Mathematics **539**, 295–324 (2011)
58. Lyons, T.: Differential equations driven by rough signals. Rev. Mat. Iberoamericana **14**(2), 215–310 (1998)
59. Lyons, T., Caruana, M.J., Lévy, T.: Differential Equations Driven by Rough Paths, Ecole d'Eté de Probabilités de Saint-Flour XXXIV-2004 1908. Springer, Berlin, Heidelberg (2007)
60. McLachlan, R.I., Modin, K., Munthe-Kaas, H., Verdier, O.: Butcher Series - A Story of Rooted Trees and Numerical Methods for Evolution Equations. Asia Pacific Mathematics Newsletter **7**(1), 1–11 (2017)
61. Magnus, W.: On the exponential solution of differential equations for a linear operator. Commun. Pure Appl. Math. **7**, 649–673 (1954)
62. Manchon, D.: A short survey on pre-Lie algebras. In: Carey, A. (ed.) E. Schrödinger Institut Lectures in Mathematics and Physics, Non-commutative Geometry and Physics: Renormalisation, Motives, Index Theory, EMS (2011)
63. Menous, F., Patras, F.: Renormalization: a quasi-shuffle approach. In: Celledoni et al. (eds.) Computation and Combinatorics in Dynamics, Stochastics and Control: The Abel Symposium 2016. Springer Abel Symposia, vol. 13 (2018)
64. F. Menous, F. Patras, Right-handed bialgebras and the Prelie forest formula, Annales I.H.P. Série D, **5**, Issue 1, (2018) 103–125
65. Mielnik, B., Plebański, J.: Combinatorial approach to Baker-Campbell-Hausdorff exponents. Ann. Inst. Henri Poincaré A XI **I**, 215–254 (1970)
66. Munthe-Kaas, H.Z., Wright, W.M.: On the Hopf algebraic structure of Lie group integrators. Found. Comput. Math. **8**(2), 227–257 (2007)
67. Murua, A.: The Hopf algebra of rooted trees, free Lie algebras, and Lie series. Found. Comput. Math. **6**, 387–426 (2006)
68. Novelli, J.-C., Patras, F., Thibon, J.-Y.: Natural endomorphisms of quasi-shuffle Hopf algebras. Bull. Soc. Math. France **141**, 107–130 (2013)
69. Novelli, J.-C., Thibon, J.-Y.: Polynomial realizations of some trialgebras. In: Proceedings of Formal Power Series and Algebraic Combinatorics, San Diego, California (2006)
70. Oudom, J.-M., Guin, D.: On the Lie enveloping algebra of a pre-Lie algebra. Journal of K-theory **2**(1), 147–167 (2008)
71. Protter, P.E.: Stochastic integration and differential equations, Version 2.1, 2nd Edn. Springer, Berlin (2005)
72. Ree, R.: Lie elements and an algebra associated With shuffles. Ann. Math. Second Series **68**(2), 210–220 (1958)
73. Reutenauer, C.: Free Lie Algebras. Oxford University Press (1993)
74. Rota, G.-C.: Baxter algebras and combinatorial identities. I, II, Bull. Amer. Math. Soc. **75**, 325–329 (1969); ibid. **75**, 330–334 (1969)
75. Rota, G.-C., Smith, D.: Fluctuation theory and Baxter algebras, Istituto Nazionale di Alta Matematica **IX**, 179 (1972)
76. Rota, G.-C.: Baxter operators, an introduction. In: Gian-Carlo Kung, J.P.S. (ed.) Rota on Combinatorics, Introductory Papers and commentaries. Contemporary Mathematicians, Birkhäuser Boston, Boston, MA (1995)

77. Rota, G.-C.: Ten mathematics problems I will never solve, Invited address at the joint meeting of the American Mathematical Society and the Mexican Mathematical Society, Oaxaca, Mexico, Dec. 6 (1997). DMV Mittellungen Heft **2**, 45 (1998)
78. Strichartz, R.S.: The Campbell-Baker-Hausdorff-Dynkin formula and solutions of differential equations. J. Func. Anal. **72**, 320–345 (1987)

Higher Order Derivatives of Heat Semigroups on Spheres and Riemannian Symmetric Spaces

K. David Elworthy

Abstract As a very special case of a more general procedure a formula is derived for the Hessian of the solutions $P_t f$ of the heat equation for functions on the sphere S^n. The formula demonstrates that for higher order derivatives there can be a spectrum of decay/growth rates, unlike the generic situation for first derivatives which is fundamental for Bakry-Emery theory. The method used is then applied for higher derivatives for spheres, and could be used for compact Riemannian symmetric spaces.

Keywords Stochastic analysis · Stochastic flows · Symmetric spaces · Heat semigroup · Bakry-Emery · Diffusion of symmetric tensors · Semi-group domination

Mathematics subject classification 58J65 · (58J70 · 60H30 60J60 · 43A85)

1 Introduction

A well known and fundamental result concerning the heat-semigroup $\{P_t\}_{t \geq 0}$ of a complete Riemannian manifold M is that of Bakry-Emery theory, [1]

$$|\nabla P_t(f)| \leq e^{-ct} P_t(|\nabla f|) \quad \text{iff} \quad c|v|^2 \leq \text{Ric}(v, v) \text{ all } v \in TM \tag{1}$$

where $\text{Ric} : TM \bigoplus TM \to \mathbf{R}$ is the Ricci curvature of M. Bakry-Emery theory, [2, 3], shows how to extend it to much more general classes of heat semi-groups, and it can then be used to define the notion of generalised Ricci curvature bounded below in much more general situations than Riemannian geometry. An obvious question is whether similar expressions hold for higher derivatives of $P_t f$ with exponential rates given in terms of the geometry of the Riemannian manifold. With this in mind we obtain expressions for the second and third derivatives of $P_t f$ when M is a sphere

---

K. D. Elworthy (✉)
Mathematics Institute, University of Warwick, Coventry, England CV4 7AL, UK
e-mail: K.D.Elworthy@warwick.ac.uk

with its standard Riemannian structure, Theorems 4.1 and 5.5 respectively, and also give expressions for all symmetrised derivatives in Theorem 5.1. These suggest that the situation is more complicated, and possibly more interesting, than expected. The approach we give, based on earlier work with Yves LeJan and Xue-Mei Li [13], can be extended to arbitrary compact Riemannian symmetric spaces, and should give similar formulae. However we have not done this.

The second derivative, or Hessian, is symmetric. This is not true in general for higher derivatives; see Sect. 5.1 below. It is simpler to compute the symmetrised versions. For the symmetrised versions the exponential rate is controlled by a Weitzenböck term, in the sense of [8, 13], which for spheres turns out to be essentially the Weitzenböck term for the Lichnerowicz Laplacian, eg see [5]. For general M, the latter has been shown by Bettiol and Mendes, [6], to characterise sectional curvature bounds. However our exponential rate can be expected to involve derivatives of curvature for general M; for spheres these vanish.

A relevant result by James Thompson, [27], is that if M is compact then for each $p \in N$ and $\epsilon > 0$ there is a constant $C_p(\epsilon) > 0$ such that for all C^1 functions $f : M \to \mathbf{R}$

$$|\nabla^p P_t f|_\infty \leq C_p(\epsilon) e^{-\lambda t} |\nabla f|_\infty \text{ for all } t > \epsilon$$

where $\lambda > 0$ is the spectral gap of M. This involves the smoothing behaviour of the semigroup when $p > 1$, which is why t needs be kept away from 0. It demonstrates that there is uniform rate of decay for all derivatives as $t \to \infty$, but our formulae suggests that a more detailed analysis involving the directions of the derivatives could be rewarding. Indeed for spheres our estimate (68) shows that for derivatives taken in orthonormal directions the rate of decay increases with $1 \leq p \leq n$. For $p = 2$ it is bounded above by e^{-nt}, (51). For S^n with $\frac{1}{2}\Delta$ the spectral gap, is $\frac{n}{2}$.

An excellent survey of work on higher order derivative formulae can be found in the introduction to Xue-Mei Li's article, [19]. Much of this concerns the technically harder problem of considering derivatives of the heat kernels. Usually just the first and second derivatives are discussed, though a notable early example giving path integral formulae is Norris's work, [25]. See Sect. 55 for the result of applying [19] to our situation on S^n.

Our treatment here of S^n is as a very special illustrative example of the more general situation described in [14]. We give the necessary geometric background, and give the proof of a simple case concerning the expectation of representations of diffusing Lie group elements. As pointed out in Sect. 4.1, below, there are alternative methods for S^n, and a purely algebraic one could be the most economical.

2 Brownian Motion on Spheres as Symmetric Spaces

2.1 The Sphere as a Symmetric Space

Consider the sphere S^n as the set of unit vectors of $\mathbf{R}^n$ with its induced topology, differential structure, and Riemannian metric. It is acted on transitively and smoothly by the special orthogonal group $SO(n+1)$. Let x_0 be a given point in S^n; we can take it to be the North Pole, $(0, 0, 0..., 1)$. This identifies the subgroup $SO(n+1)_{x_0}$, of those $\theta \in SO(n+1)$ which fix x_0, with $SO(n)$. We have the projection

$$p : SO(n+1) \to S^n \qquad p(k) = k(x_0) \quad k \in SO(n+1) \qquad (2)$$

which identifies S^n with the quotient space $SO(n+1)/SO(n)$. It is a principal bundle with group $SO(n)$. For us the main import of that will be that there is the right action of $SO(n)$

$$SO(n+1) \times SO(n) \to SO(n+1) \qquad (k, g) \mapsto k.g$$

with $p(k.g) = p(k)$.

We want p to be a Riemannian submersion. This means that we have an inner product $\langle -, - \rangle_k$ on each tangent space $T_k SO(n+1)$ such that $T_k p : T_k SO(n+1) \to T_{p(k)} S^n$, the derivative of p at k is an orthogonal projection. We also want this Riemannian structure to be bi-invariant and so it suffices to take

$$\langle A, B \rangle_{Id} = -\frac{1}{2} \text{trace } AB^* \qquad A, B \in \mathfrak{so}(n+1) \cong T_e SO(n+1).$$

With this choice, if $\{k_t\}_{t \geq 0}$ is a Brownian motion on $SO(n+1)$ starting at the identity Id, then $\{x_t\}_{t \geq 0}$ with $x_t = p(k_t) = k_t.x_0$ is a Brownian motion on S^n from x_0. Moreover if we define $\xi_t : S^n \to S^n$ by $\xi_t(y) = k_t.y$ we have a stochastic flow of Brownian motions on the sphere. For example see [7] or [13]. In particular if P_t denotes the heat semi-group acting on continuous functions on S^n then

$$P_t f(y) = \mathbf{E} f(\xi_t(y)) \qquad f : S^n \to \mathbf{R} \quad y \in S^n \qquad (3)$$

Recall that $f_t = P_t f : S^n \to \mathbf{R}$, $t \geq 0$ is the classical solution to the heat equation $\frac{df_t}{dt} = \frac{1}{2}\Delta f_t$, $f_0 = f$ on R^n. Here Δ is the Laplace Beltrami operator, $\Delta = div$ grad, on S^n.

2.2 Derivatives of the Heat Semigroup

Assume now that f is C^∞, then we can differentiate equation (3) in the direction of some $v \in T_{x_0} S^n$ to give

$$d(P_t f)(v) = \mathbf{E}\{df_{x_t}(T_{x_t} \xi_t(v))\}. \tag{4}$$

Recall that the derivative of f gives a differential one-form $df_y : T_y S^n \to \mathbf{R}$, and the derivative of the flow gives, random, linear isomorphisms, $T_y \xi_t : T_y S^n \to T_{\xi_t(y)} S^n$, for $y \in S^n$.

2.2.1 Aside on Calculus on Spheres

In order to differentiate again we need a connection on S^n. This gives a covariant derivative operator ∇ with which tensor fields such as df can be differentiated in tangent directions. Equivalently it gives a differentiation operator $\frac{D}{dt}$ of tensor fields along C^1 curves σ, and a parallel translation operator $//_t : T_{\sigma(0)} S^n \to T_{\sigma(t)} S^n$ of tangent vectors, or of other tensors. These are related, for example by

$$\frac{D}{dt} V_t = //_t \frac{d}{dt} //_t^{-1} V_t \qquad V_t \in T_{\sigma(t)} S^n, \tag{5}$$

and if $v = \dot\sigma(0)$

$$\nabla_v(df) = \frac{D}{dt}(df_{\sigma(t)})|_{t=0} = \frac{d}{dt}(df_{\sigma(t)} //_t)|_{t=0} \tag{6}$$

Stratonovich calculus allows these operations to be extended, almost surely, to the situation where σ is a continuous semi-martingale, such as our Brownian motion $\{x_t\}_t$. Also we can differentiate our stochastic flow successively, for example to get $\nabla_{u_0} T \xi_t : T_{x_0} S^n \to T_{x_t} S^n$ for $u_0 \in T_{x_0} S^n$, given by

$$\nabla_{u_0} T \xi_t(v_0) = \frac{D}{ds}(T_{\sigma(s)} \xi_t(//_s v_0))|_{s=0} \qquad u_0, v_0 \in T_{x_0} S^n \quad \dot\sigma(0) = u_0. \tag{7}$$

All this holds for any Riemannian manifold, and there is a unique connection, the Levi-Civita connection, for which parallel translations consist of orthogonal transformations and also

$$\frac{D}{\partial s} \frac{\partial}{\partial t} f(\sigma(s,t)) = \frac{D}{\partial t} \frac{\partial}{\partial s} f(\sigma(s,t)) \tag{8}$$

for a two parameter $\sigma(s,t)$ and $f : M \to \mathbf{R}$, both smooth.

We will use this. For S^n it has the natural definition that $\frac{D}{dt} V_t$ is obtained by considering the vector field V_t along σ as having values in $\mathbf{R}^{n+1}$, differentiating this in

t as usual and projecting the result back to $T_{\sigma(t)}S^n$.

Note: for $\{e_j\}_{j=1}^n$ an orthonormal base for T_yS^n, $y \in S^n$

$$\Delta f(y) = \text{trace}(\nabla(df))_y = \Sigma_{j=1}^n \nabla_{e_j}(df)e_j. \tag{9}$$

The *Hessian*, Hess(f), of f is just the second derivative considered as a bilinear form

$$\text{Hess}(f)_y = \nabla_-(df)(-) : T_yS^n \times T_yS^n \to \mathbf{R}. \tag{10}$$

By equation (8), the Hessian is symmetric and so determines a linear map on the symmetric tensor product $T_yS^n \odot T_yS^n$ by

$$\text{Hess}(f)(u \odot v) = \nabla_u(df)(v) \qquad u, v \in T_yS^n. \tag{11}$$

2.3 Higher Derivatives of $P_t f$

Using the Levi-Civita connection we can differentiate equation (4) again to obtain, for $u_0, v_0 \in T_{x_0}S^n$:

$$\text{Hess}(P_t f)(u_0 \odot v_0) = \mathbf{E}\{\text{Hess}(f)(T_{x_0}\xi_t u_0 \odot T_{x_0}\xi_t v_0) + df_{x_t}\nabla_{u_0}(T\xi_t)(v_0)\}. \tag{12}$$

An important simplification arises since our flow is a flow of isometries. In this situation covariant second order derivatives of the flow vanish, see [5]. Thus for $u_0, v_0 \in T_{x_0}S^n$:

$$\text{Hess}(P_t f)_{x_0}(u_0 \odot v_0) = \mathbf{E}\{\text{Hess}(f)_{x_t}(T_{x_0}\xi_t u_0 \odot T_{x_0}\xi_t v_0)\}. \tag{13}$$

and repeating the differentiaton, for $k = 1, 2, \ldots$ and $u_0^1, \ldots, u_0^k, v_0 \in T_{x_0}S^n$:

$$\nabla^{(k)}d(P_t f)(u_0^k, \ldots, u_0^1, v_0) = \mathbf{E}\{\nabla^k(df)(T_{x_0}\xi_t u_0^k, \ldots, T_{x_0}\xi_t u_0^1, T_{x_0}\xi_t v_0)\} \tag{14}$$

But the derivatives are not symmetric when $k \geq 2$ and $n \geq 2$; the curvature intervenes. See Subsection 5.1 below.

We can get a more precise formula from formula (12) by computing the conditional expectation of

$$T_{x_0}\xi_t \odot T_{x_0}\xi_t : T_{x_0}S^n \odot T_{x_0}S^n \to T_{x_t}S^n \odot T_{x_t}S^n$$

with respect to the σ-algebra $\mathcal{F}_t$ generated by the Brownian motion $\{x_s : 0 \leq s \leq t\}$. This technique, of filtering out the redundant noise, has been a basic tool for looking at first derivatives since [12]. It is described in detail in [10]. In essence the conditional expectation is obtained by parallel translation back to the initial point:

Write $u_t = T_{x_0}\xi_t(u_o)$ and $v_t = T_{x_0}\xi_t(v_o)$ and set $\overline{u_t \odot v_t} = \mathbf{E}\{u_t \odot v_t | \mathcal{F}_t\}$; then, essentially by definition,

$$\overline{u_t \odot v_t} = (//_t \odot //_t)\mathbf{E}\{//_t^{-1}u_t \odot //_t^{-1}v_t | \mathcal{F}_t\}. \tag{15}$$

Since $//_t^{-1}u_t \odot //_t^{-1}v_t$ lies in a fixed vector space its conditional expectation makes classical sense. There is no problem about integrability, and any choice of parallel translation in $TS^n \odot TS^n$ will do, [10].

We proceed to calculate this conditional expectation using techniques from [13], see also [8].

3 Decomposition and Conditioning of $T\xi_t \odot T\xi_t$

3.1 Decomposition of the Flow

Remember ξ_t is just the action of the Brownian motion $\{k_t\}_t$, on $SO(n+1)$, on our sphere. Also the Brownian motion $\{x_t\}_t$, from x_0 on the sphere, is given by $x_t = p(k_t) = k_t.x_0$. From [9] we have a skew product decomposition:

$$k_t = \tilde{x}_t.g_t \tag{16}$$

where $\{g_t\}_t$ is a Brownian motion on $SO(n)$ from the identity, independent of $\{\mathcal{F}_t\}_{t\geq 0}$, and $\{\tilde{x}_t\}_t$ is a diffusion process adapted to $\{\mathcal{F}_t\}_{t\geq 0}$ with $p(\tilde{x}_t) = x_t$ for $t \geq 0$. In fact $\{\tilde{x}_t\}_t$ is the "horizontal lift" of Brownian motion on S^n from the identity, and is the conditioned process of $\{k_t\}_t$ given $\{\mathcal{F}_t\}_{t\geq 0}$. Moreover if we write $\tilde{\xi}_t : S^n \to S^n$ for $y \mapsto \tilde{x}_t.y$ then parallel translation $\{//_t\}_{t\geq 0}$ along $\{x_t\}_t$ is given by

$$//_t = T_{x_0}\tilde{\xi}_t : T_{x_0}S^n \to T_{x_t}S^n. \tag{17}$$

See [11] or [13] for more.

Identifying $g \in SO(n)$ with its action on S^n let $\rho^{\odot^2}$ denote the representation of $SO(n)$ on $T_{x_0}S^n \odot T_{x_0}S^n$ given by

$$\rho^{\odot^2}(g)(u_0 \odot v_0) = T_{x_0}L_g u_0 \odot T_{x_0}L_g v_0. \tag{18}$$

From above, using the independence of g_t from $\mathcal{F}_t$, we have:

Lemma 3.1 *For a C^2 function $f : S^n \to \mathbf{R}$ and $u_0, v_0 \in T_{x_0}S^n$*

$$\operatorname{Hess}(P_t f)(u_0, v_0) = \mathbf{E}\left\{\operatorname{Hess}(f)_{x_t}\left((//_t \odot //_t)\mathbf{E}\left\{\rho^{\odot^2}(g_t)(u_0 \odot v_0)\right\}\right)\right\}. \tag{19}$$

We go on to compute the second expectation appearing above.

3.2 Expectations of Representations of Random Matrices: An Elementary Lemma

The following is a very special case of a similar result for finite dimensional representations of certain, possibly time inhomogenous, diffusions on possibly infinite dimensional groups. It is essentially Theorem 3.4.1 of [13], but see [8], or below, for a corrected sign in equation (3.19) of [13]. For completeness the simple proof is given here for Brownian motions on finite dimensional Lie groups. The more general cases will be discussed in [14]. The integrability of $\rho(g_t)v$ was proved by Baxendale, [4], for Wiener processes on Polish groups acting on Banach spaces.

Let G be a finite dimensional Lie group with right invariant metric, determined by an inner product $\langle -, - \rangle_e$ on its Lie algebra $\mathfrak{g}$ identified with the tangent space $T_e G$ at the identity $e \in G$. We will use the Maurer-Cartan form determined by *right translations* R_g, rather than the more usual left translations L_g. It is the $\mathfrak{g}$-valued one-form ϖ given by:

$$\varpi_g := T_g(R_g)^{-1} : T_g \mathcal{G} \to \mathfrak{g} := T_e \mathcal{G} \qquad g \in G.$$

The co-differential, the adjoint d^* of d, maps one-forms to functions. It acts on ϖ component wise: let $\{\alpha^j\}_j$ be an orthonormal base for $\mathfrak{g}$ and define the scalar one forms ϖ^j by $\varpi^j(v) = \langle \varpi(v), \alpha^j \rangle_\mathfrak{g}$. Then $d^* \varpi(g) := \sum_j d^* \varpi^j(g) \alpha^j \in \mathfrak{g}$ for $g \in G$.

Note that

$$\varpi^j(v) = \langle A^{\alpha^j}(g), v \rangle_g \qquad g \in G, v \in T_g G$$

for $A^{\alpha^j}(g) = T R_g(\alpha^j)$, the right invariant vector field corresponding to α^j. Therefore

$$d^* \varpi(g) = - \sum_j \operatorname{div} A^{\alpha^j}(g) \alpha^j \in \mathfrak{g} \qquad g \in G. \tag{20}$$

The divergence of a vector field measures the infinitesimal rate of change of Riemannian volume μ, say, under its flow. For us the Riemannian volume is a right Haar measure. However the flow of a right invariant vector field is left translation by its 1-parameter subgroup i.e. $L_{e^{t\alpha^j}}$ for A^{α^j}. It follows that if the right Haar measure is also left invariant, in other words if G is unimodular then $d^* \varpi = 0$. This holds in particular for G a compact Lie group; the situation of our main present interest. In general $(L_g)_* \mu$ is again right invariant and so a multiple $m(g)$ say of μ. This version $m : G \to \mathbf{R}(> 0)$ of the modular function of G is a group homomorphism. Since μ corresponds to a right invariant top dimensional form it is given by

$$m(g) = |\det \operatorname{Ad}_g|$$

for the adjoint action $\operatorname{Ad}_g = (T R_g)^{-1} T L_g : \mathfrak{g} \to \mathfrak{g}$.
Thus,

$$d^* \varpi^j(g) = -div A^{\alpha^j}(g) = -\frac{d}{dt}\frac{d\left((L_{e^{-t\alpha^j}})_*(\mu)\right)}{d\mu}\bigg|_{t=0} \quad (21)$$

$$= -\frac{d}{dt}|\det \mathrm{Ad}_{e^{-t\alpha^j}}|_{t=0} \quad (22)$$

$$= \mathrm{trace}\, \mathrm{ad}_{\alpha^j} = -\sum_k \langle \mathrm{ad}^*_{\alpha_k}\alpha_k, \alpha_j\rangle \quad (23)$$

for $\mathrm{ad}: \mathfrak{g} \to \mathbf{L}(\mathfrak{g}; \mathfrak{g})$ the adjoint representation, $\mathrm{ad}_\alpha(\beta) = [\alpha, \beta]$.

Lemma 3.2 *Let $\rho: G \to GL(V)$ be a smooth representation of G on a real finite dimensional vector space V and denote by $\rho_*: \mathfrak{g} \to \mathbf{L}(V; V)$ the derivative of ρ at the identity element.*

Let $\{g_t\}_t$ be Brownian motion on G from the identity.

Then $\rho(g_t)v$ is integrable for each $v \in V$ and $t \geq 0$ and its expectation is differentiable in t with

$$\frac{d}{dt}\mathbf{E}\{\rho(g_t)v\} = \lambda^\rho\left(\mathbf{E}\{\rho(g_t)v\}\right) \quad (24)$$

where $\lambda^\rho \in \mathbf{L}(V; V)$ is given by

$$\lambda^\rho = \frac{1}{2}\mathrm{Comp}\sum_j (\rho_*(\alpha^j) \otimes \rho_*(\alpha^j)) + \frac{1}{2}\sum_k \mathrm{ad}^*_{\alpha^k}\alpha^k \quad (25)$$

with

$$\mathrm{Comp}: \mathbf{L}(V; V) \otimes \mathbf{L}(V; V) \to \mathbf{L}(V; V)$$

the composition map $A \otimes B \mapsto AB$. For unimodular groups, and in particular for compact Lie groups, the term $\sum_k \mathrm{ad}^*_{\alpha^k}\alpha^k$ vanishes.

Proof By Itô's formula, as in equation (4.1) of [13],

$$\rho(g_t)(v) = v + M_t^{d\rho v} + \int_0^t \frac{1}{2}\Delta(\rho(-)v)(g_s)\, ds, \quad (26)$$

where $\{M_t^{d\rho v}\}_t$ is the continuous local martingale in V

$$M_t^{d\rho v} = \int_0^t d\rho_{g_s}(T_e R_{g_s} \circ dB_s)_v \quad (27)$$

where $\{B_s\}_{s\geq 0}$ is the Brownian motion on $\mathfrak{g}$ given by $dB_s := \varpi_{g_s} \circ dg_s$.

Now since $\rho: G \to GL(V)$ is a group homomorphism we see,

$$(d\rho)_k = \rho_* \circ \varpi_k(-)\rho(k): T_k G \to \mathbf{L}(V; V) \quad \text{for any } k \in G. \quad (28)$$

Thus

ns Higher Order Derivatives of Heat Semigroups ...

$$M_t^{d\rho v} = \int_0^t (\rho_*(dB_s)\rho(g_s)v). \tag{29}$$

Also, using the right invariance of the Laplacian,

$$\Delta(\rho)(g_s) = \Delta(\rho \circ R_{g_s})(e) = \Delta(\rho)(e)\rho(g_s) \in \mathbf{L}(V; V).$$

From (28) and (23) we see

$$\begin{aligned}\Delta(\rho)(e) &= -d^*(d\rho)(e) = -d^*(\rho(-)\rho_* \circ \varpi_-)(e) \\ &= \sum_j \rho_*(\alpha^j)\rho_*(\alpha^j) - d^*(\rho_* \circ \varpi)(e) \\ &= \sum_j \rho_*(\alpha^j)\rho_*(\alpha^j) + \rho_* \sum_k (\mathrm{ad})^*_{\alpha^k}\alpha^k \\ &= 2\lambda^\rho.\end{aligned}$$

Thus equation (26) reduces to the linear equation with constant coefficients

$$d\rho(g_t)(v) = \rho_*(dB_t)\rho(g_t)v + \lambda^\rho \rho(g_t)v\, dt \tag{30}$$

For compact Lie groups the result is immediate since the local martingale $\{M_t^{d\rho v}\}_t$ will be bounded and so a martingale. In general we can use a stopping time argument or the basic existence theorems for equations with Lipschitz coefficients to see that $\{\rho(g_t)(v)\}_{0 \le t \le T}$ is bounded in L^2 for each $T \ge 0$, so the local martingale has integrable quadratic variation and so is a martingale [26]. □

3.3 Calculation for S^n

We must calculate $\mathbf{E}\left\{\rho^{\odot^2}(g_t)(u_0 \odot v_0)\right\}$ to make use of our Hessian formula (19) for S^n. By Lemma 3.2 we have

$$\mathbf{E}\left\{\rho^{\odot^2}(g_t)(u_0 \odot v_0)\right\} = \mathbb{W}_t(u_0 \odot v_0) \tag{31}$$

where $\mathbb{W}_t = \mathbb{W}_t^{\rho^{\odot^2}} : T_{x_0}S^n \odot T_{x_0}S^n \to T_{x_0}S^n \odot T_{x_0}S^n$ satisfies $\mathbb{W}_0(u_0 \odot v_0) = u_0 \odot v_0$ and

$$\frac{d}{dt}\mathbb{W}_t(u_0 \odot v_0) = \lambda^{\rho^{\odot^2}}(\mathbb{W}_t(u_0 \odot v_0)) \quad t \ge 0.$$

Here $2\lambda^{\rho^{\odot^2}} = \sum_j \rho_*^{\odot^2}(\alpha^j)\rho_*^{\odot^2}(\alpha^j)$.

We will use a more algebraic formulation of our representation $\lambda^{\rho^{\otimes 2}}$ defined in (18):

3.3.1 Identification of $\mathfrak{m}$ with $T_{x_0} S^n$

As for any smooth left action of a Lie group we have a linear map $\alpha \mapsto X^\alpha$ from $\mathfrak{so}(n+1)$ to smooth vector fields on S^n. It is given by

$$X^\alpha(y) = \frac{d}{ds}((\exp s\alpha).y)|_{s=0}.$$

In particular the derivative $T_e p : \mathfrak{so}(n+1) \to T_{x_0} S^n$ at the identity of our projection $p : SO(n+1) \to S^n$ has $T_e p(\alpha) = X^\alpha(x_0)$. It is important, [5] page 182, or [18] page 469, to note the minus sign in the identity

$$[X^\alpha, X^\beta] = -X^{[\alpha,\beta]} \qquad \alpha, \beta \in \mathfrak{so}(n+1). \tag{32}$$

Let $\mathfrak{m}$ be the orthogonal complement of $\mathfrak{so}(n)$ in $\mathfrak{so}(n+1)$. A fundamental symmetric space property is that $\mathfrak{m}$ is invariant under the adjoint action, Ad, of $SO(n)$ on $\mathfrak{so}(n+1)$, and so under ad $: \mathfrak{so}(n) \to GL(\mathfrak{so}(n+1))$, its derivative at the identity e. There is the following important standard lemma with versions for more general symmetric spaces:

Lemma 3.3 *There are the commutative diagrams:*

1. For $g \in SO(n)$

$$\begin{array}{ccc} \mathfrak{m} & \xrightarrow{T_e p} & T_{x_0} S^n \\ \uparrow \mathrm{Ad}_g & & \uparrow TL_g \\ \mathfrak{m} & \xrightarrow{T_e p} & T_{x_0} S^n. \end{array}$$

2. For $\alpha \in \mathfrak{so}(n)$

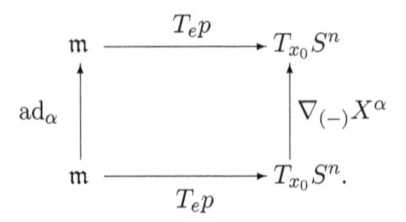

$$\begin{array}{ccc} \mathfrak{m} & \xrightarrow{T_e p} & T_{x_0} S^n \\ \uparrow \mathrm{ad}_\alpha & & \uparrow \nabla_{(-)} X^\alpha \\ \mathfrak{m} & \xrightarrow{T_e p} & T_{x_0} S^n. \end{array}$$

Proof For 1. :

$$TL_gT_ep(\alpha) = TL_gX^\alpha(x_0) = \frac{d}{ds}g\exp(s\alpha).x_0|_{s=0}$$
$$= \frac{d}{ds}g\exp(s\alpha)g.x_0|_{s=0} = X^{\mathrm{Ad}_g\alpha}(x_0)$$
$$= T_ep(\mathrm{Ad}_g\alpha).$$

For 2. : if $v \in \mathfrak{m}$

$$T_ep(\mathrm{ad}_\alpha(v)) = T_ep([\alpha, v]) = -[X^\alpha, X^v]$$
$$= -\nabla_{X^\alpha(0)}X^v + \nabla_{X^v(0)}X^\alpha$$
$$= \nabla_{X^v(0)}X^\alpha$$
$$= \nabla_{(-)}X^\alpha \circ T_ep(v).$$

$\square$

We will identify $\mathfrak{m}$ with $T_{x_0}S^n$ by T_ep. By the lemma the representation $\rho^{\odot^2}$: $SO(n) \to GL(T_{x_0}S^n \odot T_{x_0}S^n)$ gets identified with $\mathrm{Ad} \otimes \mathrm{Ad} : SO(n) \to GL(\mathfrak{m} \odot \mathfrak{m})$ using the restriction of the adjoint action. Then we have $\rho_*^{\odot^2}(\alpha) = \mathrm{ad}_\alpha \otimes \mathrm{Id} + \mathrm{Id} \otimes \mathrm{ad}_\alpha$, and so

$$\lambda^{\rho^{\odot^2}} = \frac{1}{2}\sum_j\{\mathrm{ad}_{\alpha^j} \circ \mathrm{ad}_{\alpha^j} \otimes \mathrm{Id} + \mathrm{Id} \otimes \mathrm{ad}_{\alpha^j} \circ \mathrm{ad}_{\alpha^j} + 2\mathrm{ad}_{\alpha^j} \otimes \mathrm{ad}_{\alpha^j}\}. \tag{33}$$

3.3.2 Curvature Identities

Let $R : TM \oplus TM \to L(TM; TM)$ denote the curvature tensor, with Kobayashi & Nomizu's convention, so for tangent vectors u, v, w at a point z we have a tangent vector $R(u, v)w$ at z, and for S^n :

$$R(u, v)w = \langle v, w\rangle u - \langle u, w\rangle v. \tag{34}$$

The Ricci curvature, Ric : $TM \oplus TM \to \mathbf{R}$ is given as the trace, $\mathrm{Ric}(u, v) = \mathrm{trace}\, R(-, u)v$, with $\mathrm{Ric}^\sharp : TM \to TM$ given by

$$\mathrm{Ric}^\sharp(u) = \sum_j R(u, e_j)e_j$$

for a suitable o.n. base.
For S^n:
$$\mathrm{Ric}(u, v) = (n-1)\langle u, v\rangle. \tag{35}$$

In our situation, from [18] page 231, and [5] page 193 taking account of Besse's different sign convention for R:

for $u, v, w \in \mathfrak{m}$ with $\mathfrak{m}$ identified with $T_{x_0} S^n$

$$R(u, v)w = -[[u, v], w]. \tag{36}$$

Noting that $\text{ad}_\alpha : \mathfrak{k} \to \mathfrak{k}$ is skew-symmetric for $\alpha \in \mathfrak{so}(n+1)$ we see from this that if also $a \in \mathfrak{m}$,

$$\langle R(u, v)w, a \rangle = \langle \text{ad}_w([u, v]), a \rangle \tag{37}$$
$$= -\langle \text{ad}_u v, \text{ad}_w a \rangle. \tag{38}$$

From this, for $u, v \in \mathfrak{m}$,

$$\text{Ric}(u, v) = -\text{trace}_\mathfrak{m} \text{ad}_u \text{ad}_v \tag{39}$$

Here we have written $\text{trace}_\mathfrak{m}$ to emphasise that the trace is taken for $\text{ad}_u \text{ad}_v : \mathfrak{m} \to \mathfrak{m}$. Indeed there are the fundamental relations:

$$[\mathfrak{g}, \mathfrak{g}] \subset \mathfrak{g}, \quad [\mathfrak{g}, \mathfrak{m}] \subset \mathfrak{m}, \quad [\mathfrak{m}, \mathfrak{m}] \subset \mathfrak{g} \tag{40}$$

where for us $\mathfrak{g} = \mathfrak{so}(n)$. See for example [5] page 193, or [18] page 226. Therefore ad_u interchanges $\mathfrak{g}$ and $\mathfrak{m}$ so

$$\text{Ric}(u, v) = -\text{trace}_\mathfrak{g} \text{ad}_u \text{ad}_v = -\frac{1}{2} \text{trace } \text{ad}_u \text{ad}_v \tag{41}$$

as in [5] page 194.

3.3.3 Decomposition of $V \odot V$

To go further we shall decompose $\mathfrak{m} \odot \mathfrak{m}$ into irreducible components for $\rho^{\odot^2}$.

For a real, n-dimensional, inner product space $V, \langle, \rangle$, the inner product, being symmetric and bilinear, determines a linear map $\langle - \rangle : V \otimes V \to \mathbf{R}$ given by

$$\langle u \otimes v \rangle = \langle u, v \rangle.$$

It is invariant under the action

$$u \odot v \mapsto Uu \odot Uv : U \in O(V)$$

of the orthogonal group $O(V)$ of V. Its kernel in $V \odot V$, denoted by $\mathcal{H}$, is therefore also invariant. It has codimension one and its elements are sometimes called "traceless" or "harmonic"; the latter because of the representation of symmetric tensors as homogeneous polynomials, [6, 16]. The space $V \otimes V$ has a distinguished element $\Xi := \sum_j e_j \odot e_j$ for $\{e_j\}_j$ an orthonormal basis of V. It corresponds to the identity when $V \otimes V$ is identified with $\mathbf{L}(V; V)$ using the inner product. Using the inner

Higher Order Derivatives of Heat Semigroups ...

product of $V \odot V$ inherited from that of $V \otimes V$ we see

$$\langle \Xi, u \odot v \rangle = \sum_j \langle u, e_j \rangle \langle v, e_j \rangle = \langle u \odot v \rangle.$$

Thus Ξ is the Riesz representative of $\langle - \rangle$ and so orthogonal to the kernel $\mathcal{H}$ and invariant under our orthogonal action. We write

$$V \odot V = \mathbf{R}\Xi \oplus \mathcal{H} \quad \text{with} \tag{42}$$

$$u \odot v = \frac{1}{n}\langle u, v \rangle \Xi \oplus \left(u \odot v - \frac{1}{n}\langle u, v \rangle \Xi \right). \tag{43}$$

Let $\mathcal{P}_\mathcal{H} : V \odot V \to V \odot V$ be the orthogonal projection onto $\mathcal{H}$, so

$$\mathcal{P}_\mathcal{H}(u \odot v) = u \odot v - \frac{1}{n}\langle u, v \rangle \Xi. \tag{44}$$

3.3.4 Computations

From (36), starting to compute $\lambda^{\odot^2}$ from formula (33), with our orthonormal base $\{\alpha^j\}_j$ for $\mathfrak{so}(n)$, and $u, v, a, b \in \mathfrak{m}$,

$$\sum_j \langle \mathrm{ad}_{\alpha^j} u \otimes \mathrm{ad}_{\alpha^j} v, a \odot b \rangle = \sum_j \langle \mathrm{ad}_u \alpha^j \otimes \mathrm{ad}_v \alpha^j, a \odot b \rangle$$

$$= \frac{1}{2} \sum_j \{\langle \mathrm{ad}_u a, \alpha^j \rangle \langle \mathrm{ad}_v b, \alpha^j \rangle + $$

$$+ \langle \mathrm{ad}_v a, \alpha^j \rangle \langle \mathrm{ad}_u b, \alpha^j \rangle\}$$

$$= -\frac{1}{2}\{\langle R(u, a)v, b \rangle + \langle R(v, a)u, b \rangle\}$$

$$= -\frac{1}{2}\{\langle R(u, a)v, b \rangle + \langle R(u, b)v, a \rangle\}$$

$$= -\langle R^\sharp(u, -)v, - \rangle(a \odot b), \tag{45}$$

where $R^\sharp(u, -)v \in \mathfrak{m} \odot \mathfrak{m}$ is the dual to $a \odot b \mapsto \frac{1}{2}\langle R(u, a)v + R(v, a)u, b \rangle$. For S^n using (34)

$$\langle R(u, a)v + R(v, a)u, b \rangle = \langle a, v \rangle \langle u, b \rangle + \langle b, v \rangle \langle u, a \rangle - 2\langle u, v \rangle \langle a, b \rangle$$
$$= 2\langle u \odot v, a \odot b \rangle - 2\langle u, v \rangle \langle a, b \rangle$$

whence

$$\sum_j \mathrm{ad}_{\alpha^j} u \otimes \mathrm{ad}_{\alpha^j} v = -u \odot v + \langle u, v \rangle \Xi. \tag{46}$$

Furthermore, using (41), for any $w \in \mathfrak{m}$

$$\sum_j \langle \mathrm{ad}_{\alpha^j} \circ \mathrm{ad}_{\alpha^j} u, w \rangle = -\sum_j \langle \mathrm{ad}_u \alpha^j, \mathrm{ad}_w \alpha^j \rangle$$

$$= \sum_j \langle \mathrm{ad}_w \circ \mathrm{ad}_u \alpha^j, \alpha^j \rangle$$

$$= -\operatorname{Ric}(u, w). \qquad (47)$$

We now see from (33), (47), (45)

$$\lambda^{\rho^{\odot^2}}(u \odot v) = -\frac{1}{2}\{\operatorname{Ric}^\sharp u \odot v + u \odot \operatorname{Ric}^\sharp(v)\} - R^\sharp(u, -)v. \qquad (48)$$

Using the explicit expressions, (35) and (46), for S^n this yields:

$$\lambda^{\rho^{\odot^2}}(u \odot v) = -(n-1)(u \odot v) - u \odot v + \langle u, v \rangle \Xi$$

$$= -n \mathcal{P}_\mathcal{H}(u \odot v). \qquad (49)$$

4 Main Result for S^n

Theorem 4.1 *For $x_0 \in S^n$ and u_0, v_0 in the tangent space $T_{x_0} S^n$ and a C^2 map $f : S^n \to \mathbb{R}$ the second derivative Hess $P_t f$ of the solution to the heat equation*

$$\frac{d}{dt} P_t f = \frac{1}{2} \Delta P_t f$$
$$P_0 f = f$$

is given by

$$\operatorname{Hess} P_t f(u_0, v_0) = \frac{1}{n}(1 - e^{-nt}) \langle u_0, v_0 \rangle P_t(\Delta f)(x_0) + e^{-nt} \mathbf{E}\{\operatorname{Hess}(f)_{x_t}(/\!/_t u_0, /\!/_t v_0)\}$$

$$= \frac{1}{n} \langle u_0, v_0 \rangle P_t(\Delta f)(x_0) + e^{-nt} \mathbf{E}\{\operatorname{Hess}(f)_{x_t} \mathcal{P}_\mathcal{H}(/\!/_t u_0, /\!/_t v_0)\}. \qquad (50)$$

In particular if u_0 and v_0 are orthogonal,

$$\|\operatorname{Hess} P_t f(u_0, v_0)\| \le e^{-nt} P_t(\|\operatorname{Hess} f\|)(x_0) \|u_0\| \|v_0\| \qquad t \ge 0. \qquad (51)$$

Proof From (19) and from (31),

$$\operatorname{Hess}(P_t f)(u_0, v_0) = \mathbf{E}\left\{\operatorname{Hess}(f)_{x_t}((/\!/_t \otimes /\!/_t)\mathbb{W}_t(u_0 \odot v_0))\right\}$$

where, using (31),

$$\frac{d}{dt}\mathbb{W}_t(u_0 \odot v_0) = \lambda^{\rho^{\odot^2}}(\mathbb{W}_t(u_0 \odot v_0)) \quad t \geq 0 \tag{52}$$

$$= -n\mathcal{P}_\mathcal{H}(\mathbb{W}_t(u_0 \odot v_0)). \tag{53}$$

Thus

$$\mathbb{W}_t(u_0 \odot v_0) = \frac{1}{n}\langle u_0, v_0\rangle \Xi + e^{-nt}\left(u_0 \odot v_0 - \frac{1}{n}\langle u_0, v_0\rangle \Xi\right)$$

$$= \frac{1}{n}(1 - e^{-nt})\langle u_0, v_0\rangle \Xi + e^{-nt}(u_0 \odot v_0).$$

Write $\Xi_t := (/\!/_t \otimes /\!/_t)\Xi$. We now see

$$\mathrm{Hess}(P_t f)(u_0, v_0) = \frac{1}{n}\langle u_0, v_0\rangle \mathbf{E}\{\mathrm{Hess}(f)_{x_t}(\Xi_t)\}$$

$$+ e^{-nt}\mathbf{E}\{\mathrm{Hess}(f)\mathcal{P}_\mathcal{H}(/\!/_t u_0 \odot /\!/_t v_0))\}$$

equivalently

$$\mathrm{Hess}(P_t f)(u_0, v_0) = \frac{1}{n}(1 - e^{-nt})\langle u_0, v_0\rangle \mathbf{E}\{\mathrm{Hess}(f)_{x_t} \Xi_t\}$$

$$+ e^{-nt}\mathbf{E}\{\mathrm{Hess}(f)_{x_t}(/\!/_t u_0 \odot /\!/_t v_0)\}.$$

Since $\mathrm{Hess}(f)_{x_t}\Xi_t = \Delta f(x_t)$ the results follow. □

4.1 Two Alternative Approaches

4.1.1 Algebraic

The form of formula (50) is not surprising given the symmetries of the sphere, and the decomposition of our representation of $\mathfrak{so}(n)$ into irreducible components. Indeed for $g \in \mathfrak{so}(n+1)$ and $y \in S^n$, we have $P_t f(gy) = P_t(f \circ g)(y)$. Using the fact that for $g \in \mathfrak{so}(n)$ left translation by g preserves the law of Brownian motion from x_0, we see that $\lambda^{\rho^{\odot^2}}$ must be invariant under the action of $\rho^{\odot^2}(\mathfrak{so}(n))$. It follows that it must be constant on the irreducible components $\mathcal{H}$ and $\mathbf{R}\Xi$ of $\mathfrak{m} \odot \mathfrak{m}$ for that action. Since we must have $\mathrm{Hess}(P_t f)(\Xi) = \Delta(P_t f)(x_0) = P_t(\Delta f)(x_0)$ we see the second constant must be zero. To compute the first constant we could proceed as in [13] Corollary 3.4.4, page 50 and relate $\sum_j \rho_*^{\odot^2}(\alpha_j) \circ \rho_*^{\odot^2}(\alpha_j)$ with the Casimir element of our representation, [17] 6.2. That way there need be no mention of curvature. However we have preferred to introduce curvature since it gives a geometric interpretation of the constants, and also our approach applies in greater generality.

4.1.2 Doubly Damped Parallel Translation

In [20] and [19] Xue-Mei Li obtains second derivative formulae on rather general Riemannian manifolds M by differentiating the standard first derivative formula $dP_t(f)(v_0) = \mathbf{E}\{df_{\xi_t(x_0)}W_t(v_0)\}$ with $\{W_t\}$ damped, or *Dohrn -Guerra*, parallel translation, and $\{\xi_t\}_t$, a gradient stochastic flow. This gives a term under the expectation of the form $df_{\xi_t(x_0)}\nabla_{u_0}W_t(v_0)$. If we filter out the redundant noise, i.e. condition, $\nabla_{u_0}W_t(v_0)$, this term becomes $df_{\xi_t(x_0)}W_t^{(2)}(u_0,v_0)$ for a certain process $W_t^{(2)}(u_0,v_0) \in T_{x_t}M$ which she calls the *doubly damped parallel translation*. For our sphere

$$W_t^{(2)}(u_0,v_0) = e^{-\frac{1}{2}(n-1)t}/\!/_t \int_0^t e^{-(n-1)s}\left(\langle u_0, v_0\rangle dB_s - \langle u_0, dB_s\rangle v_0\right) \tag{54}$$

for $\{B_t\}_t$ the stochastic anti-development of our Brownian motion on S^n, and her formula gives:

$$\begin{aligned}\text{Hess}(P_t f)(u_0, v_0) = e^{-(n-1)t}\,&\mathbf{E}\{\text{Hess}(f)(/\!/_t u_0, /\!/_t v_0)\}\\ &+\mathbf{E}\{df(W_t^{(2)}(u_0,v_0))\}.\end{aligned} \tag{55}$$

5 Extensions

5.1 Higher Order Derivatives

To consider 3rd order, or higher derivatives $\nabla^{(k)}d(P_t f)(u_0^k, ..., u_0^1, v_0)$, we have formula (14) but have to recall that the higher derivatives are not symmetric. To deal with this we could look at the representation theory of $so(n)$ on the full tensor algebra $\bigotimes^k \mathfrak{m}$ but this will involve sub-representations such as on $\wedge^k \mathfrak{m}$ which are not relevant to us. It seems easier to keep to the symmetric tensor products and then adjust with curvature terms as done for third derivatives below. For S^n or other symmetric spaces this is much helped by the vanishing of the covariant derivatives of the curvature.

5.1.1 Symmetrised Derivatives

For the symmetrised version, for each $p = 2, 3, ...$ we use the map

$$\mathcal{C}: \overset{p}{\bigotimes}\mathbf{R}^n \to \overset{p-2}{\bigodot}\mathbf{R}^n$$

given by

$$C(u^1 \otimes \ldots \otimes u^p) = C(u^1 \odot \ldots \odot u^p) = \sum_{i<j} \langle u^i, u^j \rangle \overset{k \neq i,j}{\underset{\cdot}{\odot}} u^k.$$

Let $\odot^p(\mathbf{R}^n)_\mathcal{H}$ denote the kernel of $\mathcal{C}$ in $\odot^p \mathbf{R}^n$. These are the *traceless* or *harmonic* elements. It is invariant under the representation $\rho^\odot$ of $SO(n)$:

$$\rho^\odot(g)(u^1 \odot \ldots \odot u^p) = (\rho(g)u^1 \odot \ldots \odot \rho(g)u^p)$$

for any given orthogonal representation ρ of $SO(n)$ on $\mathbf{R}^n$.

If ρ is irreducible we have the decomposition of $\odot^p \mathbf{R}^n$ into irreducible factors under $\rho^\odot$:

$$\overset{p}{\odot}\mathbf{R}^n = \overset{p}{\odot}(\mathbf{R}^n)_\mathcal{H} + \overset{p-2}{\odot}(\mathbf{R}^n)_\mathcal{H} \odot \Xi + \ldots + \overset{p-2k}{\odot}(\mathbf{R}^n)_\mathcal{H} \odot (\overset{k}{\odot} \Xi) + \cdots \quad (56)$$

For example see [6] or [16].

For $p = 3$ the decomposition is

$$u \odot v \odot w = \left(u \odot v \odot w - \frac{1}{n+2} \mathcal{C}(u \odot v \odot w) \odot \Xi \right)$$
$$\oplus \frac{1}{n+2} \mathcal{C}(u \odot v \odot w) \odot \Xi \quad (57)$$

We can give a precise formula for arbitrarily high symmetric derivatives:

Theorem 5.1 *For $p = 1, 2, \ldots$ and smooth $f : S^n \to \mathbf{R}$ the symmetrised p-th covariant derivative of the solution $P_t f$ to the heat equation*

$$\frac{\partial}{\partial t} P_t f = \frac{1}{2} \Delta P_t f \qquad P_0 f = f$$

is given by

$$\nabla^p(P_t f)(u_0^1 \odot \ldots \odot u_0^p) = \mathbf{E}\{\nabla^p(f) W_t^{[p]}(u_0^1 \odot \ldots \odot u_0^p)\} \quad (58)$$

where the damped parallel translation $W_t^{[p]} : \odot^p T_{x_0} M \to \odot^p T_{x_t} M$ *is given in terms of the decomposition (56) of $\odot^p T_{x_0} M$ by*

$$W_t^{[p]} = W_{\mathcal{H},t}^{[p]} + W_{\mathcal{H},t}^{[p-2]} \odot //_t|_\Xi + W_{\mathcal{H},t}^{[p-4]} \odot //_t|_\Xi \odot //_t|_\Xi + \ldots \quad (59)$$

where $//_t|_\Xi$ *refers to parallel translation restricted to Ξ, and*

$$W_{\mathcal{H},t}^{[q]} : \overset{q}{\odot}(T_{x_0} M)_\mathcal{H} \to \overset{q}{\odot}(T_{x_t} M)_\mathcal{H}$$

is the restriction of $W_t^{[q]}$ to the harmonic tensors and is given by

$$W_{\mathcal{H},t}^{[q]} U_0 = e^{-\frac{q}{2}(n+q-2)t} /\!/_t U_0 \qquad U_0 \in \overset{q}{\odot}(T_{x_0} S^n)_{\mathcal{H}}. \qquad (60)$$

Proof The same argument that gave formula (19) yields

$$\nabla^p (P_t f)(U_0) = \mathbf{E}\{\nabla^p (f)(/\!/_t \mathbf{E}\{\rho^\odot(g_t) U_0\})\} \qquad U_0 \in \overset{p}{\odot}(T_{x_0} S^n).$$

Now

$$\rho^\odot(V \odot \Xi) = \rho^\odot(V) \odot \Xi \qquad V \in \overset{q}{\odot}(T_{x_0} S^n)$$

so formulae (58) and (59) hold with $W_{\mathcal{H},t}^{[q]}$ the restriction of $\mathbf{E}\{\rho^\odot(g_t)\}$ to $\odot^q (T_{x_0} S^n)_{\mathcal{H}}$. To calculate this we have, by Lemma 3.2

$$\frac{d}{dt}\mathbf{E}\{\rho^\odot(g_t) U_0\} = \lambda^{\rho^\odot} \mathbf{E}\{\rho^\odot(g_t) U_0\}$$

for $\lambda^{\rho^\odot} = \frac{1}{2} \mathrm{Comp} \sum_r (\rho_*^\odot(\alpha^r) \otimes \rho_*^\odot(\alpha^r))$.

Since $\rho_*^\odot(\alpha^r)(u^1 \odot \ldots \odot u^q) = \sum_\ell \rho_*(\alpha^r) u^\ell \odot^{j \neq \ell} u^j$ we have

$$\rho_*^\odot(\alpha^r) \rho_*^\odot(\alpha^r)(u^1 \odot \ldots \odot u^q) = A^r + B^r$$

where

$$A^r = \sum_\ell \rho_*(\alpha^r)^2 u^\ell \odot^{j \neq \ell} u^j$$

and

$$B^r = 2 \sum_{j<k} \rho_*(\alpha^r) u^j \odot \rho_*(\alpha^r) u^k \odot^{\ell \neq j,k} u^\ell.$$

From formula (47) and the fact that $\mathrm{Ric}^\sharp(u) = (n-1)u$ for $u \in TS^n$ we have

$$\sum_r A^r(U_0) = -q(n-1)(U_0) \qquad U_0 \in \overset{q}{\odot}(T_{x_0} S^n),$$

while by (46), for $U_0 \in \odot^q (T_{x_0} S^n)_{\mathcal{H}}$,

$$\sum_r B^r(U_0) = -q(q-1) U_0 + 2\mathcal{C}(U_0) \odot \Xi = -q(q-1) U_0,$$

giving (60), to complete the proof. □

In particular for $p = 3$, using the decomposition (57) we obtain:

Corollary 5.2 *For $U_0 = u_0^1 \odot u_0^2 \odot u_0^3 \in \bigodot^3 T_{x_0}M$, with parallel translate $U_t \in \bigodot^3 T_{x_t}M$ along the Brownian paths, we have*

$$\nabla^3(P_t f)(U_0) = e^{-\frac{3}{2}(n+1)t} \mathbf{E}\{\nabla^3 f(U_t)\}$$
$$+ \frac{e^{-\frac{1}{2}(n-1)t}}{n+2}(1 - e^{-(n+2)t})\mathbf{E}\{\nabla^3 f(\mathcal{C}(U_t) \odot \Xi_t)\}. \quad (61)$$

5.1.2 The Full Derivative, P=3

Recall that with our use of Kobayashi & Nomizu's sign conventions the curvature $R^{\nabla^E} : TM \times TM \to \mathcal{L}(E; E)$ of a connection ∇^E on a vector bundle E over a manifold M is given by definition by

$$R^{\nabla^E}(u, v)S(x) = \nabla^E_u \nabla^E_V S - \nabla^E_v \nabla^E_U S - \nabla^E_{[U,V](x)} S \quad (62)$$
$$= (\nabla^E)^2 S(u, v) - (\nabla^E)^2 S(v, u) \quad (63)$$

for $u = U(x)$, $v = V(x)$ some $x \in M$, with U, V vector fields and S a section of E. To define $(\nabla^E)^2$ a torsion free connection on M is used.

For $E = TM$ with M Riemannian and $\nabla^E = \nabla$ the Levi-Civita connection, we write $R = R^{\nabla^E}$ as before. It is important to note that with the induced Levi-Civita connection on the cotangent bundle

$$R^{\nabla^{T^*M}}(u, v)\ell = -\ell \circ R(u, v) \quad \text{for } \ell \in T_x^*M.$$

This can be seen by computing the Hessian of the function $\phi(W(-))$ for ϕ a one-form and W a vector field, and using its symmetry. More generally, if S is a section of $(\bigotimes^p TM)^*$ then

$$\nabla^2 S(u, v)(w^1 \otimes \ldots \otimes w^p) - \nabla^2 S(v, u)(w^1 \otimes \ldots \otimes w^p) =$$
$$- S(x) \left(R(u, v)w^1 \otimes w^2 \otimes \ldots + \cdots + w^1 \otimes w^2 \otimes \ldots \otimes R(u, v)w^p \right) \quad (64)$$

for $u, v, w^1, \ldots w^p \in T_xM$.

Lemma 5.3 *For $u, v, w \in T_xM$ and $f : M \to \mathbf{R}$*

$$\nabla^2 df(u, v, w) = \nabla^2 df(u \odot v \odot w) + \frac{1}{3}df\left(R(v, u)w + R(w, u)v\right). \quad (65)$$

Proof By the symmetry of Hessians, $\nabla^2 df(u, v, w)$ is symmetric in v, w. Therefore

$$\nabla^2 df(u \odot v \odot w) = \frac{1}{3!}\nabla^2 df\left(2u \otimes v \otimes w + 2v \otimes u \otimes w + 2w \otimes u \otimes v\right).$$

Taking $S = df$ in (64)

$$\nabla^2 df(v \otimes u \otimes w) = \nabla^2 df(u \otimes v \otimes w) - df(R(v,u)w)$$

and

$$\nabla^2 df(w \otimes u \otimes v) = \nabla^2 df(u \otimes w \otimes v) - df(R(w,u)v),$$

giving the result by the symmetry of $\nabla^2 df$ in the last two variables. $\square$

As an example

Example 5.4

$$d\triangle f(x_0)(u) = \nabla^2 df(u \otimes \Xi) = \nabla^2 df(u \odot \Xi) - \frac{2}{3} df\left(\text{Ric}^\sharp(u)\right) \quad (66)$$

which enables us to rewrite (61) as

$$\nabla^3(P_t f)(U_0) = e^{-\frac{3}{2}(n+1)t} \mathbf{E}\{\nabla^3 f(U_t)\} +$$
$$\frac{e^{-\frac{1}{2}(n-1)t}}{n+2}(1 - e^{-(n+2)t}) \mathbf{E}\left\{\left(d\triangle f(\mathcal{C}(U_t)) + \frac{2n-2}{3} df(\mathcal{C}(U_t))\right)\right\}.$$

Theorem 5.5 *For $u_0, v_0, w_0 \in T_{x_0} S^n$ write $U_t = //_t u_0 \otimes //_t v_0 \otimes //_t w_0$ for parallel translation along a Brownian motion from x_0. Then for a C^3 function $f : S^n \to \mathbf{R}$*

$$\nabla^3 P_t f(U_0) = \mathbf{E}\{e^{-\frac{3}{2}(n+1)t} \nabla^3 f(U_t)$$
$$+ \frac{1}{n+2} e^{-\frac{1}{2}(n-1)t}(1 - e^{-(n+2)t}) d\triangle f(\mathcal{C}(U_t))$$
$$+ e^{-\frac{1}{2}(n-1)t}\left(1 - e^{-(n+2)t}\right) df(\frac{n}{n+2} \mathcal{C}(U_t) - \langle v_0, w_0\rangle //_t u_0)\}.$$

In particular if u_0, v_0, w_0 are mutually perpendicular,

$$\nabla^3 P_t f(U_0) = e^{-\frac{3}{2}(n+1)t} \mathbf{E}\{\nabla^3 f(U_t)\}. \quad (67)$$

Proof By formula (65)

$$\nabla^2 d P_t f(u_0, v_0, w_0) = \nabla^2 d P_t f(u_0 \odot v_0 \odot w_0) + \frac{1}{3} d P_t f(z_0)$$

for

$$z_0 = R(v_0, u_0)w_0 + R(w_0, u_0)v_0$$
$$= \langle u_0, w_0 \rangle v_0 + \langle u_0, v_0 \rangle w_0 - 2 \langle v_0, w_0 \rangle u_0$$
$$= \mathcal{C}(U_0^{\odot}) - 3 \langle v_0, w_0 \rangle u_0$$

for our sphere.

Now $d(P_t f)(z_0) = \mathbf{E}\{e^{-\frac{1}{2}(n-1)t} df(/\!/_t z_0)\}$ and

$$\frac{1}{3}/\!/_t z_0 + \frac{2(n-1)}{3(n+2)}(1 - e^{-(n+2)t}) \mathcal{C}(U_t^{\odot})$$
$$= -\langle v_0, w_0 \rangle /\!/_t u_0 + \left(\frac{n}{n+2} - \frac{2(n-1)e^{-(n+2)t}}{3(n+2)} \right) \mathcal{C}(U_t^{\odot}).$$

The result follows by using the formula in Example 5.4 with $U_0 = u_0 \odot v_0 \odot w_0$, together with (65) to go back again to our non-symmetrised U_t. □

Remark 5.6 Note that for all $p = 2, 3, ..n$, for a sphere S^n, formula (64) can be applied inductively to show that if $u^1, ..., u^p$ are mutually orthogonal then for any C^p function f

$$\nabla^p f(u^1, u^2, ..., u^p) = \nabla^p df(u^1 \odot u^2 ... \odot u^p)$$

For $y \in S^n$ set

$$\|\nabla^p f\|_{o.n.}(y) = \sup \{|\nabla^p f(v^1, ..., v^p)|, \text{ orthonormal } v^1, ..., v^p \in T_y S^n\}.$$

Applying (60) we obtain the pointwise semi-group domination,

$$\|\nabla^p P_t f\|_{o.n.} \leq e^{-\frac{p}{2}(n+p-2)t} P_t(\|\nabla^p f\|_{o.n.}) \qquad (68)$$

5.2 More General Diffusion Semi-groups

5.2.1 Heat Semigroups on Functions and Forms on Compact Riemannian Symmetric Spaces

The method described here for spheres should go over directly for the heat semi-group of a compact Riemannian symmetric space. The representation theory involved may be more complicated. It should extend similarly to formulae for derivatives of heat semigroups for forms. See also the alternative approach suggested in Sect. 4.1.

An additional first order term can be included by combining this method with the more standard method of filtering out redundant noise, but the formulae will be more complicated unless the term comes from a Killing vector field.

5.2.2 General Diffusions on Manifolds; Derivatives of Induced Semigroups on Functions, Forms, Jets Etc

For a heat equations on a general compact Riemannian manifold M a similar approach can be followed but replacing our bundle $p : SO(n+1) \to S^n$ by a bundle $p : \text{Diff}(M) \to M$ where $\text{Diff}(M)$ is a suitable group of diffeomorphisms of M and p the evaluation map at a base point $x_0 \in M$. This can be considered as a principal bundle with group those elements $\text{Diff}_{x_0}(M)$ of $\text{Diff}(M)$ which fix x_0. Our stochastic flow can be considered as a process on $\text{Diff}(M)$ and has a skew product decomposition generalising that described in Sect. 3.1. See [11], [13]. This gives rise to formulae like (19) and its higher order analogues, but with lower order derivative terms because the second derivative of the flow will not generally vanish: for symmetric spaces it vanished because we had a flow of isometries.

In this case we should look at k-jets of $P_t f$ rather than k-th order covariant derivatives. However the parallel translation in our formula may now not be metric preserving for any metric on the bundle of k-th order tangent vectors (essentially k-th order differential operators), or its dual, the k-jets. This makes uniform estimates difficult to obtain. For tensor bundles, associated to the frame bundle of M, such as $\bigotimes^k TM$, an SDE can be chosen for our diffusion so that its conditioned flow determines Levi-Civita parallel translation, and so is metric preserving, see [10]. For k-th order tangent vectors the question is crucial but open.

A result by Mendes & Redeschi, [24], shows that we cannot have the Levi-Civita connection on tensor bundles induced by an SDE for Brownian motion which has a solution flow of isometries except in the case we have been discussing for symmetric spaces. They call an SDE for Brownian motion which induces the Levi-Civita connection a *virtual immersion*. Ming Liao [22] has somewhat related negative results; in particular there are no isometric stochastic flows of Brownian motions on a Riemannian symmetric space of non-compact type. However Liao shows in [23] that for $n > 3$ there are a continuum of them on S^n.

In [14], in preparation, this set up is extended to a wide class of semi-groups induced on sections of natural bundles, such as jet bundles, by a sum of squares representation of a diffusion operator on M. The operator need not be elliptic or hypo-elliptic but for the method to work smoothly it should be *cohesive* in the sense of [13]. However, the crucial question of finding stochastic flows inducing metric connections on natural bundles remains open, to my knowledge.

5.3 Questions

- [Berger's spheres.] It would be interesting to see how the derivatives of the heat semigroup change as the sphere gets smoothly deformed, for example for Berger's spheres which still retain a lot of symmetry; see [15] and for a stochastic analytical discussion and more references [21].

- [Different symmetric space structures.] The same Riemannian manifold can have different symmetric space structures. For example the 3-sphere is also a Lie group and so has the symmetric space structure with group $S^3 \times S^3$ acting by $(g^1, g^2).a = g^1 a (g^2)^{-1}$ for $a \in S^3$. Can such different structures give different derivative formulae?
- [Non-compact type] Are there corresponding formulae for symmetric spaces of non-compact type? In particular for hyperbolic space. As remarked above, [22] and [24] imply that the use of isometric flows as here does not go over immediately to the non-compact case.

Acknowledgements This was written for the 80th Birthday of Sergio Albeverio, and I am very happy to be able to record my appreciation of Sergio as a mathematical colleague, and my enjoyment at having known him personally for close to half of those 80 years. Thanks also to the organisers of the joyous and stimulating workshop in the beautiful city of Verona in honour of that birthday.

References

1. Bakry, D.L.: On Sobolev and logarithmic Sobolev inequalities for Markov semigroups. In: New Trends in Stochastic Analysis (Charingworth, 1994), pp. 43–75. World Scientific Publishing, River Edge, NJ (1997)
2. Bakry, D., Michel Émery: Propaganda for Γ_2. In: From Local Times to Global Geometry, Control and Physics (Coventry, 1984/85), Pitman Research Notes in Mathematics Series, vol. 150, pp. 39–46. Longman Scientific and Technical, Harlow (1986)
3. Bakry, D., Gentil, I., Ledoux, M.: Analysis and geometry of Markov diffusion operators. Grundlehren der Mathematischen Wissenschaften [Fundamental Principles of Mathematical Sciences], vol. 348. Springer, Cham (2014)
4. Baxendale, Peter: Brownian motions in the diffeomorphism group. I. Compos. Math. **53**(1), 19–50 (1984)
5. Besse, A.L.: Einstein manifolds. Classics in Mathematics. Springer, Berlin (2008). Reprint of the 1987 edition
6. Bettiol, R.G., Mendes, R.A.E.: Sectional curvature and Weitzenböck formulae, Aug. 2017. arXiv:1708.09033
7. Elworthy, K.D.: Stochastic Differential Equations on Manifolds. LMS Lecture Notes Series, vol. 70. Cambridge University Press (1982)
8. Elworthy, K.D.: Generalised Weitzenböck formulae for differential operators in Hörmander form. In: Stochastic Partial Differential Equations and Related Fields, volume 229 of Springer Proceedings in Mathematics and Statistics, pp. 319–331. Springer, Cham (2018)
9. Elworthy, K.D., Kendall, W.S.: Factorization of harmonic maps and Brownian motions. In: From Local Times to Global Geometry, Control and Physics (Coventry, 1984/85), Pitman Research Notes in Mathematics Series, vol. 150, pp. 75–83. Longman Scientific and Technical, Harlow (1986)
10. Elworthy, K.D., Le Jan, Y., Li, X.-M.: On the geometry of diffusion operators and stochastic flows. Lecture Notes in Mathematics, vol. 1720. Springer (1999)
11. Elworthy, K.D., Le Jan, Y., Li, X.-M.: Equivariant diffusions on principal bundles. In: Stochastic Analysis and Related Topics in Kyoto, volume 41 of Advanced Studies in Pure Mathematics, pp. 31–47. The Mathematical Society of Japan, Tokyo (2004)
12. Elworthy, K.D., Yor, M.: Conditional expectations for derivatives of certain stochastic flows. In: Azéma, J., Meyer, P.A., Yor, M. (eds.), SEM de Problem XXVII. Lecture Notes in Mathematics, vol. 1557, pp. 159–172. Springer (1993)

13. David Elworthy, K., Le Jan, Y., Li, X.-M.: The geometry of filtering. Frontiers in Mathematics. Birkhäuser Verlag, Basel (2010)
14. Elworthy, K.D.: Generalised Weitzenböck formulae for differential operators in Hörmander form ii: natural bundles and higher order derivative formulae. In preparation (2019)
15. Gadea, P.M., Oubiña, J.A.: Homogeneous Riemannian structures on Berger 3-spheres. Proc. Edinb. Math. Soc. **48**(2), 375–387 (2005)
16. Goodman, R., Wallach, Nolan, R.: Representations and invariants of the classical groups. Encyclopedia of Mathematics and its Applications, vol. 68. Cambridge University Press, Cambridge (1998)
17. Humphreys, J.E.: Introduction to Lie algebras and representation theory. Volume 9 of Graduate Texts in Mathematics. Springer, New York (1978). Second printing, revised
18. Kobayashi, S., Nomizu, K.: Foundations of differential geometry, Vol. II. Interscience Publishers (1969)
19. Li, X.-M.: Hessian formulas and estimates for parabolic Schrödinger operators. J. Stoch. Anal. **2**(3): 7, 53 (2021). arXiv:1610.09538
20. Li, X.-M.: Doubly damped stochastic parallel translations and Hessian formulas. In: Stochastic Partial Differential Equations and Related Fields, Volume 229 of Springer Proceedings in Mathematics and Statistics, pp. 345–357. Springer, Cham (2018)
21. Li, X.-M.: Homogenisation on homogeneous spaces. J. Math. Soc. Jpn. **70**(2), 519–572 (2018). With an appendix by Dmitriy Rumynin
22. Liao, M.: The existence of isometric stochastic flows for Riemannian Brownian motions. In: Diffusion processes and related problems in analysis, vol. II (Charlotte, NC, 1990), volume 27 of Progress in Probability, pp. 95–109. Birkhäuser Boston, Boston, MA (1992)
23. Liao, Ming: Isometric stochastic flows on spheres. J. Theoret. Probab. **12**(2), 475–488 (1999)
24. Mendes, R.A.E., Radeschi, M.: Virtual immersions and a characterization of symmetric spaces. Ann. Global Anal. Geom. **55**(1), 43–53 (2019)
25. Norris, J.R.: Path integral formulae for heat kernels and their derivatives. Probab. Theory Related Fields **94**(4), 525–541 (1993)
26. Revuz, D., Yor, M.: Continuous Martingales and Brownian Motion, Volume 293 of Grundlehren der Mathematischen Wissenschaften [Fundamental Principles of Mathematical Sciences], 3rd edn. Springer, Berlin (1999)
27. Thompson, J.: Approximation of Riemannian measures by Stein's method, Jan. 2020. arXiv:2001.09910

Rough Homogenisation with Fractional Dynamics

Johann Gehringer and Xue-Mei Li

Dedicated to Sergio Albeverio on the occasion of his 80th birthday

Abstract We review recent developments of slow/fast stochastic differential equations, and also present a new result on Diffusion Homogenisation Theory with fractional and non-strong-mixing noise and providing new examples. The emphasise of the review will be on the recently developed effective dynamic theory for two scale random systems with fractional noise: Stochastic Averaging and 'Rough Diffusion Homogenisation Theory'. We also study the geometric models with perturbations to symmetries.

Keywords Homogenisation · Stochastic averaging · Fractional Brownian motion · Multi-scale · Rough paths

1 Introduction

When we study the evolution of a variable/quantity, which we denote by x, we often encounter other interacting variables which either have the same scale as x and are therefore treated equally, or are much smaller in size or slower in speed

---

J. Gehringer research supported by an EPSRC-Roth scholarship.
X.-M. Li research partially supported by EPSRC grant CDT EP/S023925/1.

J. Gehringer · X.-M. Li (✉)
Imperial College London, London, UK
e-mail: xue-mei.li@imperial.ac.uk

J. Gehringer
e-mail: johann.gehringer18@imperial.ac.uk

© Springer Nature Switzerland AG 2021
S. Ugolini et al. (eds.), *Geometry and Invariance in Stochastic Dynamics*,
Springer Proceedings in Mathematics & Statistics 378,
https://doi.org/10.1007/978-3-030-87432-2_8

and are essentially negligible, or they might evolve in a microscopic scale ε, such variables are called the fast variables which we denote by y. It happens often that y is approximately periodic or has chaos behaviour or exhibits ergodic properties, then its effect on the x-variables can be analysed. During any finite time on the natural scale of x, the y-variable will have explored everywhere in its state space. As $\varepsilon \to 0$, the persistent effects from the fast variables will be encoded in the 'averaged' slow motions through adiabatic transformation. We then expect that $\lim_{\varepsilon \to 0} x_t^\varepsilon$ exists; its limit will be autonomous, not depending on the y-variables. In other words, the action of y is transmitted adiabatically to x and the evolution of x can be approximated by that of an autonomous system called the effective dynamics of x.

We explain this theory with the two scale slow/fast random evolution equation

$$dx_t^\varepsilon = f(x_t^\varepsilon, y_t^\varepsilon)\,dt + g(x_t^\varepsilon, y_t^\varepsilon)\,dB_t. \tag{1.1}$$

Here $\varepsilon > 0$ is a small parameter, y_t^ε is a fast oscillating noise, and B_t is another noise. The stochastic processes B_t and y_t^ε will be set on a standard probability space $(\Omega, \mathcal{F}, \mathcal{P})$ with a filtration of σ-algebra $\mathcal{F}_t$. Typically the sample paths of the stochastic processes $t \mapsto B_t(\omega)$, and $t \mapsto y_t^\varepsilon(\omega)$ are not differentiable but have Hölder regularities. The equation is then interpreted as the integral equation:

$$x_t^\varepsilon = x_0 + \int_0^t f(x_s^\varepsilon, y_s^\varepsilon)\,ds + \int_0^t g(x_s^\varepsilon, y_s^\varepsilon)\,dB_s. \tag{1.2}$$

We take the initial values to be the same for all ε. The x-variables are usually referred to as the slow variables. If the fast dynamics depend on the slow variables, we refer this as the 'feedback dynamics'. If the fast variables do not depend on the slow variables, we have the 'non-feedback dynamics'.

If (B_s) is a Brownian motion, the integral is an Itô integral. The solutions of a Itô stochastic differential equation (SDE) are Markov processes, they have continuous sample paths and therefore diffusion processes. Within the Itô calculus realm, the study of two scale systems began in the 1960s, almost as soon as a rigorous theory of Itô stochastic differential equations was established, and has been under continuous exploration. In the averaging regime, the effective dynamics are obtained by averaging the original system in the y-variable. This non-trivial dynamical theory is related to the Law of Large Numbers (LLN) and the ergodic theorems. The averaged dynamics for the Markovian system is expected to be again the Markov process whose Markov generator is obtained by averaging the y-components in the family of Markov generators $\mathcal{L}^y$ of the slow variables with a parameter y. Stochastic Averaging for Markovian ordinary differential equations was already studied in the 1960s and 1970s in [52, 86, 105, 106], [12, 39, 68, 104]. See also [7, 39, 40, 57, 97, 102, 113]. Stochastic averaging with periodic and stationary vector fields from the point of view of dynamical systems is a related classic topic, see [61]. For more recent work, see [2, 10, 62, 85]. Stochastic averaging on manifolds for Markovian systems are studied in [1, 9, 17, 73, 74, 100]. In the homogenisation regime, this theory is linked to Functional Central Limit Theorems (CLTs). In the classical setting this falls within the theory of diffusion creation, we therefore refer it as the diffusive homogenisation theory. A meta functional CLT is as follows [54, 69]: Let f be such that $\int f d\pi = 0$ where π is the unique invariant probability measure of a Markov process Y_s, then $\frac{1}{\sqrt{t}} \int_0^t f(Y_s) ds$ converges to a Markov process. Such limit theorems are the foundation for diffusive homogenisation theorems and for studying weak interactions. If the dynamics is Markovian, we naturally expect the limit of the slow variable to be another Markov process. This theory is known as homogenisation; let us call this 'diffusive homogenisation' to distinguish it from the settings where the dynamics is fractional. For Markovian systems, there are several well developed books, see e.g. [63, 102], see also [11]. Diffusive homogenisation was studied in [16, 24, 58, 96], see also [10, 79].

An Itô type stochastic differential equation is a good model if the randomness is obtained under the assumptions that there are a large number of independent (or weakly correlated) components. However, long range dependence (LRD) is prevalent in mathematical modelling and observed in time series data such as economic cycles and data networks. One of the simplest LRD noise is the fractional noise, which is the 'derivative' of fractional Brownian motions (fBM). Fractional Brownian motions was popularised by Mandelbrot and Van Ness [84] for modelling the long range dependent phenomenon observed by H. Hurst [53]. This is a natural process to use. Within the Gaussian class, those with stationary increments and the covariance structure $\mathbb{E}(B_t - B_s)^2 = (t-s)^{2H}$ are necessarily fractional Brownian motions with similarity exponent/Hurst parameter H. Fractional Gaussian fields and strongly correlated fields are used to study critical phenomena in mathematical physics, see e.g. R. L. Dobrushin, G. Jona-Lasinio, G. Pavoliotti [26, 42, 59], and Ja. G. Sinai [103].

If B_t is a fractional BM of Hurst parameter H, the stochastic integral in (1.2) is a Riemann-Stieljes integral if $H > \frac{1}{2}$. Otherwise this can be understood in the sense of rough path integration or fractional calculus. We explain the essence for this using the basics in the rough path theory. The instrument for this is the 'Young bound' [114]: if $F \in \mathcal{C}^\alpha$ and $b \in \mathcal{C}^\beta$ with $\alpha + \beta > 1$

$$\left| \int_r^t (F_s - F_0) db_s \right| \lesssim |F|_\alpha |b|_\beta (t-r)^{\alpha+\beta}.$$

With this, Young showed that the map $(F, b) \mapsto \int_0^t F_s db_s$ is continuous where

$$\int_0^t F_s db_s = \lim_{|\mathcal{P}| \to 0} \sum_{[u,v] \subset \mathcal{P}} F_u (b_u - b_v) \in \mathcal{C}^\beta,$$

a Riemannan Stieljes integral/Young integral. This argument also established the continuity of the solutions to Young equations. The solution x_t inherits the regularity of the driver and is in $\mathcal{C}^\beta$. This means if $H > \frac{1}{2}$, the equation

$$dx_t = b(x_t) dt + F(x_t) dB_t$$

below can be interpreted as a Young integral equation. It is well posed if F, b are in BC^3 [81]. In general this type of SDE can be made sense of if the Hurst parameter of the fBM B_t is greater than $\frac{1}{4}$, see [23, 27, 48] and also [70, 82]. The study of stochastic evolution equations with fractional Brownian motions has since become popular, see e.g. [3, 15, 20, 36, 41, 56, 88] and [22, 91] for their study in mathematical finance and in economics. See also [46].

Despite of this popularity of stochastic equations with fractional noise, there had not been much activity on the effect dynamics of a multi-scale systems. A stochastic averaging with LRD fractional dynamics is obtained in [55] for the without feedback case and also for a feedback Markovian dynamics y_t^ε which solves the following interacting SDE:

$$dy_t^\varepsilon = \frac{1}{\sqrt{\varepsilon}} \sum_{k=1}^{m_2} Y_k(x_t^\varepsilon, y_t^\varepsilon) \circ dW_t^k + \frac{1}{\varepsilon} Y_0(x_t^\varepsilon, y_t^\varepsilon) \, dt, \quad y_0^\varepsilon = y_0 \quad (1.3)$$

where W_t^i are independent Wiener processes. A uniform ellipticity is assumed of the equation. A priori the fast variables $y_t^\varepsilon \in C^{\frac{1}{2}-}$. Since the sample paths of B_t is in C^{H-}, for the LRD case where $H > \frac{1}{2}$, (1.1) is a Young equation. More precisely, for ω fixed,

$$\left(\int_0^t g(x_s^\varepsilon, y_s^\varepsilon) dB_s \right)(\omega) := \int_0^t g(x_s^\varepsilon(\omega), y_s^\varepsilon(\omega)) dB_s(\omega),$$

is a Riemannan Stieljes integral/Young integral. The solution x_t^ε inherits the regularity of B_t^H and is in C^{H-}. These integrals are not defined with the classic Itô stochastic calculus in any natural way, it is therefore reasonable to use this pathwise interpretation. A solution theory for the equation (1.1) and (1.3) also exist, see [47, 55].

As mentioned earlier, if B_t is a BM, the classic averaging theory states that the effective dynamics is the Markov process with its generator obtained by averaging the Markov generators of the slow variables. This is obtained within the theory of Itô calculus. However standard analysis within the integration theory does not lead to 'pathwise' estimates on x_t^ε that are uniform in ε, which means a pathwise limit theory is not to be expected. It is clear that the 'Young bounds' are totally ineffective for obtaining the essential uniform pathwise estimates for $\left(\int_0^t g(x_s^\varepsilon, y_s^\varepsilon) dB_s \right)(\omega)$. When $\varepsilon \to 0$, the Hölder norm of the y^ε is expected to blow up. If g does not depend on the fast variables, it is of course possible to obtain pathwise bounds. Indeed, the sewing lemma of Gubinelli [49] and Feyel-de la Pradelle [34], which neatly encapsulates the main analytic estimates of both the work of Young and that of Lyons [81], and has since become a fundamental tool in pathwise integration theory, does not provide the required estimates for the feedback dynamics. Without any uniform pathwise estimates, the slow variables (for a generic equation) cannot be shown to converge for fixed fBM path.

In [55], a novel approximation, of the pathwise Young integrals by Wiener integrals, was introduced with the help of the stochastic sewing lemma of Lê [71]. This approach used, paradoxically, the stochastic nature of the fractional Brownian motion in an essential way and, therefore, effectively departed the pathwise framework. Since Itô integrals and Wiener integrals are defined as an element of $L^2(\Omega)$, the uniform estimates are L^p-estimates and thus the limit theorem is an 'annealed' limit. It was shown, [55], that x_t^ε converges in joint probability to the solution of the following equation with the same initial data as x_0^ε:

$$dx_t = \bar{f}(x_t)\,dt + \bar{g}(x_t)\,dB_t.$$

where $\bar{f}$ and $\bar{g}$ are obtained by directly averaging f and g respectively. Stochastic averaging with fractional dynamics is now a fast moving area, see [94, 95, 101] [?] for more recent work see [8, 30, 80].

For the homogenisation theory the main references are [44, 45]; see also [43] which is the preliminary version of the previous two articles, equations of the form

$$\dot{x}_t^\varepsilon = h(x_t^\varepsilon, y_t^\varepsilon)$$

are studied. They can be used to model the dynamics of a passive tracer in a fractional and turbulent random environment.

In [43, 45], a functional limit theorem is obtained. The limit theorems are build upon the results in [14, 25, 87–89, 98, 109, 111, 112]. In [43, 44], a homogenisation theorem for random ODE's for fractional Ornstein-Uhlenbeck processes. See also [38, 66]. Since the tools for diffusive homogenisation do not apply, we have to rely on a theorem from the theory of rough path differential equations. This approach is close to that in [18, 19, 65, 93], see also [4, 13, 35]. However, in these references only the dynamics are Markovian and the results are of diffusive homogenisation type. In [43–45], the fast dynamics is a fractional Ornstein-Uhlenbeck process and the effective dynamics are not necessarily Markov processes and the limiting equation is a rough differential equation. We refer this theory as 'rough creation' theory and 'rough homogenisation' theory.

The study for the stochastic averaging theory and the homogenisation theory for fractional dynamics has just started. These theories departed from the classical theory both in terms of the methods of averaging, the techniques, and the effective dynamics. We will compare the methodologies and obtain the following intermediate result. for $y_t^\varepsilon = y_{t/\varepsilon}$ where y_t is a stationary stochastic process with stationary distribution μ. Let G_k be a collection of real valued $L^2(\mu)$ functions on $\mathbb{R}$ we define

$$X_t^{k,\varepsilon} = \sqrt{\varepsilon}\int_0^{\frac{t}{\varepsilon}} G_k(y_s)\,ds, \qquad X_t^\varepsilon = \left(X_t^{1,\varepsilon}, \ldots, X_t^{n,\varepsilon}\right)$$

1.1 Main Results

Our main result, c.f. Sect. 2, can be proved using the methods in [43]. The statement is as follows:

Theorem 1 *Let y_t be a real valued stationary and ergodic process with stationary measure μ. Let $G_k : \mathbb{R} \to \mathbb{R}$ be L^2 functions satisfying*

$$\int_0^\infty \|\mathbb{E}[G_k(y_s)|\mathcal{F}_0]\|_{L^2}\, ds < \infty. \tag{1.4}$$

Suppose that furthermore the following moment bounds hold

$$\|X^{k,\varepsilon}_{s,t}\|_{L^p} \lesssim |t-s|^{\frac{1}{2}}, \qquad \|\mathbb{X}^{i,j,\varepsilon}_{s,t}\|_{L^{\frac{p}{2}}} \lesssim |t-s|,$$

and the functional central limit theorem holds (i.e. X^ε_t converges jointly in finite dimensional distributions to a Wiener process $X_t = (X^1_t, \ldots, X^n_t)$). Then the following statements hold.

(i) *The canonical lift $\mathbf{X}^\varepsilon = (X^\varepsilon, \mathbb{X}^\varepsilon)$ converges weakly in C^γ for $\gamma \in (\frac{1}{3}, \frac{1}{2} - \frac{1}{p})$.*
(ii) *As $\varepsilon \to 0$, the solutions of*

$$\dot{x}^\varepsilon_t = \sum_{k=1}^N \frac{1}{\sqrt{\varepsilon}} f_k(x^\varepsilon_t)\, G_k(y^\varepsilon_t), \qquad x^\varepsilon_0 = x_0$$

converges to the solution of the equation $dx_t = \sum_{k=1}^n f_k(x_t) \circ dX^k_t$.

We will discuss examples of y^ε_t and G_k for which the above holds including those discussed, see Sect. 2.4.

2 Homogenization via Rough Continuity

With the Law of Large Number theory in place we now explain the homogenisation theory. The homogenisation problem is about fluctuations from the average. We therefore take

$$\dot{x}^\varepsilon_t = h(x^\varepsilon_t, y^\varepsilon_t), \tag{2.1}$$

where $h(x, y)$ is a function averaging to zero and $h(x, y^\varepsilon_t)$ is a fast oscillating fractional nature moving at a microscopic scale ε. The aim is to obtain an 'effective' closed equation whose solution $\bar{x}_t$ approximates x^ε_t. This effective dynamics will have taken into accounts of the persistent averaging effects from the fast oscillations. Homogenisation for random ODEs has been dominated by the diffusion creation theory, with only a handful of exceptions where the limit is a fBM. We will obtain a range of dynamics with local self-similar characteristics of the fractional Brownian motion, Brownian motion, and Hermite processes.

Recall that the fractional Brownian motion is a Gaussian process with stationary increments. Its Hurst parameter H, given by its covariance structure: $\mathbb{E}(B_t - B_s)^2 = (t-s)^{2H}$, indicates also the exponent in the power law decay of the corresponding fractional noise. Indeed, the correlation of two increments of the fractional Brownian motion of length 1 and time t apart,

$$\mathbb{E}(B_{t+s+1} - B_{t+s})(B_{s+1} - B_s) = \frac{1}{2}(t+1)^{2H} + \frac{1}{2}(t-1)^{2H} - t^{2H}$$

which is approximately $2Ht^{2H-2}$ for large t.

2.1 CLT for Stationary Processes

Given a stochastic process y_s we are concerned with the question, whether for some scaling $\alpha(\varepsilon)$ and function $G : \mathbb{R} \to \mathbb{R}$ the following term converges in the sense of finite dimensional distributions

$$X_t^\varepsilon = \alpha(\varepsilon) \int_0^{\frac{t}{\varepsilon}} G(y_s) ds.$$

Usual functional central limit theorems would set $\alpha(\varepsilon) = \sqrt{\varepsilon}$ and the limit would be a Wiener process. For Markovian noises these question has been studied a lot, see e.g. [69] and the book [63] and the references therein. The basic idea is: If $\mathcal{L}$ is the Markov generator and the Poisson equation $\mathcal{L} = G$ is solvable, with solution in the domain of $\mathcal{L}$, then the central limit theorem holds. This follows from the martingale formulation for Markov processes.

Another kind of condition often imposed on the noise is some kind of mixing condition. Processes that satisfy these conditions are in a sense nicely behaved as they obey the usual CLT. However, in this section we aim to treat cases in which non of this conditions are satisfied. For example the fractional Ornstein-Uhlenbeck process for $H > \frac{1}{2}$ is neither Markovian nor obeys usual mixing assumptions. Looking at this question from a Gaussian perspective Rosenblatt [99] gave an example of a stationary Gaussian sequence X_k such that $\frac{1}{\sqrt{N}} \sum_{k=1}^N X_k$ does not converge, however the right scaling $\frac{1}{N^\alpha} \sum_{k=1}^N X_k$, for a suitable $\alpha > \frac{1}{2}$ converges to the so called Rosenblatt process. Taqqu and Dobrushin [26, 108, 110] added to this work of so called non-central limit theorems. Philosophically, if the covariance function $\rho(j) = \mathbb{E}[X_0 X_j]$ does not decay fast enough the limiting distribution can not have independent increments. The notions of short and long range dependence capture this idea. We say that a sequence is short range dependent if $\sum_{j=1}^\infty |\rho(j)| < \infty$ and long range dependent otherwise.

We now want to discuss a method to conclude convergence to a Wiener process for these kind of processes. In case of a Gaussian noise y_s the rich toolboxes of Malliavin calculus, in particular the fourth moment theorem enables one to conclude limit theorems for a wide variety of situations, c.f. [88–90, 92]. However, the method we are going to explore further relies on a martingale approximation method, see [60] Theorem 3.79.

The idea is to impose a condition on the functional instead on the noise in order to obtain a decomposition, as in the Markovian case, into a martingale and coboundary term.

Furthermore, the method of martingale approximation can be used to obtain convergence of the lifted process via Theorem 2.2 in [67], cf. [6, 64]. Given a function U satisfying Assumption 2.7 we may define the following L^2 martingale

$$M_t = \int_0^\infty \mathbb{E}\left[U(y_r)|\mathcal{F}_{t+}\right] - \mathbb{E}\left[U(y_r)|\mathcal{F}_{0+}\right] dr.$$

Now we may decompose X^ε as follows

$$X_t^\varepsilon = \sqrt{\varepsilon} \int_0^{\frac{t}{\varepsilon}} U(y_r) dr = \sqrt{\varepsilon} M_{\frac{t}{\varepsilon}} + \sqrt{\varepsilon} \left(Z_{\frac{t}{\varepsilon}} - Z_0\right),$$

where $Z_t = \int_t^\infty \mathbb{E}\left[U(y_r)|\mathcal{F}_{t+}\right] dr$. As by assumption $\|Z_t\|_{L^2}$ is uniformly bounded we may drop the coboundary term and apply the Martingale central limit theorem.

2.2 Fractional Ornstein Uhlenbeck as Fast Dynamics

To illustrate the type of theorem we are seeking, we review the recently obtained result for the fractional Ornstein-Uhlenbeck process. In [44], we studied equation (2.4) where y_t^ε is the fractional Ornstein-Uhlenbeck process. A fluctuation theorem from the average was obtained. We showed furthermore that the effective dynamics is the solution of (2.5).

Definition 2.1 A function $G \in L^2(\mu)$, $G = \sum_{l=0}^\infty c_l H_l$, is said to satisfy the fast chaos decay condition with parameter $q \in \mathbb{N}$, if

$$\sum_{l=0}^\infty |c_l| \sqrt{l!} \, (2q-1)^{\frac{l}{2}} < \infty.$$

The lowest index l with $c_l \neq 0$ is called the Hermite rank of G. If m is the Hermite rank we define $H^*(m) = m(H-1) + 1$.

Remark 2.2 [43, 44] Let y_t be the stationary fractional Ornstein-Uhlenbeck process with $H \in (0, 1) \setminus \{\frac{1}{2}\}$. Then, for any real valued functions $U \in L^2(\mu)$ with Hermite rank $m > 0$ and $H^*(m) = m(H-1) + 1 < 0$, we have

$$\int_0^\infty \|\mathbb{E}\left[U(y_s)|\mathcal{F}_0\right]\|_{L^2}\, ds < \infty. \tag{2.2}$$

Note that, without the conditioning on $\mathcal{F}_0$ the integral would be infinite due to the stationarity of y_s.

Let $\alpha(\varepsilon, H^*(m))$ be positive constants as follows, they depend on m, H and ε and tend to ∞ as $\varepsilon \to 0$,

$$\alpha\left(\varepsilon, H^*(m)\right) = \begin{cases} \frac{1}{\sqrt{\varepsilon}}, & \text{if } H^*(m) < \frac{1}{2}, \\ \frac{1}{\sqrt{\varepsilon|\ln(\varepsilon)|}}, & \text{if } H^*(m) = \frac{1}{2}, \\ \varepsilon^{H^*(m)-1}, & \text{if } H^*(m) > \frac{1}{2}. \end{cases} \tag{2.3}$$

For $H^*(m) \neq \frac{1}{2}$ this is equivalent to $\alpha(\varepsilon, H^*(m)) = \varepsilon^{(H^*(m_k) \vee \frac{1}{2})-1}$. The following is proved in [43] for $H > \frac{1}{2}$, see [44] for also the $H < \frac{1}{2}$ case.

Theorem 2.3 [43, 44] *Let y_t be the fractional Ornstein-Uhlenbeck process with stationary measure μ. Let G_k be $L^2 \cap L^{p_k}$ functions with satisfies the fast chaos decay condition with parameter $q \geqslant 4$ for p_k sufficiently large (see blow). We order the functions $\{G_k\}$ so that their Hermite rank m_k does not increase with k. We also assume either $H^*(m_k) < 0$ for $k \leqslant n$ or $H^*(m_k) > \frac{1}{2}$ otherwise. Then, the solutions of*

$$\begin{cases} \dot{x}_t^\varepsilon = \sum_{k=1}^N \varepsilon^{(H^*(m_k)\vee\frac{1}{2})-1} f_k(x_t^\varepsilon)\, G_k(y_t^\varepsilon), \\ x_0^\varepsilon = x_0, \end{cases} \tag{2.4}$$

converges, as $\varepsilon \to 0$ to the solution of the following equation with the same initial data:

$$dx_t = \sum_{k=1}^n f_k(x_t) \circ dX_t^k + \sum_{k=n+1}^N f_k(x_t)\, dX_t^k. \tag{2.5}$$

Here X_t^k is a Wiener process for $k \leqslant n$, and otherwise a Gaussian or a non-Gaussian Hermite process. The covariances between the processes are determined by the functions G_k, for which there are explicit formulas. In these equations, the symbol $\circ$ denotes the Stratonovich integral and the other integrals are in the sense of Young integrals.

The conditions for p_k are:

Assumption 2.4 *If G_k has low Hermite rank, assume $H^*(m_k) - \frac{1}{p_k} > \frac{1}{2}$; otherwise assume $\frac{1}{2} - \frac{1}{p} > \frac{1}{3}$. Furthermore,*

$$\min_{k \leqslant n}\left(\frac{1}{2} - \frac{1}{p_k}\right) + \min_{n < k \leqslant N}\left(H^*(m_k) - \frac{1}{p_k}\right) > 1. \tag{2.6}$$

2.3 The Rough Path Topology

To explain the methodology we first explain the necessities from the rough path theory. If X and Y are Hölder continuous functions on $[0, T]$ with exponent α and β respectively, such that $\alpha + \beta > 1$, then by Young integration theory

$$\int_0^T Y dX = \lim_{\mathcal{P} \to 0} \sum_{[u,v] \in \mathcal{P}} Y_u(X_v - X_u).$$

where $\mathcal{P}$ denotes a partition of $[0, T]$. Furthermore $(X, Y) \mapsto \int_0^T Y dX$ is a continuous map. Thus, for $X \in C^{\frac{1}{2}+}$ and $f \in C_b^2$, one can make sense of a solution Y to the Young integral equation $dY_s = f(Y_s)dX_s$. Furthermore the solution is continuous with respect to both the driver X and the initial data, see [114]. If X has Hölder continuity less or equal to $\frac{1}{2}$, this fails and one cannot define a pathwise integration for $\int X dX$ by the above Riemann sum anymore. Rough path theory provides us with a machinery to treat less regular functions by enhancing the process with a second order process, giving a better local approximation, which then can be used to enhance the Riemann sum and show it converges.

A rough path of regularity $\alpha \in (\frac{1}{3}, \frac{1}{2})$, is a pair of process $\mathbf{X} = (X_t, \mathbb{X}_{s,t})$ where $(\mathbb{X}_{s,t}) \in \mathbb{R}^{d \times d}$ is a two parameter stochastic processes satisfying the following algebraic conditions: for $0 \leqslant s < u < t \leqslant T$,

$$\mathbb{X}_{s,t} - \mathbb{X}_{s,u} - \mathbb{X}_{u,t} = X_{s,u} \otimes X_{u,t}, \qquad \text{(Chen's relation)}$$

where $X_{s,t} = X_t - X_s$, and $(X_{s,u} \otimes X_{u,t})^{i,j} = X_{s,u}^i X_{u,t}^j$ as well as the following analytic conditions,

$$\|X_{s,t}\| \lesssim |t-s|^\alpha, \qquad \|\mathbb{X}_{s,t}\| \lesssim |t-s|^{2\alpha}. \tag{2.7}$$

The set of such paths will be denoted by $\mathcal{C}^\alpha([0, T]; \mathbb{R}^d)$. The so called second order process $\mathbb{X}_{s,t}$ can be viewed as a possible candidate for the iterated integral $\int_s^t X_{s,u} dX_u$.

Given a path X, which is regular enough to define its iterated integral, for example $X \in \mathcal{C}^1([0, T]; \mathbb{R}^d)$, we define its natural rough path lift to be given by

$$\mathbb{X}_{s,t} := \int_s^t X_{s,u} dX_u.$$

It is now an easy exercise to verify that $\mathbf{X} = (X, \mathbb{X})$ satisfies the algebraic and analytic conditions (depending on the regularity of X), by which we mean Chen's relation and (2.7). Given two rough paths $\mathbf{X}$ and $\mathbf{Y}$ we may define, for $\alpha \in (\frac{1}{3}, \frac{1}{2})$, the following defines a complete metric on $C^\alpha([0, T]; \mathbb{R}^d)$, called the in-homogenous α-Hölder rough path metric:

$$\rho_\alpha(\mathbf{X}, \mathbf{Y}) = \sup_{s \neq t} \frac{\|X_{s,t} - Y_{s,t}\|}{|t - s|^\alpha} + \sup_{s \neq t} \frac{\|\mathbb{X}_{s,t} - \mathbb{Y}_{s,t}\|}{|t - s|^{2\alpha}}. \tag{2.8}$$

We are also going to make use of the norm like object

$$\|\mathbf{X}\|_\alpha = \sup_{s \neq t \in [0,T]} \frac{\|X_{s,t}\|}{|t - s|^\alpha} + \sup_{s \neq t \in [0,T]} \frac{\|\mathbb{X}_{s,t}\|^{\frac{1}{2}}}{|t - s|^\alpha}, \tag{2.9}$$

where we denote for any two parameter process $\mathbb{X}$ a semi-norm:

$$\|\mathbb{X}\|_{2\alpha} := \sup_{s \neq t \in [0,T]} \frac{\|\mathbb{X}_{s,t}\|}{|t - s|^{2\alpha}}.$$

Our proof will be based on the following results:

Lemma 2.5 *Let $\mathbf{X}^\varepsilon$ be a sequence of rough paths with $\mathbf{X}(0) = 0$ and*

$$\sup_{\varepsilon \in (0,1]} \mathbb{E}\left(\|\mathbf{X}^\varepsilon\|_\gamma\right)^p < \infty,$$

for some $\gamma \in (\frac{1}{3}, \frac{1}{2} - \frac{1}{p})$, then $\mathbf{X}^\varepsilon$ is tight in $C^{\gamma'}$ for every $\frac{1}{3} < \gamma' < \gamma$.

Theorem 2.6 [37] *Let $Y_0 \in \mathbb{R}^m$, $\beta \in (\frac{1}{3}, 1)$, $f \in C_b^3(\mathbb{R}^m, L(\mathbb{R}^d, \mathbb{R}^m))$ and $\mathbf{X} \in C^\beta([0, T], \mathbb{R}^d)$. Then, the differential equation*

$$Y_t = Y_0 + \int_0^t f(Y_s) d\mathbf{X}_s \tag{2.10}$$

has a unique solution which belongs to C^β. Furthermore, the solution map $\Phi_f : \mathbb{R}^d \times C^\beta([0, T], \mathbb{R}^d) \to \mathcal{D}_X^{2\beta}([0, T], \mathbb{R}^m)$, where the first component is the initial condition and the second component the driver, is continuous.

2.4 Homogenization via Rough Continuity

2.4.1 Main Idea of the Method

Theorem 2.6 has an interesting application to our homogenisation problem as weak convergence is preserved under continuous operations. A simple equation for the demonstration is

$$dx_t^\varepsilon = \alpha(\varepsilon) f(x_t^\varepsilon) G(y_t^\varepsilon) dt,$$

for a suitable choice of α and stochastic process y_t^ε, we may rewrite this equation into a rough differential equation. To do so set $X_t^\varepsilon = \alpha(\varepsilon) \int_0^t G(y_s^\varepsilon)\,ds$ thus,

$$dx_t^\varepsilon = f(x_t^\varepsilon)dX_t^\varepsilon.$$

To obtain a rough differential equation we now define the canonical lift of X^ε as $\mathbf{X}^\varepsilon = (X_t^\varepsilon, \mathbb{X}_{s,t}^\varepsilon)$, where $\mathbb{X}_{s,t}^\varepsilon = \int_s^t X_{s,r}^\varepsilon\,dX_r^\varepsilon$ and study

$$dx_t^\varepsilon = f(x_t^\varepsilon)d\mathbf{X}_t^\varepsilon.$$

As stated above to conclude weak convergence in a Hölder space of the solutions x^ε it is sufficient to obtain weak convergence of $\mathbf{X}^\varepsilon$ in a rough path space $\mathcal{C}^\gamma$ for some $\gamma \in (\frac{1}{3}, \frac{1}{2})$. This can be done by following a two step approach. Firstly, proving convergence in finite dimensional distributions for $\mathbf{X}^\varepsilon$ and secondly tightness via moment bounds. See [6, 18, 35, 44, 65].

Overall we have reduced the question to proving a functional central limit for $X_t^\varepsilon = \alpha(\varepsilon) \int_0^t G(y_s^\varepsilon)\,ds$, as well as for

$$\mathbb{X}_{s,t}^\varepsilon = \int_s^t X_{s,r}^\varepsilon dX_r^\varepsilon = \alpha(\varepsilon)^2 \int_s^t \int_s^r G(y_r)G(y_u)du\,dr,$$

in a suitable path space. In one dimensions, by symmetry, $\mathbb{X}_{s,t}^\varepsilon = \frac{1}{2}\left(X_{s,t}^\varepsilon\right)^2$, hence a continuous functional of X^ε. This makes the one dimensional case quite simple. However, when dealing with equations of the form

$$dx_t^\varepsilon = \sum_{k=1}^N \alpha_k(\varepsilon) f_k(x_t^\varepsilon) G_k(y_t^\varepsilon)dt,$$

where the fast motions are channelled through functions of different scales, or

$$dx_t^\varepsilon = \alpha(\varepsilon) f(x_t^\varepsilon, y_t^\varepsilon),$$

the canonical lift is more involved.

2.4.2 Proof of Theorem 1

We apply the rough path theory and CLT theorem to diffusion creation. To deal with components converging to a Wiener process, to prove convergence of the second order process as mentioned above one usually relies on a martingale coboundary decomposition with suitable regularity, cf. [6, 65]. Then, one can use the stability of weak convergence of the Itô integral, Theorem 2.2 [67], for L^2 martingales.

Assumption 2.7 Given a stochastic process y_t with stationary distribution μ, a function U in $L^2(\mu)$ is said to satisfy the 'conditional memory loss' condition:

$$\int_0^\infty \|\mathbb{E}[U(y_s)|\mathcal{F}_0]\|_{L^2} \, ds < \infty. \tag{2.11}$$

Assumption 2.8 A vector valued process

$$X_t^\varepsilon = \left(X_t^{1,\varepsilon}, \ldots, X_t^{n,\varepsilon}\right),$$

is said to satisfy the rough moment condition if the following moment bounds hold

$$\|X_{s,t}^{k,\varepsilon}\|_{L^p} \lesssim |t-s|^{\frac{1}{2}}$$
$$\|\mathbb{X}_{s,t}^{i,j,\varepsilon}\|_{L^{\frac{p}{2}}} \lesssim |t-s|.$$

Given a stationary and ergodic process y_t and functions G_k, we define $X_t^{k,\varepsilon} = \sqrt{\varepsilon}\int_0^{\frac{t}{\varepsilon}} G_k(y_s)ds$. Suppose that there exists a Wiener process $X_t = (X_t^1, \ldots, X_t^n)$ such that $X_t^\varepsilon \to X_t$ in finite dimensional distributions. To prove Theorem 1, by the continuity theorem, it is sufficient to show that the canonical lift $\mathbf{X}^\varepsilon = (X^\varepsilon, \mathbb{X}^\varepsilon)$ converges weakly in $\mathcal{C}^\gamma$ for $\gamma \in (\frac{1}{3}, \frac{1}{2} - \frac{1}{p})$.

Remark 2.9 In case the assumptions of Theorem 1 are satisfied one has in particular,

$$\mathbb{E}[X_t^j X_s^l] = 2(t \wedge s) \int_0^\infty \mathbb{E}[G_j(y_r)G_l(y_0)]dr.$$

Proof of Theorem 1. To prove Theorem 1 one may argue similarly as in Sect. 3.3 in [44]. Recall that we assume that y_t and G_k satisfy the conditional memory loss condition in Assumption 2.7, and X^ε satisfies Assumption 2.8.

Firstly, due to Assumption 2.7 we may decompose each $X^{k,\varepsilon}$ as follows,

$$X_t^{k,\varepsilon} = \sqrt{\varepsilon}\int_0^{\frac{t}{\varepsilon}} G_k(y_r)dr = \sqrt{\varepsilon}M_{\frac{t}{\varepsilon}}^k + \sqrt{\varepsilon}\left(Z_{\frac{t}{\varepsilon}}^k - Z_0^k\right),$$

where $Z_t^k = \int_t^\infty \mathbb{E}\left[G_k(y_r)|\mathcal{F}_{t+}\right]dr$ and $M_t^k = \int_0^\infty \mathbb{E}\left[G_k(y_r)|\mathcal{F}_{t+}\right] - \mathbb{E}\left[G_k(y_r)|\mathcal{F}_{0+}\right]dr$. By the construction, M_t^k is a martingale and $\|Z_t^k\|_{L^2}$ is bounded by Assumption 2.7. Hence, the term $\sqrt{\varepsilon}\left(Z_{\frac{t}{\varepsilon}}^k - Z_0^k\right)$ converges to 0 in L^2. Thus, also the multidimensional martingales

$$(\sqrt{\varepsilon}M_{\frac{t}{\varepsilon}}^1, \ldots, \sqrt{\varepsilon}M_{\frac{t}{\varepsilon}}^n)$$

converge jointly to the n-dim Wiener process,

$$X_t = (X_t^1, \ldots, X_t^n).$$

When computing the iterated integrals $\mathbb{X}_{s,t}^{i,j,\varepsilon} = \int_s^t X_{s,r}^{i,\varepsilon} dX_r^{j,\varepsilon}$, we obtain for the diagonal entries $\mathbb{X}_{s,t}^{i,i,\varepsilon} = \frac{1}{2}\left(X_{s,t}^{i,\varepsilon}\right)^2$ by symmetry as above. However, in case $i \neq j$ we need to argue differently. Using the same martingale coboundary decomposition as above one can show the following lemma.

Lemma 2.10 (See the Appendix) *For $L(\varepsilon) = \lfloor \frac{t}{\varepsilon} \rfloor$,*

$$\varepsilon \int_0^{\frac{t}{\varepsilon}} \int_0^s G_i(y_s) G_j(y_r) dr = \varepsilon \sum_{k=0}^{L(\varepsilon)} (M_{k+1}^i - M_k^i) M_k^j + t A^{i,j} + \mathbf{Er}(\varepsilon),$$

where $A^{i,j} = \int_0^\infty G_i(y_s) G_j(y_0) ds$ and $\mathbf{Er}(\varepsilon)$ converges to 0 in probability.

Now, defining the cadlag martingales $M_t^{i,\varepsilon} = \sqrt{\varepsilon} M_{\lfloor \frac{t}{\varepsilon} \rfloor}^i$ one may identify the sum above as Itô integral. Furthermore, by Assumption 2.7 these martingales are bounded in L^2, hence, using Theorem 2.2 in [67] we obtain,

$$\left(M_t^{k,\varepsilon}, \int_0^t M_s^{i,\varepsilon} dM_s^{j,\varepsilon}\right) \to (X_t, \mathbb{X}_{0,t}),$$

where $\mathbb{X}$ is given by Itô integrals, in the sense of finite dimensional distributions. Additionally, the moment bounds guarantee that the convergence actually takes place $\mathcal{C}^\gamma$ for $\gamma \in (\frac{1}{3}, \frac{1}{2} - \frac{1}{p})$.

An example that satisfies above conditions is given by choosing y_t as the fractional Ornstein Uhlenbeck process and functions G_k that satisfy,

Assumption 2.11 Each G_k belongs to $L^{p_k}(\mu)$, where $p_k > 2$, and has Hermite rank $m_k \geq 1$. Furthermore,

(1) Each G_k satisfies the fast chaos decay condition with parameter $q \geq 4$.
(2) For each k, $\frac{1}{2} - \frac{1}{p_k} > \frac{1}{3}$ and $H^*(m_k) < 0$.

where $H^*(m) = (H-1)m + 1$, the Hermite rank is as defined in Sect. 2.5.3.

2.5 Examples Satisfying Assumption 2.7

2.5.1 Strong Mixing Environment

Using this method one can also treat the classical setup. Given a stationary stochastic process y_t and assume it is strong mixing with mixing rate $\alpha(s)$.

$$\alpha(t) = \sup_{A \in \mathcal{F}_\infty^0, B \in \mathcal{F}_t^\infty} |\mathcal{P}(A \cap B) - \mathcal{P}(A)\mathcal{P}(B)|.$$

By Lemma 3.102 in [60], given $G : \mathbb{R} \to \mathbb{R}$ such that $\|G(y_0)\|_{L^1} < \infty$ the following inequality holds,

$$\|\mathbb{E}[G(y_t)|\mathcal{F}_0] - \mathbb{E}[G(y_t)]\|_{L^q} \leqslant 2\left(2^{\frac{1}{q}} + 1\right)\alpha(t)^{\frac{1}{q}-\frac{1}{r}}\|G(y_t)\|_{L^r},$$

where $1 \leqslant q \leqslant r \leqslant \infty$.

Lemma 2.12 *Let y_t be a stationary process with mixing rate $\alpha(t)$ and $G \in L^r(\mu)$ is centred. Then Assumption 2.7 holds if*

$$\int_0^\infty \alpha(t)^{\frac{1}{2}-\frac{1}{r}} dt < \infty.$$

2.5.2 Volterra Kernel Moving Averages

Let B denote a fBm of Hurst parameter H and set $y_t = \int_{-\infty}^t K(t-s) dB_s$, where K denotes a kernel K such that $\|y_t\|_{L^2} = 1$. Using the decomposition $B_t - B_k = \tilde{B}_t^k + \bar{B}_t^k$, where $\bar{B}_t^k = \int_{-\infty}^k (t-k)^{H-\frac{1}{2}} - (k-r)^{H-\frac{1}{2}} dr$ and $\tilde{B}_t^k = \int_k^t (t-r)^{H-\frac{1}{2}} dW_r$, we may also decompose

$$y_t = \int_{-\infty}^t K(t-r) dB_r = \int_{-\infty}^k K(t-r) dB_r + \int_k^t K(t-r) d(B_r - B_k)$$
$$= \left(\int_{-\infty}^k K(t-r) dB_r + \int_k^t K(t-r) d\bar{B}_r^k\right) + \int_k^t K(t-r) d\tilde{B}_r^k$$
$$= \bar{y}_t^k + \tilde{y}_t^k.$$

It was shown in [50] that the term $\tilde{B}_t^k$ is independent of $\mathcal{F}_k$ and $\bar{B}_t^k$ is $\mathcal{F}_k$ measurable, hence $\tilde{y}_t^k$ is independent of $\mathcal{F}_k$ and $\bar{y}_t^k$ is $\mathcal{F}_k$ measurable. Moreover, both terms are Gaussians. We set $y_t^k = y_t$ for $k \geqslant t$. Using an expansion into Hermite polynomials one can show the following:

Lemma 2.13 *Let $y_t = \int_{-\infty}^t K(t-s) dB_s$ such that $\|y_t\|_{L^2} = 1$ be given. If the kernel K is such that*

$$\int_{k-1}^\infty \left(\mathbb{E}\left[(\bar{y}_s^k)^2\right]\right)^{\frac{m}{2}} dt,$$

then Assumption 2.7 hold.

An example of this is the fractional Ornstein-Uhlenbeck process.

2.5.3 Effective Dynamics Driven by Non-wiener Process

As mentioned above for Gaussian noises one may obtain a variety of limits. The Hermite rank plays a central role in the analysis of limit theorems with Gaussian noises. Let γ denote the standard normal distribution and y_t a stationary process with distribution γ, then each $G \in L^2(\gamma)$ admits an expansion into Hermite polynomials, $G(y_s) = \sum_{k=0}^{\infty} c_k H_k(y_s)$. The Hermite rank is now defined as the rank of the smallest non zero Hermite polynomial, thus, a function with Hermite rank m can be written as $G(y_s) = \sum_{k=m}^{\infty} c_k H_k(y_s)$ with $c_m \neq 0$. Hermite polynomials are mutually orthogonal and satisfy $\mathbb{E}\left(H_k(y_s)H_j(y_r)\right) = \delta(k-j)\rho(|s-r|)^k$, where ρ denotes the correlation function of y_s. In case ρ admits only an algebraic decay, i.e. the fractional OU process satisfies $\rho(t) \lesssim 1 \wedge t^{2H-2}$, higher order polynomials accelerate the correlation decay. Thus, the picture one obtains is that there is a critical m such that $H_k(y_s)$ is short range dependent if $k \geqslant m$ and long range dependent otherwise. In case $(H-1)k+1 > \frac{1}{2}$ the process is long range dependent and short range dependent if $(H-1)k+1 < \frac{1}{2}$, the border line case is long range dependent as well, however the sum of the correlations only diverges logarithmically, which leads to a separate behaviour which will not be discussed here. Thus, if a function G has Hermite rank m such that $(H-1)k+1 > \frac{1}{2}$ one obtains convergence to Hermite processes for $\alpha(\varepsilon) = \varepsilon^{(H-1)m+1}$. As this processes have Hölder regularity greater than $\frac{1}{2}$ they can be treated within the Young framework.

Remark 2.14 To combine both frameworks in [44] the assumption was made that the Hölder regularity of Wiener components plus the ones for Hermite components is bigger than 1, hence the joint lifts are irrelevant in the limit and well defined as Young integrals.

3 Recent Progress on Slow/Fast Markovian Dynamics

In order to compare the methods, we will explain the classical Stochastic Averaging Theory and the Diffusion Homogenisation theory with Markovian Dynamics, which have been continuously re-inventing itself since the 1960s and 1970s. They are typically a system of two-scale stochastic equations as follows

$$\begin{cases} dx_t^\varepsilon = \sum_{k=1}^{m_1} X_k(x_t^\varepsilon, y_t^\varepsilon) \circ d\tilde{W}_t^k + X_0(x_t^\varepsilon, y_t^\varepsilon)\, dt, & x_0^\varepsilon = x_0; \\ dy_t^\varepsilon = \frac{1}{\sqrt{\varepsilon}} \sum_{k=1}^{m_2} Y_k(x_t^\varepsilon, y_t^\varepsilon) \circ dW_t^k + \frac{1}{\varepsilon} Y_0(x_t^\varepsilon, y_t^\varepsilon)\, dt, & y_0^\varepsilon = y_0. \end{cases} \quad (3.1)$$

where $W_t^i, \tilde{W}_t^i$ are independent Brownian motions. Such models have the flavour of the following multi-scale system in dynamical system

$$\dot{x}_t^\varepsilon = f(x_t^\varepsilon, y_t^\varepsilon), \qquad \dot{y}_t^\varepsilon = \frac{1}{\varepsilon} Y_0(x_t^\varepsilon, y_t^\varepsilon).$$

In dynamical system, the fast dynamics is often periodic or has chaotic behaviour.

Because these are well known, we will focus on the newer developments that applies to non-linear state spaces which is sufficient to illustrate the underlying ideas and the differences between the theories within the Markovian dynamics and that with the fractional dynamics. Since the scale in a multi-scale system are note naturally separated, we will need to use geometric methods to separate them, the slow and fast variables so obtained often lives in a non-linear space. For example if we take an approximately integrable Hamiltonian system the natural state space is a torus. If the slow-fast system is obtained by using symmetries, the state space of the fast motions are Lie groups. SDEs with symmetries are popular topics, see [5, 28, 31–33, 107]. See also [74, 78] for perturbation to symmetries.

3.1 The Basic Averaging Principal

For $y \in \mathbb{R}^n$, let $\sigma_i(\cdot, y) : \mathbb{R}^d \to \mathbb{R}^d$, and for $x \in \mathbb{R}^d$, $Y_i(x, \cdot) : \mathbb{R}^n \to \mathbb{R}^n$. The averaging principle states that if $(x_t^\varepsilon, y_t^\varepsilon)$ is the solution of the following stochastic differential equation, with Itô integrals,

$$\begin{cases} dx_t^\varepsilon = \sum_{k=1}^{m_1} \sigma_k(x_t^\varepsilon, y_t^\varepsilon) d\tilde{W}_t^k + \sigma_0(x_t^\varepsilon, y_t^\varepsilon) dt, & x_0^\varepsilon = x_0; \\ dy_t^\varepsilon = \frac{1}{\sqrt{\varepsilon}} \sum_{k=1}^{m_2} Y_k(x_t^\varepsilon, y_t^\varepsilon) dW_t^k + \frac{1}{\varepsilon} Y_0(x_t^\varepsilon, y_t^\varepsilon) dt, & y_0^\varepsilon = y_0, \end{cases} \qquad (3.2)$$

then x_t^ε converges to a Markov process. This principle requires technical verification.

The popular format and assumptions for the Stochastic Averaging Principle is as follows, see [52] and [113]. Let Y_s^x be the solution to the equation below with frozen slow variable (we assume sufficient regularity assumption so that there equation has a unique strong solution,)

$$dy_t^x = \sum_{k=1}^{m_2} Y_k(x, y_t^x) dW_t^k + Y_0(x, y_t^x) dt, \quad y_0^x = y_0.$$

suppose that there exists a unique invariant probability measure μ^x. Assume furthermore that the coefficients of the equations are Lipschitz continuous in both variables and suppose that there exist functions $\bar{a}_{i,j}$ and $\bar{b}$ such that on $[0, T]$,

$$\left|\frac{1}{t}\mathbf{E}\int_0^t b(x, Y_s^x)ds - \bar{b}(x)\right| \lesssim (|x|^2 + |y_0|^2 + 1),$$
$$\left|\frac{1}{t}\mathbf{E}\int_0^t \sum_k \sigma_k^i \sigma_k^j(x, Y_s^x)ds - \bar{a}_{i,j}(x)\right| \lesssim (|x|^2 + |y_0|^2 + 1). \quad (3.3)$$

Then, x_t^ε converges weakly to the Markov process with generator

$$\bar{\mathcal{L}} = \frac{1}{2}\bar{a}_{i,j}(x)\frac{\partial^2}{\partial x_i \partial x_j} + \bar{b}_k(x)\frac{\partial}{\partial x_k}.$$

The notation $\bar{h}$ denotes the average of a function h, i.e. $\bar{h}(x) = \int h(y)\mu^x(dy)$. In [52], boundedness in y is assumed in (3.3). This is replaced by the quadratic growth in [113], where a uniform ellipticity is also used to replace the regularity assumption needed on $x \mapsto \mu^x$.

In the next section we outlined conditions posed directly on the coefficients of the equations, under which these assumptions hold.

3.2 Quantitative Locally Uniform LLN

The law of large numbers is the foundation for stochastic averaging theory. Let y_t^ε be an ergodic Markov process on a state space $\mathcal{Y}$ with invariant measure π, then the Birkhoff's ergodic theorem holds, i.e.

$$\lim_{t \to \infty} \left|\frac{1}{t}\int_0^t h(x_s)ds - \int h(y)\pi(dy)\right| \to 0.$$

If $\mathcal{L}$ is a strictly elliptic operator, this is seen easily by the martingale inequalities and Schauder's estimates. There is typically a rate of convergence of the order $\frac{1}{\sqrt{t}}$. Such results are classic for elliptic diffusions on $\mathbb{R}^n$. For Brownian motion on a compact manifolds this is proven in [72], see also [29], we generalise this result to non-elliptic operators and obtain a quantitative estimate. Furthermore, we obtain locally uniform estimates for diffusion operators $\mathcal{L}_x$ where $x \in \mathcal{X}$ is a parameter. Our main contribution is to obtain quantitative estimates that are locally uniform in x. We indicate one such result below.

Definition 3.1 Let $X_0, X_1, \ldots, X_k$ be smooth vector fields.

(i) The differential operator $\sum_{k=1}^m (X_i)^2 + X_0$ is said to satisfy *Hörmander's condition* if $\{X_k, k = 0, 1, \ldots, m\}$ and their iterated Lie brackets generate the tangent space at each point.
(ii) The differential operator $\sum_{k=1}^m (X_i)^2 + X_0$, is said to satisfy *strong Hörmander's condition* if $\{X_k, k = 1, \ldots, m\}$ and their iterated Lie brackets generate the tangent space at each point.

Let $s \geq 0$, let dx denote the volume measure of a Riemannian manifold G and let Δ denote the Laplacian. If f is a C^∞ function we define its Sobolev norm to be

$$\|f\|_s = \left(\int_G f(x)(I + \Delta)^{s/2} f(x) \, dx \right)^{\frac{1}{2}}$$

If the Strong Hörmander's condition holds and $\mathcal{Y}$ is compact, then the Markov process with generator $\sum_{k=1}^m (X_i)^2 + X_0$ has a unique invariant probability measure. We state a theorem for $\mathcal{Y}$ compact, a version with $\mathcal{Y}$ not compact can also be found in [78].

Theorem 3.2 [78] *Let $\mathcal{Y}$ be a compact manifold. Suppose that $Y_i(x, \cdot)$ are bounded vector fields on $\mathcal{Y}$ with bounded derivatives and C^∞ in both variables. Suppose that for each $x \in \mathcal{X}$,*

$$\mathcal{L}_x = \frac{1}{2} \sum_{i=1}^m Y_i^2(x, \cdot) + Y_0(x, \cdot)$$

satisfies Hörmander's condition and has a unique invariant probability measure which we denote by μ_x. Then the following statements hold.

(a) *$x \mapsto \mu_x$ is locally Lipschitz continuous in the total variation norm.*
(b) *For every $s > 1 + \frac{\dim(\mathcal{Y})}{2}$ there exists a positive constant $C(x)$, depending continuously in x, such that for every smooth function $f : \mathcal{Y} \to \mathbb{R}$,*

$$\left| \frac{1}{T} \int_t^{t+T} f(z_r^x) \, dr - \int_G f(y) \mu_x(dy) \right|_{L_2(\Omega)} \leq C(x) \|f\|_s \frac{1}{\sqrt{T}}, \quad (3.4)$$

where z_r denotes an $\mathcal{L}_x$-diffusion.

The proof for this follows from an application of Itô's formula, applied to the solution of the Poisson equation $\mathcal{L}_x h = f(x, \cdot)$ where $\int f(x, y) \mu_x(dy) = 0$. For such functions,

$$\frac{1}{T} \int_0^T f(x, z_r^x) dr = \frac{1}{T} \left(g(x, z_T^x) - g(x, y_0) \right) - \frac{1}{T} \left(\sum_{k=1}^{m_2} \int_0^T dg(x, \cdot)(Y_k(x, z_r^x)) dW_r^k \right).$$

(We take $t = 0$ for simplicity.) It then remain to bound the supremum norm of $dg(x, \cdot)$, which is a consequence of the sub-elliptic estimates of Hörmander.

3.3 Averaging with Hörmander's Conditions

Let $\mathcal{X}, \mathcal{Y}$ be smooth manifolds. We take a family of vector fields $\sigma_i(\cdot, y)$ on $\mathcal{X}$ indexed by $y \in \mathcal{Y}$ and a family of vector fields $Y_i(x, \cdot)$ with parameter $x \in \mathcal{X}$. The vector field $\sigma_i(\cdot, y)$ acts on a real function $h : \mathcal{X} \to \mathbb{R}$ so that $\sigma_i(\cdot, y)h$ is the derivative of h in the direction of $\sigma_i(\cdot, y)$. If X is a vector field we denote by $Dh(X(x))$ the derivative of the function h in the direction of $X(x)$ at the point x. The function so obtained is also denoted by $\mathcal{L}_X h$, or Xh, or $Dh(X)$.

The assumption (3.3) can in fact be verified with ergodicity and regularity conditions on the coefficients. We state such a result on manifolds. We switch to Stranovich integrals and denote this by $\circ$. We also denote by $\tilde{\sigma}_0$ and $\tilde{Y}_0$ the effective drifts (including the Stratonovich corrections). Consider

$$\begin{cases} dx_t^\varepsilon = \sum_{k=1}^{m_1} \sigma_k(x_t^\varepsilon, y_t^\varepsilon) \circ d\tilde{W}_t^k + \sigma_0(x_t^\varepsilon, y_t^\varepsilon) dt, & x_0^\varepsilon = x_0; \\ dy_t^\varepsilon = \frac{1}{\sqrt{\varepsilon}} \sum_{k=1}^{m_2} Y_k(x_t^\varepsilon, y_t^\varepsilon) \circ dW_t^k + \frac{1}{\varepsilon} Y_0(x_t^\varepsilon, y_t^\varepsilon) dt, & y_0^\varepsilon = y_0. \end{cases} \quad (3.5)$$

Suppose that, for each x, $\mathcal{L}_x = \frac{1}{2} \sum_{i=1}^m Y_i^2(x, \cdot) + Y_0(x, \cdot)$ has a unique invariant probability measure which we denote by μ^x. For $h : \mathcal{X} \to \mathbb{R}$ a smooth function with compact support, we define

$$\bar{\mathcal{L}}h(x) = \frac{1}{2} \sum_k \int_{\mathcal{Y}} \left(D^2 h \right)_x (\sigma_k(x, y), \sigma_k(x, y)) (y) \mu^x(dy) + \int_{\mathcal{Y}} (Dh)_x \left(\sigma_0(x, y) \right) \mu^x(dy).$$

The key condition for the convergence of the slow motion is the Hörmander's condition. More precisely we will make use of the conclusion of Theorem 3.2, from which we obtain the following estimates on a compact subset D of $\mathcal{X}$:

$$\mathbb{E} \sum_{i=0}^{N-1} \left| \int_{t_i}^{t_{i+1}} f\left(x_{t_i}^\varepsilon, y_r^{x_{t_i}^\varepsilon}\right) ds - \Delta t_i \, f\left(x_{t_i}^\varepsilon\right) \right| \leqslant c T \, \lambda\left(\frac{\Delta t_i}{\varepsilon}\right) \sup_{x \in D} \left\| f(x, \cdot) - \bar{f}(x) \right\|_s. \quad (3.6)$$

Here $\lambda(t)$ is a function converging to zero as $t \to \infty$. With this we obtain the dynamical Law of Large Number as follows.

Theorem 3.3 [78] *Let $\mathcal{Y}$ be compact and Y_i are in $C^2 \cap BC^1$. Suppose that $\tilde{\sigma}_0$ and σ_i are C^1, where $i = 1, \ldots, m$, and $\mathcal{L}_x$ satisfies Hörmander's condition and that it has a unique invariant probability measure. Suppose that **one** of the following two statements holds.*

(i) *Let ρ denote the distance function on $\mathcal{Y}$. Suppose that ρ^2 is smooth and*

$$\frac{1}{2} \sum_{i=1}^m \nabla d\rho^2(\sigma_i(\cdot, y), \sigma_i(\cdot, y)) + d\rho^2(\tilde{\sigma}_0(\cdot, y)) \leqslant K(1 + \rho^2(\cdot)), \quad \forall y \in G.$$

(ii) *The sectional curvature of $\mathcal{Y}$ is bounded. There exists a constant K such that*

Rough Homogenisation with Fractional Dynamics

$$\sum_{i=1}^{m} |\sigma_i(x, y)|^2 \leqslant K(1 + \rho(x)), \quad |\sigma_0(x, y)| \leqslant K(1 + \rho(x)), \quad \forall x \in N, \forall y \in G.$$

Then $\{x_t^\varepsilon, \varepsilon > 0\}$ converges weakly, on any compact time intervals, to the Markov process with the Markov generator $\tilde{\mathcal{L}}$.

This theorem is useful when we have an stochastic differential equation on a manifold which is invariant under a group action, and when we study its small perturbations.

3.3.1 Geometric Models

The LLN can be used to study the following models.

Example 3.4 (1) **Approximately Integrable Hamiltonian systems.** If $H : M \to \mathbb{R}$ is a smooth Hamiltonian function on a symplectic manifold or on $\mathbb{R}^{2n}$ with its standard symplectic structure, we use X_H to denote its Hamiltonian vector fields. Let k be a smooth vector field. Let $\{H_i\}$ be a family of Poisson commuting Hamiltonian functions on a symplectic manifold.

$$dz_t^\varepsilon = \frac{1}{\sqrt{\varepsilon}} \sum_{i=1}^{n} X_{H_i}(z_t^\varepsilon) \circ dW_t^i + \frac{1}{\varepsilon} X_{H_0}(z_t^\varepsilon) dt + k(z_t^\varepsilon) dt.$$

Let $x_t^\varepsilon = (H_1(z_t^\varepsilon), \ldots, H_1(z_t^\varepsilon))$ and let y_t^ε denote the angle components. It was shown in [73] that an averaging principle holds if $\{H_1, \ldots, H_n\}$ forms a completely integrable systems and x_t^ε converging to the solution of and ODE. Furthermore if k is also a Hamiltonian vector field, then x^ε converges on the scale $[0, \frac{1}{\varepsilon}]$ to a Markov process whose generator can be explicitly computed.

(2) **Stirring geodesics.** Let H_0 is a horizontal vector field on the orthonormal frame bundle OM of a manifold M. The orthonormal frames over $x \in M$E are the set of directions, i.e. linear maps from $\mathbb{R}^n$ to $T_x M$. Then the equation $\dot{u}_t = H_0(u_t)$ on OM is the equation for geodesics, its projection to the base manifold is a geodesic with unit speed, it solves a second order differential equation on M. Let $\{A_k\}$ be a collection of skew symmetric $d \times d$ matrices where d is the dimension of the manifold. Let A_k denote also the vertical fundamental fields obtained on OM by rotating an initial tangent vector in the direction of the exponential map of A_k. Consider the equation on OM:

$$du_t^\varepsilon = \frac{1}{\varepsilon} H_0(u_t^\varepsilon) dt + \frac{1}{\varepsilon} \sum_{k=1}^{\frac{n(n-1)}{2}} A_k(u_t^\varepsilon) \circ dW_t^k.$$

These are large oscillatory perturbations on the the geodesic equation, considered on the scale $[0, \frac{1}{\varepsilon}]$.

Let π denote the projection of a frame in $O_x M \triangleq \mathcal{L}(\mathbb{R}^n, T_x M)$ to x. We are only concerned with the projection of u_t^ε to the base manifold M and we set $x_t^\varepsilon = \pi(u_t^\varepsilon)$. Then our equation reduces to the following:

$$\dot{x}_t^\varepsilon = \frac{1}{\varepsilon} x_t^\varepsilon g_t^\varepsilon$$

on the manifold M, where g_t^ε is a fast diffusion on $SO(d)$. If $\{A_k\}$ is an o.n.b. of $\mathfrak{so}(d)$, it was shown in [75] that $\pi(u_t^\varepsilon)$ converges as $\varepsilon \to 0$. Theorem 3.3 allow this theorem to be extended to a set of $L = \{A_k\}$, which is not necessarily an o.n.b. of $\mathfrak{so}(n)$, but the elements of L and their Lie bracket has dimension n.

(3) **Perturbation to equi-variant diffusions.** Perturbed equi-variant SDEs on principal bundles were studied in [74, 78]. See also [31–33] for the study of equivariant diffusions.

(4) **Inhomogeneous scaling of Riemannian metric and collapsing of manifolds.** In [77], a singularly perturbed model on a lie group G with a compact subgroup H was studied. Let G be endowed with an Ad_H-invariant left invariant Riemannian metric, this exists if H is compact. Let X_i be elements of the Lie algebra of H which we denote by $\mathfrak{h}$. Let Y be an element of its orthogonal complement. We identify an element of the Lie algebra with the left invariant vector fields generated by it. Consider

$$dg_t^\varepsilon = \frac{1}{\varepsilon} \sum_{k=1}^p X_k(g_t^\varepsilon) \circ dW_t^k + \frac{1}{\varepsilon} Y(g_t^\varepsilon) \, dt.$$

Then, if $\{X_k\}$ and their brackets generates $\mathfrak{h}$, it was shown in [77] that $\pi(g_t^\varepsilon)$ converges to a diffusion process on the orbit space G/H.

In these examples, a geometric reduction method is used, see also [21, 78].

3.4 Diffusion Homogenisation Theory

The models in [2] and [4] above can be reduced to random ODEs with a random right hand sides. The effective dynamic theory falls into the diffusive homogenisation theory. The first example also falls within this theory when the perturbation vector field k is again a Hamiltonian vector field. According to [51], the study of ODEs with a random right hand side was already considered in an article by Stratonovich in 1961. See [63, 83].

A diffusive homogenisation theory, in its simplest form, is about a family of random ordinary differential equation of the form

$$\dot{x}_t^\varepsilon = \frac{1}{\sqrt{\varepsilon}} V(x_t^\varepsilon, y_t^\varepsilon)$$

where V is a function with $\int V(x,y)\mu(dy) = 0$ and y_t is an ergodic Markov process or a stationary strong mixing stochastic process, and y_t^ε is distributed as $y_{\frac{t}{\varepsilon}}$, with stationary measure μ. The zero averaging assumption is about the 'oscillatory property' on the y process and also restrictions on the functions V, without this x_t^ε may blow up as $\varepsilon \to 0$ even if V is bounded and smooth. Assuming the square root scaling is the correct scaling for a non-trivial effective limit, then we expect the effective dynamics to be a Markov process in which case we employ the martingale method.

Suppose that y_t is a Markov process with generator $\mathcal{L}_0$, then y_t^ε is a Markov process with generator $\frac{1}{\varepsilon}\mathcal{L}_0$ with invariant measure π. To proceed further let F be a real valued function, then

$$F(x_t^\varepsilon) = F(x_0^\varepsilon) + \frac{1}{\sqrt{\varepsilon}} \sum_k \int_0^t (DF)_{x_s^\varepsilon}\left(V(x_s^\varepsilon, y_s^\varepsilon)\right)ds.$$

We take $x_0^\varepsilon = x_0$ for simplicity. Using a semi-martingale decomposition,

$$\int_0^t dF((V(x_s^\varepsilon, y_s^\varepsilon))ds = M_t^\varepsilon + A_t^\varepsilon,$$

it remains to show that

(i) $(x_\cdot^\varepsilon, \varepsilon \in (0,1])$ is a relatively compact set of stochastic processes;
(ii) any of its limit point $\bar{x}_t$ solves a martingale problem with a generator $\bar{\mathcal{L}}$ which comes down to show that conditioning on the past;
(iii) the drift part A_t^ε converges to the conditional process of $\int_0^t \bar{\mathcal{L}} f(x_s)ds$.

If y_t is a Markov process with generator $\mathcal{L}$, then the semi-martingale decomposition comes from solving the Poisson equation:

$$\mathcal{L}\beta(\cdot, y) = dF \circ V(\cdot, y).$$

This equation is solvable, for each y, precisely under the center condition on V. Of course we will need to assume some technical conditions, for example $\mathcal{L}_0$ is elliptic and the state space for y is compact. Then one applies the Itô formula to obtain:

$$\beta(x_t^\varepsilon, y_t^\varepsilon) = \beta(x_0, y_0) + \frac{1}{\sqrt{\varepsilon}}\int_0^t D_x\beta\left(V(x_s^\varepsilon, y_{\frac{s}{\varepsilon}})\right)ds + \frac{1}{\varepsilon}\int_0^t \mathcal{L}_y\beta(x_s^\varepsilon, y_{\frac{s}{\varepsilon}})ds + N_t^\varepsilon$$

$$= \beta(x_0, y_0) + \frac{1}{\sqrt{\varepsilon}}\int_0^t D_x\beta\left(V(x_s^\varepsilon, y_{\frac{s}{\varepsilon}})\right)ds + \frac{1}{\varepsilon}\int_0^t dF(V(x_s^\varepsilon, y_{\frac{s}{\varepsilon}}))ds + N_t^\varepsilon.$$

the subscript y in $\mathcal{L}_y$ indicates applying the operator to the second variable y, and similarly $D_x\beta$ indicates differentiation in the first variable. Also, $N_t^\varepsilon = \int_0^t D_y\beta(x_s^\varepsilon, y_s^\varepsilon))d\{y_s^\varepsilon\}$, where $\{y_s^\varepsilon\}$ denotes the martingale part of y^ε, is a local martingale. This means

$$F(x_t^\varepsilon) = F(x_0^\varepsilon) + \sqrt{\varepsilon}\big(\beta(x_t^\varepsilon, y_t^\varepsilon) - \beta(x_0, y_0)\big) - \int_0^t D_x\beta\left(V(x_s^\varepsilon, y_{\frac{s}{\varepsilon}})\right) ds - N_t^\varepsilon. \tag{3.7}$$

This identity can now be used to show $\{x^\varepsilon, \varepsilon \in (0, 1]\}$ is tight. Let us examine the terms in the above equation. We expect that that $\sqrt{\varepsilon}\big(\beta(x_t^\varepsilon, y_t^\varepsilon) - \beta(x_0, y_0)\big)$ is negligible when ε is small and $\int_0^t D_x\beta\left(V(x_s^\varepsilon, y_{\frac{s}{\varepsilon}})\right) ds$ converges if x^ε converges as a process. Indeed if V does not depend on x, the ergodic assumption will imply that $\int_0^t D_x\beta\left(V(x_s^\varepsilon, y_{\frac{s}{\varepsilon}})\right) ds$ converges to the spatial average $\int V(y)\pi(dy)$. Let us define

$$\bar{\mathcal{L}} = \int D_x\beta(V(x, y))\pi(dy).$$

It remains to show that any limit x_t satisfies that $F(x_t) - F(x_s) - \int \bar{\mathcal{L}} F(x_s) ds$ is a martingale, which can be obtained from (3.7) by ergodic theorems.

If both the x and y-variables taking values in a manifold and $V(x, y) = \sum_{k=1}^N Y_k(x) G_k(y)$ where $\bar{G}_k = \int G_k(y)\pi(dy) = 0$, a diffusive homogenisation theorem is proved in [76], from which we extract a simple version and state it below.

Theorem 3.5 [76] *Suppose that $\mathcal{L}_0$ is a smooth and elliptic operator on a smooth compact manifold $\mathcal{Y}$ with invariant measure π and G_k smooth functions with $\int_\mathcal{Y} G_k d\pi = 0$. Let Y_i be smooth vector fields on $\mathbb{R}^N$, growing at mostly linearly, and have bounded derivatives of all order. Consider*

$$\dot{x}_t^\varepsilon = \frac{1}{\sqrt{\varepsilon}} \sum_{k=1}^N Y_k(x_t^\varepsilon) G_k(y_t^\varepsilon), \qquad x_0^\varepsilon = x_0.$$

Then, the following statements hold.

(1) *As $\varepsilon \to 0$, the solution x^ε converges to a diffusion measure $\bar{\mu}$ on any bounded time intervals. Furthermore for every $r < \frac{1}{4}$, their Wasserstein distance is bounded above by ε^r, i.e.*

$$\sup_{t \leqslant T} d(\hat{P}_{y_{\frac{t}{\varepsilon}}^\varepsilon}, \bar{\mu}_t) \lesssim \varepsilon^{\frac{1}{4}-}.$$

(2) *Let β_j be the solutions to $\mathcal{L}_0 \beta_j = G_j$, $\overline{Y_i \beta_j} = \int_G (Y_i \beta_j) d\pi$, and P_t the Markov semigroup for*

$$\bar{\mathcal{L}} = -\sum_{i,j=1}^m \overline{Y_i \beta_j}\, Y_i(Y_j f),$$

where $Y_i F = d F(Y_i)$. Then, for every $F \in BC^4$ and any T,

$$\sup_{t \leqslant T} \left|\mathbb{E} F(x_t^\varepsilon) - P_t F(x_0)\right| \lesssim \varepsilon \sqrt{|\log \varepsilon|}(1 + |x_0|^2)(1 + |F|_{C^4}).$$

Remark 3.6 In [76] a more general theorem is proved without assuming the uniqueness of the invariant probability measures for the fast variables. The compactness of the state space for y_t is not important, what we really used is the exponential rate of convergence of the diffusion at time t to the invariant measure in the total variation norm.

4 Appendix

Here we give a proof for Lemma 2.10, which is completely analogue to that of [44, Prop. 4.10], where the theorem is stated for the fractional OU-process, but the proof is also valid for the processes specified in Lemma 2.10. The proof is added here for reader's the convenience. We will however omit lengthy algebraic manipulations identical to that for the proof of [44, Prop. 4.10], pointing out only where to find them.

Proof of Lemma 2.10. We write $U = G_i$ and $V = G_j$ and denote by $\mathcal{F}_k$ the filtration generated by the fractional Brownian motion defining our fOU process. For $k \in \mathbb{N}$, we define the $\mathcal{F}_k$-adapted processes:

$$I(k) = \int_{k-1}^{k} U(y_s)ds, \quad J(k) = \int_{k-1}^{k} V(y_s)ds,$$

$$\hat{U}(k) = \int_{k-1}^{\infty} \mathbb{E}[U(y_s)|\mathcal{F}_k]ds, \quad \hat{V}(k) = \int_{k-1}^{\infty} \mathbb{E}[V(y_s)|\mathcal{F}_k]ds,$$

$$M_k = \sum_{l=1}^{k} \hat{U}(l) - \mathbb{E}\left[\hat{U}(l-1)|\mathcal{F}_{l-1}\right], \quad N_k = \sum_{l=1}^{k} \hat{V}(l) - \mathbb{E}\left[\hat{V}(l-1)|\mathcal{F}_{l-1}\right].$$

In particular, M_k and N_k are $\mathcal{F}_k$ adapted L^2 martingales. There are the following useful identities. For $k \in \mathbb{N}$

$$\hat{U}(k) = I(k) + \mathbb{E}[\hat{U}(k+1)|\mathcal{F}_k],$$

$$M_{k+1} - M_k = I(k) + \hat{U}(k+1) - \hat{U}(k)$$

$$\sum_{j=1}^{k} I(j) = \int_{0}^{k} U(y_r)dr = M_k - \hat{U}(k) + \hat{U}(1) - M_1,$$

and similarly for J, $\hat{V}$ and N, where the function U is replaced by V.

In a nutshell, the proof is as follows. Combining Lemma 4.1 below, and using $\mathbb{E}(U(y_s)V(y_r)) = \mathbb{E}(U(y_{s-r})V(y_0))$, we see that

$$\varepsilon \int_0^{\frac{L}{\varepsilon}} \int_0^s U(y_s) V(y_r) dr ds$$

$$= \varepsilon \sum_{k=1}^{L} (M_{k+1} - M_k) N_k + t \int_0^\infty \mathbb{E}\left(U(y_v)V(y_0)\right) du + \mathbf{Er}_1(\varepsilon) + \mathbf{Er}_2(\varepsilon).$$

The proof of Lemma 2.10 is then concluded with Lemma 4.2 and the identity

$$\left(\int_0^1 \int_0^s + \int_1^\infty \int_0^1\right) \mathbb{E}\left(U(y_s)V(y_r)\right) dr ds = -\frac{1}{2} \int_0^\infty \int_u^{u-2} \mathbb{E}\left(U(y_v)V(y_0)\right) du dv$$

$$= \int_0^\infty \mathbb{E}\left(U(y_v)V(y_0)\right) dv.$$

Henceforth in this section we set $L = L(\varepsilon) = [\frac{t}{\varepsilon}]$.

Lemma 4.1 *There exists a function* $\mathbf{Er}_1(\varepsilon)$, *which converges to zero in probability as* $\varepsilon \to 0$, *such that*

$$\varepsilon \int_0^{\frac{L}{\varepsilon}} \int_0^s U(y_s)V(y_r) dr ds = \varepsilon \sum_{k=1}^{L} I(k) \sum_{l=1}^{k-1} J(l) + t \int_0^1 \int_0^s \mathbb{E}\left(U(y_s)V(y_r)\right) dr ds + \mathbf{Er}_1(\varepsilon) \tag{4.1}$$

The proof for this follows exactly the same way as that of Lemma 4.13 in [44] where the proof is for the fractional Ornstein-Uhlenbeck process, but the proof used only the stationary property and the ergodicity of the process.

Lemma 4.2 *The following converges in probability:*

$$\lim_{\varepsilon \to 0} \left(\varepsilon \sum_{k=1}^{L} I(k) \sum_{l=1}^{k-1} J(l) - \varepsilon \sum_{k=1}^{L} (M_{k+1} - M_k) N_k\right) = t \int_1^\infty \int_0^1 \mathbb{E}\left(U(y_s)V(y_r)\right) dr ds.$$

This lemma can be proved in the same way as that of [44, Lemma 4.14], again that was for the fractional Ornstein-Uhlenbeck process. The idea is to use the above identities to write the summation on the left hand side as follows:

$$\varepsilon \left(\sum_{k=1}^{L} -I(k)\hat{V}(k) + \sum_{k=1}^{L} I(k)(\hat{V}(1) - N_1) - \sum_{k=1}^{L} (\hat{U}(k+1) - \hat{U}(k)) N_k\right) = I_1^\varepsilon + I_2^\varepsilon + I_3^\varepsilon.$$

To the first term we apply shift invariance and ergodic theorem to obtain

$$I_1^\varepsilon \to -t \, \mathbb{E}[I(1)\hat{V}(1)] = (-t)\mathbb{E}\left(\int_0^1 U(y_r) dr \int_0^\infty V(y_s) ds\right). \tag{4.2}$$

Also,

$$I_2^\varepsilon = \mathbb{E}\left|\varepsilon \sum_{k=1}^{L} I(k)(\hat{V}(1) - N_1)\right|^2 \lesssim \varepsilon^2\, \mathbb{E}[\hat{V}(1)]^2 \int_0^L \int_0^L \mathbb{E}[U(y_r)U(y_s)]\,ds\,dr \to 0,$$

Since $\int_0^L \int_0^L \mathbb{E}[U(y_r)U(y_s)]\,ds\,dr \sim \frac{t}{\varepsilon}$. We then change the order of summation to obtain the following decomposition

$$-I_3^\varepsilon = -\varepsilon \sum_{j=1}^{L-1}(N_{j+1} - N_j)\hat{U}(L+1) + \varepsilon \sum_{j=1}^{L-1}(N_{j+1} - N_j)\hat{U}(j+1) - \varepsilon\left(\hat{U}(L+1) - \hat{U}(1)\right)N_1$$
$$= J_1^\varepsilon + J_2^\varepsilon + J_3^\varepsilon.$$

By Birkhoff's ergodic theorem, $J_1^\varepsilon \to 0$ a.s. and similarly,

$$J_2^\varepsilon \to t\,\mathbb{E}\left(\hat{U}(2)(N_2 - N_1)\right). \tag{4.3}$$

Since $\hat{U}(j)$ is bounded in $L^2(\Omega)$ by the assumption (for the fractional OU-process, this assumption is proved to hold), $|J_3^\varepsilon|_{L^2(\Omega)} \lesssim \varepsilon \to 0$. This concludes the limit of the left hand side to be

$$t\,\mathbb{E}\left[\hat{U}(2)(N_2 - N_1) - I(1)\hat{V}(1)\right], \tag{4.4}$$

which we need to rewrite. Firstly,

$$(N_2 - N_1) = \hat{V}(2) - \hat{V}(1) + \int_0^1 V(y_s)\,ds, \qquad I(1) = \hat{U}(1) - \mathbb{E}[\hat{U}(2)|\mathcal{F}_1].$$

Secondly,

$$\mathbb{E}\left(\hat{U}(2)(N_2 - N_1) - I(1)\hat{V}(1)\right)$$
$$= \int_1^\infty \int_0^1 \mathbb{E}(U(y_s)V(y_r))\,drds + \mathbb{E}\left(\hat{U}(2)\left(\hat{V}(2) - \hat{V}(1)\right) - \left(\hat{U}(1) - \mathbb{E}[\hat{U}(2)|\mathcal{F}_1]\right)\hat{V}(1)\right).$$

Since $\hat{V}(1)$ is $\mathcal{F}_1$ measurable and by the shift covariance of $\hat{U}(k)\hat{V}(k)$,

$$\mathbb{E}\left(\hat{U}(2)\left(\hat{V}(2) - \hat{V}(1)\right) - \left(\hat{U}(1) - \mathbb{E}[\hat{U}(2)|\mathcal{F}_1]\right)\hat{V}(1)\right) = 0.$$

This concludes the proof of the lemma.

References

1. Angst, J., Bailleul, I., Tardif, C.: Kinetic Brownian motion on Riemannian manifolds. Electron. J. Probab. **20**(no. 110), 40 (2015)
2. Arnold, Ludwig, Imkeller, Peter, Yonghui, Wu.: Reduction of deterministic coupled atmosphere–ocean models to stochastic ocean models: a numerical case study of the Lorenz-Maas system. Dyn. Syst. **18**(4), 295–350 (2003)
3. Albeverio, S., Jorgensen, P.E.T., Paolucci, A.M.: On fractional Brownian motion and wavelets. Complex Anal. Oper. Theory **6**(1), 33–63 (2012)
4. Al-Talibi, H., Hilbert, A.: Differentiable approximation by solutions of Newton equations driven by fractional Brownian motion. Preprint (2012)
5. Albeverio, S., De Vecchi, F.C., Morando, P., Ugolini, S.: Weak symmetries of stochastic differential equations driven by semimartingales with jumps. Electron. J. Probab. **25** (2020)
6. Bailleul, I., Catellier, R.: Rough flows and homogenization in stochastic turbulence. J. Differ. Equ. **263**(8), 4894–4928 (2017)
7. Berglund, N., Gentz, B.: Noise-induced phenomena in slow-fast dynamical systems. Probability and its Applications (New York). Springer London, Ltd., London (2006). A sample-paths approach
8. Bourguin, S., Gailus, S., Spiliopoulos, K.: Discrete-time inference for slow-fast systems driven by fractional brownian motion (2020)
9. Birrell, Jeremiah, Hottovy, Scott, Volpe, Giovanni, Wehr, Jan: Small mass limit of a Langevin equation on a manifold. Ann. Henri Poincaré **18**(2), 707–755 (2017)
10. Bakhtin, Victor, Kifer, Yuri: Diffusion approximation for slow motion in fully coupled averaging. Probab. Theory Relat. Fields **129**(2), 157–181 (2004)
11. Bensoussan, A., Lions, J.-L., Papanicolaou, G.: Asymptotic analysis for periodic structures. AMS Chelsea Publishing, Providence, RI (2011)
12. Borodin, A.N.: A limit theorem for the solutions of differential equations with a random right-hand side. Teor. Verojatnost. i Primenen. **22**(3), 498–512 (1977)
13. Boufoussi, Brahim, Tudor, Ciprian A.: Kramers-Smoluchowski approximation for stochastic evolution equations with FBM. Rev. Roumaine Math. Pures Appl. **50**(2), 125–136 (2005)
14. Bai, Shuyang, Taqqu, Murad S.: Multivariate limit theorems in the context of long-range dependence. J. Time Ser. Anal. **34**(6), 717–743 (2013)
15. Brzeźniak, Z., van Neerven, J., Salopek, D.: Stochastic evolution equations driven by Liouville fractional Brownian motion. Czechoslov. Math. J. **62**(137)(1), 1–27 (2012)
16. Barret, Florent, von Renesse, Max: Averaging principle for diffusion processes via Dirichlet forms. Potential Anal. **41**(4), 1033–1063 (2014)
17. Catuogno, P.J., da Silva, F.B., Ruffino, P.R.: Decomposition of stochastic flows in manifolds with complementary distributions. Stoch. Dyn. **13**(4), 1350009, 12 (2013)
18. Chevyrev, I., Friz, P.K., Korepanov, A., Melbourne, I., Zhang, H.: Multiscale systems, homogenization, and rough paths. In: Probability and Analysis in Interacting Physical Systems (2019)
19. Catellier, R., Gubinelli, M.: Averaging along irregular curves and regularisation of ODEs. Stoch. Process. Appl. **126**(8), 2323–2366 (2016)
20. Cass, Thomas, Hairer, Martin, Litterer, Christian, Tindel, Samy: Smoothness of the density for solutions to Gaussian rough differential equations. Ann. Probab. **43**(1), 188–239 (2015)
21. Ciccotti, Giovanni, Lelievre, Tony, Vanden-Eijnden, Eric: Projection of diffusions on submanifolds: application to mean force computation. Comm. Pure Appl. Math. **61**(3), 371–408 (2008)
22. Cont, R.: Long range dependence in financial markets. In: Fractals in Engineering. Springer, London (2005)
23. Coutin, Laure, Qian, Zhongmin: Stochastic analysis, rough path analysis and fractional Brownian motions. Probab. Theory Relat. Fields **122**(1), 108–140 (2002)
24. Dolgopyat, Dmitry, Kaloshin, Vadim, Koralov, Leonid: Sample path properties of the stochastic flows. Ann. Probab. **32**(1A), 1–27 (2004)

25. Dobrushin, R.L., Major, P.: Non-central limit theorems for nonlinear functionals of Gaussian fields. Z. Wahrsch. Verw. Gebiete **50**(1), 27–52 (1979)
26. Dobrushin, R.L.: Gaussian and their subordinated self-similar random generalized fields. Ann. Probab. **7**(1), 1–28 (1979)
27. Decreusefond, L., Üstünel, A.S.: Stochastic analysis of the fractional Brownian motion. Potential Anal. **10**(2), 177–214 (1999)
28. De Vecchi, F.C., Morando, P., Ugolini, S.: Symmetries of stochastic differential equations: a geometric approach. J. Math. Phys. **57**(6), 063504, 17 (2016)
29. Enriquez, N., Franchi, J., Le Jan, Y.: Central limit theorem for the geodesic flow associated with a Kleinian group, case $\delta > d/2$. J. Math. Pures Appl. (9) **80**(2), 153–175 (2001)
30. Eichinger, K., Kuehn, C., Neamt, A.: Sample paths estimates for stochastic fast-slow systems driven by fractional brownian motion. J. Stat. Phys. (2020)
31. David Elworthy, K., Li, X.-M.: Intertwining and the Markov uniqueness problem on path spaces. In: Stochastic partial differential equations and applications—VII. Lecture Notes in Pure and Applied Mathematics, vol. 245, pp. 89–95. Chapman & Hall/CRC, Boca Raton, FL (2006)
32. David Elworthy, K., Le Jan, Y., Li, X.-M.: Equivariant diffusions on principal bundles. In: Stochastic analysis and related topics in Kyoto. Advanced Studies in Pure Mathematics, vol. 41, pp. 31–47. Math. Soc. Japan, Tokyo (2004)
33. David Elworthy, K., Le Jan, Y., Li, X.-M.: The geometry of filtering. Frontiers in Mathematics. Birkhäuser Verlag, Basel (2010)
34. Feyel, Denis, de La Pradelle, Arnaud: Curvilinear integrals along enriched paths. Electron. J. Probab. **11**(34), 860–892 (2006)
35. Friz, Peter, Gassiat, Paul, Lyons, Terry: Physical Brownian motion in a magnetic field as a rough path. Trans. Amer. Math. Soc. **367**(11), 7939–7955 (2015)
36. Flandoli, Franco, Gubinelli, Massimiliano, Russo, Francesco: On the regularity of stochastic currents, fractional Brownian motion and applications to a turbulence model. Ann. Inst. Henri Poincaré Probab. Stat. **45**(2), 545–576 (2009)
37. Friz, P.K., Hairer, M.: A course on rough paths. Universitext. Springer, Cham (2014). With an introduction to regularity structures
38. Fannjiang, Albert, Komorowski, Tomasz: Fractional Brownian motions in a limit of turbulent transport. Ann. Appl. Probab. **10**(4), 1100–1120 (2000)
39. Freĭdlin, M.I.: Fluctuations in dynamical systems with averaging. Dokl. Akad. Nauk SSSR **226**(2), 273–276 (1976)
40. Freidlin, M.I., Wentzell, A.D.: Averaging principle for stochastic perturbations of multifrequency systems. Stoch. Dyn. **3**(3), 393–408 (2003)
41. Garrido-Atienza, María J., Schmalfuss, Björn.: Local stability of differential equations driven by Hölder-continuous paths with Hölder index in $(1/3, 1/2)$. SIAM J. Appl. Dyn. Syst. **17**(3), 2352–2380 (2018)
42. Gallavotti, G., Jona-Lasinio, G.: Limit theorems for multidimensional Markov processes. Comm. Math. Phys. **41**, 301–307 (1975)
43. Gehringer, J., Li, X.-M.: Homogenization with fractional random fields (2019). arXiv:1911.12600. (This is now improved and split into 'Functional limit theorem for fractional OU' and 'Diffusive and rough homogenisation in fractional noise field')
44. Gehringer, J., Li, X.-M.: Diffusive and rough homogenisation in fractional noise field (2020). (Based on Part 2 of arXiv:1911.12600)
45. Gehringer, J., Li, X.-M.: Functional limit theorem for fractional OU. J. Theoretical Probability (2020). https://doi.org/10.1007/s10959-020-01044-7 (Based on Part 1 of arXiv:1911.12600)
46. Grahovac, Danijel, Leonenko, Nikolai N., Taqqu, Murad S.: Limit theorems, scaling of moments and intermittency for integrated finite variance supOU processes. Stoch. Process. Appl. **129**(12), 5113–5150 (2019)
47. Guerra, João., Nualart, David: Stochastic differential equations driven by fractional Brownian motion and standard Brownian motion. Stoch. Anal. Appl. **26**(5), 1053–1075 (2008)

48. Gradinaru, Mihai, Nourdin, Ivan, Russo, Francesco, Vallois, Pierre: m-order integrals and generalized Itô's formula: the case of a fractional Brownian motion with any Hurst index. Ann. Inst. H. Poincaré Probab. Statist. **41**(4), 781–806 (2005)
49. Gubinelli, M.: Controlling rough paths. J. Funct. Anal. **216**(1), 86–140 (2004)
50. Hairer, Martin: Ergodicity of stochastic differential equations driven by fractional Brownian motion. Ann. Probab. **33**(2), 703–758 (2005)
51. Hasminskiĭ, R.Z.: A limit theorem for solutions of differential equations with a random right hand part. Teor. Verojatnost. i Primenen **11**, 444–462 (1966)
52. Has′minskii, R.Z.: On the principle of averaging the Itô's stochastic differential equations. Kybernetika (Prague) **4**, 260–279 (1968)
53. Hurst, H.E., Black, R.P., Sinaika, Y.M.: Long Term Storage in Reservoirs, An Experimental Study. Constable
54. Helland, Inge S.: Central limit theorems for martingales with discrete or continuous time. Scand. J. Statist. **9**(2), 79–94 (1982)
55. Hairer, Martin, Li, Xue-Mei.: Averaging dynamics driven by fractional Brownian motion. Ann. Probab. **48**(4), 1826–1860 (2020)
56. Hu, Y., Nualart, D.: Differential equations driven by Hölder continuous functions of order greater than 1/2. In: Stochastic analysis and applications. Abel Symp., vol. 2, pp. 399–413. Springer, Berlin (2007)
57. Hairer, M., Pavliotis, G.A.: Periodic homogenization for hypoelliptic diffusions. J. Statist. Phys. **117**(1–2), 261–279 (2004)
58. Hairer, Martin, Pardoux, Etienne: Homogenization of periodic linear degenerate PDEs. J. Funct. Anal. **255**(9), 2462–2487 (2008)
59. Jona-Lasinio, G.: Probabilistic approach to critical behavior. In: New developments in quantum field theory and statistical mechanics (Proc. Cargèse Summer Inst., Cargèse, 1976), pp. 419–446. NATO Adv. Study Inst. Ser., Ser. B: Physics, 26 (1977)
60. Jacod, J., Shiryaev, A.N.: Limit theorems for stochastic processes, 2nd edn. Grundlehren der Mathematischen Wissenschaften [Fundamental Principles of Mathematical Sciences], vol. 288. Springer, Berlin (2003)
61. Kryloff, N., Bogoliuboff, N.: Introduction to Non-Linear Mechanics. Annals of Mathematics Studies, no. 11. Princeton University Press, Princeton, N. J. (1943)
62. Kifer, Yuri: Averaging in dynamical systems and large deviations. Invent. Math. **110**(2), 337–370 (1992)
63. Komorowski, T., Landim, C., Olla, S.: Fluctuations in Markov processes. Grundlehren der Mathematischen Wissenschaften [Fundamental Principles of Mathematical Sciences], vol. 345. Springer, Heidelberg (2012). Time symmetry and martingale approximation
64. Kelly, David, Melbourne, Ian: Smooth approximation of stochastic differential equations. Ann. Probab. **44**(1), 479–520 (2016)
65. Kelly, David, Melbourne, Ian: Deterministic homogenization for fast-slow systems with chaotic noise. J. Funct. Anal. **272**(10), 4063–4102 (2017)
66. Komorowski, Tomasz, Novikov, Alexei, Ryzhik, Lenya: Homogenization driven by a fractional brownian motion: the shear layer case. Multiscale Model. Simul. **12**(2), 440–457 (2014)
67. Kurtz, Thomas G., Protter, Philip: Weak limit theorems for stochastic integrals and stochastic differential equations. Ann. Probab. **19**(3), 1035–1070 (1991)
68. Kurtz, Thomas G.: A general theorem on the convergence of operator semigroups. Trans. Amer. Math. Soc. **148**, 23–32 (1970)
69. Kipnis, C., Varadhan, S.R.S.: Central limit theorem for additive functionals of reversible Markov processes and applications to simple exclusions. Comm. Math. Phys. **104**(1), 1–19 (1986)
70. Klingenhöfer, F., Zähle, M.: Ordinary differential equations with fractal noise. Proc. Amer. Math. Soc. **127**(4), 1021–1028 (1999)
71. Lê, K.: A stochastic sewing lemma and applications. Electron. J. Probab. **25**, 55 pp. (2020)
72. Ledrappier, François: Central limit theorem in negative curvature. Ann. Probab. **23**(3), 1219–1233 (1995)

73. Li, Xue-Mei.: An averaging principle for a completely integrable stochastic Hamiltonian system. Nonlinearity **21**(4), 803–822 (2008)
74. Li, X.-M.: Effective diffusions with intertwined structures (2012). arxiv:1204.3250
75. Li, Xue-Mei.: Random perturbation to the geodesic equation. Ann. Probab. **44**(1), 544–566 (2015)
76. Li, Xue-Mei.: Limits of random differential equations on manifolds. Probab. Theory Relat. Fields **166**(3–4), 659–712 (2016). https://doi.org/10.1007/s00440-015-0669-x
77. Li, Xue-Mei.: Homogenisation on homogeneous spaces. J. Math. Soc. Japan **70**(2), 519–572 (2018)
78. Li, X.-M.: Perturbation of conservation laws and averaging on manifolds. In: Computation and combinatorics in dynamics, stochastics and control. Abel Symp., vol. 13, pp. 499–550. Springer, Cham (2018)
79. Liverani, Carlangelo, Olla, Stefano: Toward the Fourier law for a weakly interacting anharmonic crystal. J. Amer. Math. Soc. **25**(2), 555–583 (2012)
80. Li, X.-M., Sieber, J.: Slow/fast systems with fractional environment and dynamics. In preparation (2020)
81. Lyons, T.: Differential equations driven by rough signals. I. An extension of an inequality of L. C. Young. Math. Res. Lett. **1**(4), 451–464 (1994)
82. Mishura, Y.S.: Stochastic calculus for fractional Brownian motion and related processes. Lecture Notes in Mathematics, vol. 1929. Springer, Berlin (2008)
83. Mathieu, P., Piatnitski, A.: Steady states, fluctuation-dissipation theorems and homogenization for reversible diffusions in a random environment. Arch. Ration. Mech. Anal. **230**(1), 277–320 (2018)
84. Mandelbrot, Benoit B., Van Ness, John W.: Fractional Brownian motions, fractional noises and applications. SIAM Rev. **10**, 422–437 (1968)
85. Neishtadt, A.I.: Averaging and passage through resonances. In: Proceedings of the International Congress of Mathematicians, Vol. I, II (Kyoto, 1990), pp. 1271–1283. Math. Soc. Japan, Tokyo (1991)
86. Nelson, E.: Dynamical Theories of Brownian Motion. Princeton University Press, Princeton, N.J. (1967)
87. Nourdin, Ivan, Nualart, David, Peccati, Giovanni: Strong asymptotic independence on Wiener chaos. Proc. Amer. Math. Soc. **144**(2), 875–886 (2016)
88. Nourdin, Ivan, Nualart, David, Zintout, Rola: Multivariate central limit theorems for averages of fractional Volterra processes and applications to parameter estimation. Stat. Inference Stoch. Process. **19**(2), 219–234 (2016)
89. Nualart, David, Peccati, Giovanni: Central limit theorems for sequences of multiple stochastic integrals. Ann. Probab. **33**(1), 177–193 (2005)
90. Nourdin, I., Peccati, G.: Normal approximations with Malliavin calculus. Cambridge Tracts in Mathematics, vol. 192. Cambridge University Press, Cambridge (2012). From Stein's method to universality
91. Neuman, E., Rosenbaum, M.: Fractional Brownian motion with zero Hurst parameter: a rough volatility viewpoint. Electron. Commun. Probab. **23**(Paper No. 61), 12 (2018)
92. Nualart, D.: The Malliavin calculus and related topics. Probability and its Applications (New York), 2nd edn. Springer, Berlin (2006)
93. Perruchaud, P.: Homogénísation pour le mouvement brownien cinétique et quelques résultats sur son noyau (2019). Université de Rennes 1
94. Pei, B., Inahama, Y., Xu, Y.: Averaging principles for mixed fast-slow systems driven by fractional brownian motion (2020). arXiv:Dynamical Systems
95. Pei, B., Inahama, Y., Xu, Y.: Pathwise unique solutions and stochastic averaging for mixed stochastic partial differential equations driven by fractional brownian motion and brownian motion (2020). arXiv:Probability
96. Papanicolaou, G.C., Kohler, W.: Asymptotic theory of mixing stochastic ordinary differential equations. Comm. Pure Appl. Math. **27**, 641–668 (1974)

97. Pavliotis, G.A., Stuart, A.M., Zygalakis, K.C.: Homogenization for inertial particles in a random flow. Commun. Math. Sci. **5**(3), 507–531 (2007)
98. Pipiras, Vladas, Taqqu, Murad S.: Integration questions related to fractional Brownian motion. Probab. Theory Relat. Fields **118**(2), 251–291 (2000)
99. Rosenblatt, M.: Independence and dependence. In: Proceedings of 4th Berkeley Sympos. Math. Statist. and Prob., Vol. II, pp. 431–443. Univ. California Press, Berkeley, Calif. (1961)
100. Ruffino, P.R.: Application of an averaging principle on foliated diffusions: topology of the leaves. Electron. Commun. Probab. **20**(no. 28), 5 (2015)
101. Röckner, M., Xie, L.: Averaging principle and normal deviations for multiscale stochastic systems (2020)
102. Skorokhod, A.V., Hoppensteadt, F.C., Salehi, H.: Random perturbation methods with applications in science and engineering. Applied Mathematical Sciences, vol. 150. Springer, New York (2002)
103. Sinaı, Ja.G.: Self-similar probability distributions. Teor. Verojatnost. i Primenen. **21**(1), 63–80 (1976)
104. Skorohod, A.V.: The averaging of stochastic equations of mathematical physics. In: Problems of the asymptotic theory of nonlinear oscillations (Russian), pp. 196–208, 279 (1977)
105. Stratonovich, R.L.: Selected problems in the theory of fluctuations in radio engineering. Sov. Radio, Moscow (1961). In Russian
106. Stratonovich, R.L.: Topics in the theory of random noise. Vol. I: General theory of random processes. Nonlinear transformations of signals and noise. Revised English edition. Translated from the Russian by Richard A. Silverman. Gordon and Breach Science Publishers, New York-London (1963)
107. Takao, S.: Stochastic geometric mechanics for fluid modelling and mcmc, imperial college london (2020)
108. Taqqu, Murad S.: Weak convergence to fractional Brownian motion and to the Rosenblatt process. Z. Wahrscheinlichkeitstheorie und Verw. Gebiete **31**, 287–302 (1975)
109. Taqqu, Murad S.: Law of the iterated logarithm for sums of non-linear functions of Gaussian variables that exhibit a long range dependence. Z. Wahrscheinlichkeitstheorie und Verw. Gebiete **40**(3), 203–238 (1977)
110. Taqqu, Murad S.: Convergence of integrated processes of arbitrary Hermite rank. Z. Wahrsch. Verw. Gebiete **50**(1), 53–83 (1979)
111. Taqqu, M.S.: Colloquium and Workshop on Random Fields: rigorous results in statistical mechanics and quantum field theory. Cambridge Tracts in Mathematics, vol. 192. Cambridge University Press, Cambridge (1981). From Stein's method to universality
112. Üstünel, A.S., Zakai, M.: On independence and conditioning on Wiener space. Ann. Probab. **17**(4), 1441–1453 (1989)
113. Veretennikov, AYu.: On an averaging principle for systems of stochastic differential equations. Mat. Sb. **181**(2), 256–268 (1990)
114. Young, L.C.: An inequality of the Hölder type, connected with Stieltjes integration. Acta Math. **67**(1), 251–282 (1936)

Stochastic Geometric Mechanics with Diffeomorphisms

Darryl D. Holm and Erwin Luesink

Abstract Noether's celebrated theorem associating symmetry and conservation laws in classical field theory is adapted to allow for broken symmetry in geometric mechanics and is shown to play a central role in deriving and understanding the generation of fluid circulation via the Kelvin-Noether theorem for ideal fluids with stochastic advection by Lie transport (SALT).

Keywords Noether's theorem · Geometric mechanics · Diffeomorphisms · Reduction by symmetry · Symmetry breaking

1 Noether's Theorem in Geometric Mechanics

1.1 Euler-Poincaré Reduction

Geometric mechanics deals with group-invariant variational principles. In this setting, Noether's theorem [23, 24] plays a key role. Given the tangent lift action $G \times TM \to TM$ of a Lie group G on the tangent bundle TM of a manifold M[1] on which G acts transitively, Noether's theorem states that each Lie symmetry of a Lagrangian $L : TM \to \mathbb{R}$ defined in the action integral $S = \int L(q, v)dt$ for Hamilton's variational principle $\delta S = 0$ with $(q, v) \in TM$ implies a conserved quantity for the corresponding Euler-Lagrange equations defined on the cotangent bundle T^*M. The conserved quantities arising from Noether's theorem in the case where the configuration manifold M is a Lie group G were studied by Smale, in [26, 27], where

---

[1] M is called the *configuration manifold* in classical mechanics.

D. D. Holm (✉) · E. Luesink
Department of Mathematics, Imperial College London, London SW7 2AZ, UK
e-mail: d.holm@ic.ac.uk

E. Luesink
e-mail: e.luesink16@imperial.ac.uk

it was shown that the reduction procedure $TG \to TG \setminus G \simeq \mathfrak{g}$ leads to dynamics which take place on the dual $\mathfrak{g}^*$ of the Lie algebra $\mathfrak{g}$. The dynamical variable $m \in \mathfrak{g}^*$ in the dual Lie algebra is now called the momentum map (Smale called it angular momentum). In general, the configuration manifold M is not a Lie group. However, when a Lie group G acts transitively on a configuration manifold M the proof of Noether's theorem induces a cotangent-lift momentum map $J : T^*M \to \mathfrak{g}^*$. The momentum map induced this way is an infinitesimally equivariant Poisson map taking functions on the cotangent bundle T^*M of M to the dual Lie algebra $\mathfrak{g}^*$ of the Lie group G. The momentum map $J : T^*M \to \mathfrak{g}^*$ is equivariant and Poisson, even if G is not a Lie symmetry of the Lagrangian in Hamilton's principle. Momentum maps naturally lead from the Lagrangian side to the Hamiltonian side. The Hamiltonian dynamics on T^*M involves symplectic transformations. However, as we shall discuss below, for the class of Hamiltonians which can be defined as $H \circ J : \mathfrak{g}^* \to \mathbb{R}$, the momentum map induces Euler-Poincaré motion on the Lagrangian side and Lie-Poisson motion on the Hamiltonian side. To illustrate these remarks, we return to the situation in which the configuration manifold, M, is a Lie group, G.

For hyperregular Lagrangians, the Legendre transform to the Hamiltonian side is invertible and one may reconstruct the solution on G from its representation on $T^*G \setminus G \simeq \mathfrak{g}^*$. In that case, solving the equations describing the evolution of the momentum map on the dual Lie algebra $\mathfrak{g}^*$ is equivalent to solving the equations on the cotangent bundle T^*G when the configuration manifold is G. When the Lie group G acts transitively, freely and properly on the configuration manifold M, then one may reconstruct the solution on M from its representation on $T^*G \setminus G \simeq \mathfrak{g}^*$. The last statement is proven for finite-dimensional Lie groups G in, e.g., [1].

The Lie-group reduced equations defined on the dual Lie algebra $\mathfrak{g}^*$ via Smale's procedure of reduction by symmetry $T^*G \setminus G \simeq \mathfrak{g}^*$ are called Euler-Poincaré equations after [25]. Provided the Lagrangian is hyperregular, the Euler-Poincaré reduction procedure can be expressed in terms of the cube of linked commutative diagrams shown in Fig. 1.

To summarise the notation in Fig. 1, G denotes the configuration manifold which is assumed to be isomorphic to a Lie group, TG is the tangent bundle, T^*G is the cotangent bundle, $TG \setminus G \simeq \mathfrak{g}$ is the Lie algebra and $T^*G \setminus G \simeq \mathfrak{g}^*$ is the dual of the Lie algebra. The Lagrangian is a functional $L : TG \to \mathbb{R}$ and the Hamiltonian is a functional $H : T^*G \to \mathbb{R}$. Euler-Poincaré reduction takes advantage of Lie group symmetries to transform the Lagrangian and Hamiltonian into group-invariant variables, which leads to a reduced Lagrangian $\ell : \mathfrak{g} \to \mathbb{R}$ and a reduced Hamiltonian $\hbar : \mathfrak{g}^* \to \mathbb{R}$. The diagram comprising the face of the cube involving these functionals in Fig. 1 commutes if the Legendre transform is a diffeomorphism. This is guaranteed if the Lagrangian or Hamiltonian is hyperregular. The Euler-Lagrange equations and Hamilton's equations are related via a change of variables, which also holds for the Euler-Poincaré equations and the Lie-Poisson equations. Many finite dimensional mechanical systems may be described naturally in this framework. The classic example is the rotating rigid body, discussed from the viewpoint of symmetry reduction by Poincaré in [25]. In his 1901 paper, Poincaré also raised the issue of *symmetry breaking*, by introducing the vertical acceleration of gravity, which breaks the $SO(3)$ symmetry for free rotation and restricts it to $SO(2)$ for rotations about the vertical axis.

Stochasticity may also be included in the framework of Euler-Poincaré reduction by symmetry. The first attempt to include noise consistently in finite-dimensional symplectic Hamiltonian mechanics was by [5] and reduction by symmetry of stochastic systems was studied by [21].

Plan of the paper. In the present work, we will review Euler-Poincaré reduction of stochastic infinite dimensional variational systems with symmetry breaking. The infinite dimensional case is interesting because it is the natural setting for fluid dynamics, quantum mechanics and elasticity. The foundations of the finite dimensional stochastic geometric mechanics are established in [12]. We will explore the infinite dimensional case in context of fluid dynamics, where symmetry under the smooth invertible maps of the flow domain is broken by the spatial dependence of the initial mass density.

1.2 Sobolev Class Diffeomorphisms

Consider an n-dimensional compact and oriented smooth manifold M, equipped with a Riemannian metric $\langle \cdot, \cdot \rangle$. This will be the spatial domain of flow and $X \in M$ will denote the initial position of any given fluid particle. The manifold M is acted upon by a group of Sobolev class diffeomorphisms. In [13] it is shown that the space of C^∞ diffeomorphisms, defined by $\mathfrak{D} = \{g \in C^\infty(M, M) | \, g$ is bijective and $g^{-1} \in C^\infty(M, M)\}$, is not the convenient setting to study fluid dynamics, but that one should use $\mathfrak{D}^s = \{g \in H^s(M, M) | \, g$ is bijective and $g^{-1} \in H^s(M, M)\}$, the space of Sobolev class diffeomorphisms with s weak derivatives. The reason for this choice

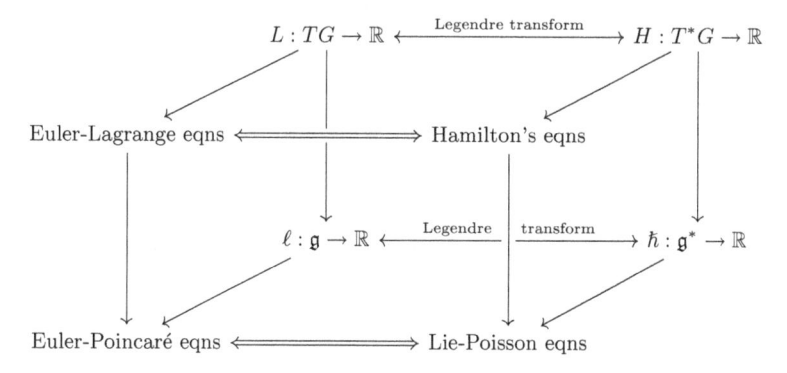

Fig. 1 The cube of commutative diagrams for geometric mechanics on Lie groups. Euler-Poincaré reduction (on the left side) and Lie-Poisson reduction (on the right side) are both indicated by the arrows pointing down. The diagrams are all commutative, provided the Legendre transformation and reduced Legendre transformation are both invertible

is that the smooth diffeomorphisms constitute a Fréchet manifold for which there is no inverse or implicit function theorem and no general solution theorem for ordinary differential equations. Each of these latter features would prohibit the study of geodesics.

The space of Sobolev class diffeomorphisms is both a Hilbert manifold and a topological group if $s > n/2 + 1$, as was shown by [14]. The Hilbert manifold structure implies the existence of function inverses and the implicit function theorem, as well as the existence of a general solution theorem for ordinary differential equations. This additional structure also implies that one can construct the tangent space of $\mathfrak{D}^s$ in the usual way and study geodesics. The space $\mathfrak{D}^s$ is the configuration space for continuum mechanics and each $g \in \mathfrak{D}^s$ is called a configuration. A fluid trajectory starting from $X \in M$ at time $t = 0$ is given by $x(t) = g_t(X) = g(X, t)$, with $\mathfrak{D}^s \ni g : M \times \mathbb{R}^+ \to M$ being a continuous one-parameter subgroup of $\mathfrak{D}^s$. In the deterministic case, computing the time derivative of this one-parameter subgroup gives rise to the *reconstruction equation*, given by

$$\frac{\partial}{\partial t} g_t(X) = u(g_t(X), t), \tag{1}$$

where $u_t(\,\cdot\,) = u(\,\cdot\,, t) \in \mathfrak{X}^s$ is a time dependent vector field with flow $g_t(\,\cdot\,) = g(\,\cdot\,, t)$. The initial data is given by $g(X, 0) = X$. Here $\mathfrak{X}^s = H^s(TM)$ denotes the space of Sobolev class vector fields on M, which is also the Lie algebra associated to the Sobolev class diffeomorphisms.

1.3 Stochastic Advection by Lie Transport (SALT)

In the setting of stochastic advection by Lie transport (SALT), which was introduced by [16], the deterministic reconstruction equation in (1) is replaced by the semimartingale

$$\mathrm{d}g(X, t) = u(g_t(X), t)dt + \sum_{i=1}^{M} \xi_i(g_t(X)) \circ dW_t^i, \tag{2}$$

where the symbol $\circ$ means that the stochastic integral is taken in the Stratonovich sense. The initial data is given by $g(X, 0) = X$. The W_t^i are independent, identically distributed Brownian motions, defined with respect to the standard stochastic basis $(\Omega, \mathcal{F}, (\mathcal{F}_t)_{t \geq 0}, \mathbb{P})$. Such a noise was shown to arise from a multi-time homogenisation argument in [10]. The $\xi_i(\,\cdot\,) \in \mathfrak{X}^s$ are called data vector fields and are prescribed. These data vector fields represent the effects of unresolved degrees of freedom on the resolved scales of fluid motion and account for unrepresented processes. They are determined by applying empirical orthogonal function analysis to appropriate numerical and/or observational data. For instance, for an application to the two dimensional

Euler equations for an ideal fluid, see [8] and for an application to a two-layer quasi-geostrophic model, see [7]. Stochastic models enable the use of a variety of methods in data assimilation, which are discussed in [9]. It is not difficult to make sense of (1), but understanding (2) is more complicated. In [4], a *stochastic chain rule* is shown to exist. This stochastic chain rule is called the *Kunita-Itô-Wentzell (KIW) formula* and helps interpret the semimartingale in (2). The KIW formula will also be used later to prove the stochastic Kelvin circulation theorem. First, however, the space $\mathfrak{D}^s$ needs to be given more structure.

The space $\mathfrak{D}^s$ inherits a *weak Riemannian structure* from the underlying manifold M in a natural way. For $g \in \mathfrak{D}^s$ and $V, W \in T_g\mathfrak{D}^s$, one can define the following bilinear form

$$(V, W) = \int_M \langle (V(X), W(X) \rangle_{g(X)} \mu(dX), \qquad (3)$$

where μ is the volume form on M induced by the metric. The Riemannian structure induced by (3) is weak because the topology is of type L^2, which is strictly weaker than the H^s topology. This bilinear form is a linear functional on the Hilbert space $T_g\mathfrak{D}^s$ and can be used to define the dual space $T_g^*\mathfrak{D}^s$. The pairing between $V \in T_g\mathfrak{D}^s$ and $\alpha \in T_g^*\mathfrak{D}^s$ is given by

$$\langle \alpha, V \rangle = \int_M \alpha(X) \cdot V(X). \qquad (4)$$

Hence the metric on M and the volume form $\mu(dX)$ can be used to construct the isomorphism between $T\mathfrak{D}^s$ and $T^*\mathfrak{D}^s$ as $V(X) \mapsto \alpha(X) = V^\flat(X)\mu(dX)$, where $\flat : TM \to T^*M$ is one of the musical isomorphisms that are induced by the metric on M. The group $\mathfrak{D}^s$ is not a Lie group; since right multiplication is smooth, but left multiplication is only continuous. Hence $\mathfrak{D}^s$ is a topological group with a weak Riemannian structure. In general, these properties are not sufficient to guarantee the existence of an exponential map. However, [13] showed that an exponential map can exist in many important cases. In particular, they showed that the geodesic spray associated to (3) (with and without forcing) is smooth.[2] The smoothness of the geodesic spray persists even though H^s diffeomorphisms are considered rather than smooth diffeomorphisms. Combined with the existence of an exponential map, the smoothness property implies a regular interpretation of the Euler-Poincaré equations on $\mathfrak{D}^s$, provided that one uses right translations and right representations of the group on itself and its Lie algebra, as shown in [17]. However, due to the presence of the volume form $\mu(dm)$, the bilinear form (3) is not right-invariant under the action of the entire H^s diffeomorphism group, although there is right-invariance under the action of the isotropy subgroup $\mathfrak{D}^s_\mu = \{g \in \mathfrak{D}^s | g_*\mu = \mu\}$. Since this subgroup is a proper subgroup, as it is smaller than $\mathfrak{D}^s$ itself. Thus, one speaks of *symmetry breaking*.

In deriving the equations of ideal deterministic fluid dynamics, one needs to keep track of the volume form as well. The appropriate mathematical setting for this is an

---

[2] The *geodesic spray* is the vector field whose integral curves are the geodesics.

outer semidirect product group. This means that one constructs a new group from two given groups with a particular type of group operation. For continuum mechanics, the ingredients are $\mathfrak{D}^s$ and V^*, where V^* is a vector space of tensor fields. This vector space is the space of *advected quantities* and it will always contain at least the volume form μ.

Definition 1.1 (*Advected quantity*) A fluid variable is said to be *advected*, if it keeps its value along Lagrangian particle trajectories. Advected quantities are sometimes called *tracers*, because the evolution histories of scalar advected quantities with different initial values (labels) trace out the Lagrangian particle trajectories of each label, or initial value, via the *push-forward* of the full diffeomorphism group, i.e., $a_t = g_{t*}a_0 = a_0 g_t^{-1}$, where g_t is a time-dependent curve on the manifold of diffeomorphisms that represents the fluid flow.

Remark 1 (*Advected quantities as order parameters*) When several advected quantities are involved, the space V^* is the direct sum of several vector spaces, where each summand space hosts a different advected quantity. In general, each additional advected quantity decreases the dimension of the isotropy subgroup. For example, consider an ideal deterministic fluid with a buoyancy variable b, then the Lagrangian corresponding to the model will depend on μ and b in a parametric manner. This Lagrangian will be right invariant under the action of the isotropy subgroup $\mathfrak{D}^s_{\mu,b} = \{g \in \mathfrak{D}^s \mid g_*\mu = \mu \text{ and } g_*b = b\}$. Hence, advected quantities are *order parameters* and each additional order parameter breaks more symmetry. For the sake of notation, one usually writes $\mathfrak{D}^s_{a_0}$ for the isotropy subgroup, no matter how many advected quantities there are. One then uses a to represent all advected quantities and a_0 to denote the initial value of the advected quantities.

1.4 Semidirect Product Group Adjoint and Coadjoint Actions

The semidirect product group action is constructed in the following way. The representation of $\mathfrak{D}^s$ on a vector space V is by push-forward, which is a left representation, as shown by [22]. The representation of the group on itself and on its Lie algebra is a right representation. In terms of analysis, this means that all representations are smooth and no derivatives need to be counted. The group action of the semidirect product group is given by

$$\bullet : (\mathfrak{D}^s \times V) \times (\mathfrak{D}^s \times V) \to (\mathfrak{D}^s \times V)$$
$$(g_1, v_1) \bullet (g_2, v_2) := (g_1 \circ g_2, v_2 + (g_2)_* v_1) \tag{5}$$

with $g_1, g_2 \in \mathfrak{D}^s$ and $v_1, v_2 \in V$. The semidirect product group is often denoted as $\mathfrak{D}^s \circledS V = (\mathfrak{D}^s \times V, \bullet)$. In the group action above, $(g_2)_* v_1$ denotes the *push-forward* of v_1 by g_2 and $\circ$ denotes composition. Note that the group affects both slots in (5), but the vector space only appears in the second slot. The identity element of the semidirect product group is $(e, 0)$ where $e \in \mathfrak{D}^s$ is the identity diffeomorphism and $0 \in V$ is the zero vector. An inverse element is given by

$$(g, v)^{-1} = (g^{-1}, -(g^{-1})_* v) = (g^{-1}, -g^* v), \tag{6}$$

where $g^* v$ denotes the pull-back of v by g. To understand how reduction works for semidirect products, it is helpful to know how the group acts on its Lie algebra and on the dual of its Lie algebra. Duality will be defined with respect to the sum of the pairing (4) and the dual linear transformation $[\cdot]^*$ on V. This pairing induces another pairing in a natural way on $\mathfrak{X}^s \times V$. Consider two at least C^1 one parameter subgroups $(g_t, v_t), (\tilde{g}_\epsilon, \tilde{v}_\epsilon) \in \mathfrak{D}^s \times V$. Using these one parameter subgroups, one can compute the inner automorphism, or adjoint action of the group on itself. This adjoint action is defined by conjugation

$$\mathrm{AD} : (\mathfrak{D}^s \times V) \times (\mathfrak{D}^s \times V) \to (\mathfrak{D}^s \times V),$$
$$\mathrm{AD}_{(g_t, v_t)}(\tilde{g}_\epsilon, \tilde{v}_\epsilon) := (g_t, v_t) \bullet (\tilde{g}_\epsilon, \tilde{v}_\epsilon) \bullet (g_t, v_t)^{-1} \tag{7}$$
$$= \left(g_t \circ \tilde{g}_\epsilon \circ g_t^{-1}, g_t^*(\tilde{v}_\epsilon - v_t + \tilde{g}_{\epsilon *} v_t)\right).$$

To see how the group acts on its Lie algebra, one can compute the derivative with respect to ϵ and evaluate at $\epsilon = 0$ in the adjoint action of the group on itself. Let $\mathfrak{X}^s \ni \tilde{u} = \frac{d}{d\epsilon}|_{\epsilon=0} \tilde{g}_\epsilon$ and $V \ni \tilde{b} = \frac{d}{d\epsilon}|_{\epsilon=0} \tilde{v}_\epsilon$. This choice for a vector field is guided by the deterministic reconstruction equation in (1). For any tensor $S_\epsilon \in T_s^r(M)$ whose dependence on ϵ is at least C^1 it holds that

$$\frac{d}{d\epsilon} \tilde{g}_{\epsilon *} S_\epsilon = \tilde{g}_{\epsilon *} \left(\frac{d}{d\epsilon} S_\epsilon - \mathcal{L}_{\tilde{u}} S_\epsilon \right). \tag{8}$$

Important here is that the Lie derivative does not commute with pull-backs and push-forwards that depend on parameters, see [1]. The adjoint action of the group on its Lie algebra can be computed as

$$\mathrm{Ad} : (\mathfrak{D}^s \times V) \times (\mathfrak{X}^s \times V) \to (\mathfrak{X}^s \times V),$$
$$\mathrm{Ad}_{(g_t, v_t)}(\tilde{u}, \tilde{b}) := \frac{d}{d\epsilon}\bigg|_{\epsilon=0} \mathrm{AD}_{(g_t, v_t)}(\tilde{g}_\epsilon, \tilde{v}_\epsilon) \tag{9}$$
$$= (g_{t*}\tilde{u}, g_t^* \tilde{b} - g_t^* \mathcal{L}_{\tilde{u}} v_t).$$

By means of the pairing on $\mathfrak{X}^s \times V$, one can compute the dual action to the adjoint action (9). This is called the coadjoint action of the group on the dual of its Lie algebra. Let $(\tilde{m}, \tilde{a}) \in (\mathfrak{X}^s \times V)^*$, then the coadjoint action is given by

$$\mathrm{Ad}^* : (\mathfrak{D}^s \times V) \times (\mathfrak{X}^s \times V)^* \to (\mathfrak{X}^s \times V)^*,$$
$$\langle \mathrm{Ad}^*_{(g_t^{-1}, -g_t^{-1} v_t)}(\tilde{m}, \tilde{a}), (\tilde{u}, \tilde{b}) \rangle := \langle (\tilde{m}, \tilde{a}), \mathrm{Ad}_{(g_t, v_t)}(\tilde{u}, \tilde{b}) \rangle, \tag{10}$$
$$\mathrm{Ad}^*_{(g_t^{-1}, -g_t^{-1} v_t)}(\tilde{m}, \tilde{a}) = (g_t^* \tilde{m} + v_t \diamond g_{t*}\tilde{a}, g_{t*}\tilde{a}).$$

Definition 1.2 (*The diamond operator*) The coadjoint action (10) features the *diamond operator*, which is defined for $a \in V^*$, $u \in \mathfrak{X}^s$ and fixed $v \in V$ as

$$\langle v \diamond a, u \rangle_{\mathfrak{X}^{s*} \times \mathfrak{X}^s} := -\langle a, \mathcal{L}_u v \rangle_{V^* \times V}. \tag{11}$$

Note that the diamond operator is the dual of the Lie derivative regarded as a map $\mathcal{L}_{(\cdot)} v : \mathfrak{X}^s \to V$, hence $v \diamond (\cdot) : V^* \to \mathfrak{X}^{s*}$. The diamond operator shows how an element from the dual of the vector space acts on the dual of the Lie algebra.

When evaluated at $t = 0$, the t-derivatives of Ad in (9) and Ad* in (10) define, respectively, the adjoint and coadjoint actions of the Lie algebra on itself and on its dual. Denote by $\mathfrak{X}^s \ni u = \frac{d}{dt}|_{t=0} g_t$ and $V \ni b = \frac{d}{dt}|_{t=0} v_t$. The adjoint action of the Lie algebra on itself is

$$\begin{aligned}
\mathrm{ad} &: (\mathfrak{X}^s \times V) \times (\mathfrak{X}^s \times V) \to (\mathfrak{X}^s \times V), \\
\mathrm{ad}_{(u,b)}(\tilde{u}, \tilde{b}) &:= \frac{d}{dt}\bigg|_{t=0} \mathrm{Ad}_{(g_t, v_t)}(\tilde{u}, \tilde{b}), \\
\mathrm{ad}_{(u,b)}(\tilde{u}, \tilde{b}) &= (-\mathcal{L}_u \tilde{u}, \mathcal{L}_u \tilde{b} - \mathcal{L}_{\tilde{u}} b) \\
&= (-[u, \tilde{u}], \mathcal{L}_u \tilde{b} - \mathcal{L}_{\tilde{u}} b),
\end{aligned} \tag{12}$$

where the bracket $[\cdot, \cdot]$ in (12) is the commutator of vector fields. The minus sign is due to fact that group acts on itself from the right. The coadjoint action of the Lie algebra on its dual can be obtained by computing the dual to (12) or by taking the derivative with respect to t and evaluate at $t = 0$ in (10). Either way, one arrives at

$$\begin{aligned}
\mathrm{ad}^* &: (\mathfrak{X}^s \times V) \times (\mathfrak{X}^s \times V)^* \to (\mathfrak{X}^s \times V)^*, \\
\langle \mathrm{ad}^*_{(u,b)}(\tilde{m}, \tilde{a}), (\tilde{u}, \tilde{b}) \rangle &:= \langle (\tilde{m}, \tilde{a}), \mathrm{ad}_{(u,b)}(\tilde{u}, \tilde{b}) \rangle, \\
\mathrm{ad}^*_{(u,b)}(\tilde{m}, \tilde{a}) &= (\mathcal{L}_u \tilde{m} + b \diamond \tilde{a}, -\mathcal{L}_u \tilde{a}),
\end{aligned} \tag{13}$$

in which (12) implies the last line in (13). Alternatively, one can also obtain (13) by taking the derivative with respect to t in (10) and evaluate at $t = 0$.

Remark 2 (*Coadjoint action and the diamond operator*) The coadjoint action is an important operator in geometric mechanics and representation theory. It was shown by [18] and in further work by [20, 28] that the coadjoint orbits of a Lie group G have the structure of symplectic manifolds and are connected with Hamiltonian mechanics. See [19] for a review. The computations of the adjoint and coadjoint actions for the semidirect product group is valuable for fluid mechanics, as they introduce the two fundamental operators that appear in the equations of motion. The Lie derivative is responsible for transport of tensors along vector fields and its dual action given by the diamond operator encodes the symmetry breaking. In particular, the diamond operator introduces the effect of symmetry breaking into the Euler-Poincaré equations of motion.

2 Deterministic Geometric Fluid Dynamics

With the adjoint and coadjoint actions defined, one can derive continuum mechanics equations with advected quantities by using symmetry reduction. Euler-Poincaré reduction for a semidirect product group $\mathfrak{D}^s \times V$ as developed in [17] is sketched below in Fig. 2.

As shown by comparison of Fig. 2 with Fig. 1, several new features arise in semidirect product Lie group reduction which differ from Euler-Poincaré reduction by symmetry when the configuration space itself is a Lie group. These differences can be conveniently explained by introducing the physical concept of an order parameter. As discussed earlier, the order parameters in continuum mechanics are the elements of V^* which are advected by the action of the diffeomorphism group $\mathfrak{D}^s$. The advection is defined simply as the semidirect product action on the elements of V^*. The introduction of each additional advected state variable (or, order parameter) into the physical problem further breaks the original symmetry $\mathfrak{D}^s$. The remaining symmetry of the Lagrangian in Hamilton's principle is the isotropy subgroup $\mathfrak{D}^s_{a_0}$ of the initial conditions, a_0, for the entire set of advected quantities, a. The action of the diffeomorphism group $\mathfrak{D}^s$ on these initial conditions then describes their advection as the action of $\mathfrak{D}^s$ on its coset space $\mathfrak{D}^s \setminus \mathfrak{D}^s_{a_0} = V^*$. Once the initial values of the order parameters, a_0, have been set, one must still define a Legendre transform from the Lagrangian formulation into the Hamiltonian formulation and vice versa. The Legendre transform in the setting of semidirect products is a partial Legendre transform, since it transforms between $T\mathfrak{D}^s$ and $T^*\mathfrak{D}^s$ or $T\mathfrak{D}^s \setminus \mathfrak{D}^s_{a_0} \simeq \mathfrak{X}^s$ and $T^*\mathfrak{D}^s \setminus \mathfrak{D}^s_{a_0} \simeq \mathfrak{X}^{s*}$ only after having fixed the value a_0 of the order parameters, which live in V^*. This coset reduction is what Fig. 2 shows. The remaining invariance of a functional under the action of the isotropy subgroup is called its *particle relabelling symmetry*.

Our exploration continues on the Lagrangian side in Fig. 2. Consider a fluid Lagrangian $L : T\mathfrak{D}^s \times V^* \to \mathbb{R}$. By fixing the value of $a_0 \in V^*$, one can construct $L_{a_0} : T\mathfrak{D}^s \to \mathbb{R}$. If this Lagrangian is right invariant under the action of the isotropy

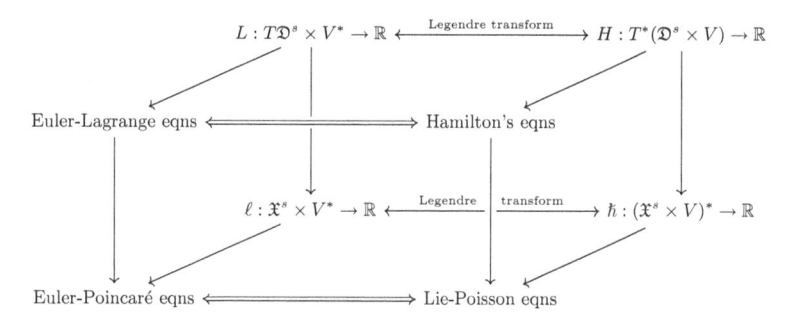

Fig. 2 The cube of continuum mechanics in the semidirect product group setting. Reduction is indicated by the arrows pointing down

subgroup $\mathfrak{D}^s_{a_0}$, then one can construct

$$L\left(\frac{d}{dt}g\circ g^{-1}, e, a_0\right) = L_{a_0}\left(\frac{d}{dt}g\circ g^{-1}, e\right)$$
$$= \ell_{a_0}\left(\frac{d}{dt}g\circ g^{-1}\right) = \ell\left(\frac{d}{dt}g\circ g^{-1}, g_*a_0\right). \quad (14)$$

Here ∘ means composition of functions. The same procedure applies to the Hamiltonian. Since the coadjoint action is known, it is straightforward to formulate the Lie-Poisson equations. The details of Hamiltonian semidirect product reduction and also more information on the Lagrangian semidirect product reduction can be found in [17].

The coadjoint action of the Lie algebra on its dual is also required for the Lagrangian semidirect product reduction. One can use the deterministic reconstruction equation to see that the argument of the Lagrangians in (14) is

$$\frac{d}{dt}g\circ g^{-1} = u. \quad (15)$$

Using this information, one can integrate the Lagrangian in time to construct the action functional. By requiring the variational derivative of the action functional to vanish, one can compute the equations of motion. However, due to the removal of symmetries, the variations are no longer free.

3 Stochastic Geometric Fluid Dynamics

In the situation where noise is present, that is, when the reconstruction equation is (2), the Euler-Poincaré variations become stochastic. Consider $g : \mathbb{R}^2 \to \mathfrak{D}^s$ with $g_{t,\epsilon} = g(t, \epsilon)$ to be a two parameter subgroup with smooth dependence on ϵ, but only continuous dependence on t. Let us denote

$$d\chi_{t,\epsilon}(X) = (dg_{t,\epsilon} \circ g_{t,\epsilon})(X) = u_{t,\epsilon}(X)dt + \sum_{i=1}^{N}\xi_i(X) \circ dW^i_t$$

and

$$v_{t,\epsilon}(X) = (\frac{\partial}{\partial \epsilon}g_{t,\epsilon} \circ g_{t,\epsilon})(X).$$

When a ∘ symbol is followed by dW_t it means Stratonovich integration and in every other context the ∘ symbol is used to denote composition. Note that the data vector fields ξ_i are prescribed and hence will not have a dependence on ϵ.

In order to compute with these stochastic subgroups and their associated vector fields, one needs a stochastic Lie chain rule. The Kunita-Itô-Wentzell (KIW) formula is the stochastic generalisation of the Lie chain rule (8). A proof of the KIW formula is given in [4] for differential k-forms and vector fields. That proof includes the technical details on regularity that will be omitted here. In the KIW formula, the k-form is allowed to be a semimartingale itself. Let K be a continuous adapted semimartingale that takes values in the k-forms and satisfies

$$K_t = K_0 + \int_0^t G_s\,ds + \sum_{i=1}^N \int_0^t H_{i\,s} \circ dB_s^i, \qquad (16)$$

where the B_t^i are independent, identically distributed Brownian motions. The drift of the semimartingale K is determined by G and the diffusion by H_i, both of which are k-form valued continuous adapted semimartingales with suitable regularity. Let g_t satisfy (2), then [4] shows that the following holds

$$\mathsf{d}(g_t^* K_t) = g_t^*\bigl(\mathsf{d}K_t + \mathcal{L}_{u_t} K_t\,dt + \mathcal{L}_{\xi_i} K_t \circ dW_t^i\bigr). \qquad (17)$$

Equation (16) helps to interpret the $\mathsf{d}K_t$ term in the KIW formula (17). This formula will be particularly useful in computing the variations of the variables in the Lagrangian. To compute these variations, one needs the variational derivative.

The variational derivative. The variational derivative of a functional $F : \mathcal{B} \to \mathbb{R}$, where $\mathcal{B}$ is a Banach space, is denoted $\delta F/\delta \rho$ with $\rho \in \mathcal{B}$. The variational derivative can be defined by the first variation of the functional

$$\delta F[\rho] := \frac{d}{d\epsilon}\bigg|_{\epsilon=0} F[\rho + \epsilon\delta\rho] = \int \frac{\delta F}{\delta \rho}(x)\delta\rho(x)\,dx = \left\langle \frac{\delta F}{\delta \rho}, \delta\rho \right\rangle. \qquad (18)$$

In the definition above, $\epsilon \in \mathbb{R}$ is a parameter, $\delta\rho \in \mathcal{B}$ is an arbitrary function and the first variation can be understood as a Fréchet derivative. A precise and rigorous definition can be found in [15]. With the definition of the functional derivative in place, the following lemma can be formulated.

Lemma 3 *With the notation as above, the variations of u and any advected quantity a are given by*

$$\delta u(t) = \mathsf{d}v(t) + [\mathsf{d}\chi_t, v(t)], \quad \delta a(t) = -\mathcal{L}_{v(t)} a(t), \qquad (19)$$

where $v(t) \in \mathfrak{X}^s$ is arbitrary.

Proof The proof of the variation of $a(t)$ is a direct application of the Kunita-Itô-Wentzell formula to $a(t, \epsilon) = g_{t,\epsilon*} a_0$. Note that the data vector fields ξ_i are prescribed and do not depend on ϵ. Denote by $x_{t,\epsilon} = g_{t,\epsilon}(X)$. Then one has

$$dg_{t,\epsilon}(X) = dx_{t,\epsilon} = u_{t,\epsilon}(x_{t,\epsilon}) \, dt + \sum_{i=1}^{N} \xi_i(x_{t,\epsilon}) \circ dW_t^i =: d\chi_{t,\epsilon}(x_{t,\epsilon}). \qquad (20)$$

The vector field associated to the ϵ-dependence of the two parameter subgroup is given by

$$\frac{\partial}{\partial \epsilon} g_{t,\epsilon} = \frac{\partial}{\partial \epsilon} x_{t,\epsilon} = v_{t,\epsilon}(x_{t,\epsilon}). \qquad (21)$$

Computing the derivative with respect to ϵ of (20) gives

$$\frac{\partial}{\partial \epsilon} dx_{t,\epsilon} = \frac{\partial}{\partial \epsilon} (d\chi_{t,\epsilon}(x_{t,\epsilon}))$$
$$= \left(\frac{\partial}{\partial \epsilon} u_{t,\epsilon} + v_{t,\epsilon} \cdot \frac{\partial}{\partial x_{t,\epsilon}} d\chi_{t,\epsilon} \right) (x_{t,\epsilon}), \qquad (22)$$

where the independence of the data vector fields ξ_i on ϵ was used. Taking the differential with respect to time of (21) gives

$$d\left(\frac{\partial}{\partial \epsilon} x_{t,\epsilon} \right) = d(v_{t,\epsilon}(x_{t,\epsilon}))$$
$$= \left(dv_{t,\epsilon}(x_{t,\epsilon}) + d\chi_{t,\epsilon} \cdot \frac{\partial}{\partial x_{t,\epsilon}} v_{t,\epsilon} \right) (x_{t,\epsilon}). \qquad (23)$$

One can then evaluate at $\epsilon = 0$ and call upon equality of cross derivative-differential to obtain the result by subtracting. Since $g_{t,\epsilon}$ depends on t in a C^0 manner, the integral representation is required. The particle relabelling symmetry permits one to stop writing the explicit dependence on space,

$$\delta u(t) \, dt = dv(t) + [d\chi_t, v(t)]. \qquad (24)$$

This completes the proof of formula (19) for the variation of $u(t)$.

The notation in (20) needs careful explanation, because it comprises both a stochastic differential equation and a definition. The symbol $d\chi_{t,\epsilon}$ is used to define a vector field, whereas $dx_{t,\epsilon}$ denotes a stochastic differential equation. This lemma makes the presentation of the stochastic Euler-Poincaré theorem particularly simple.

Theorem 4 (Stochastic Euler-Poincaré) *With the notation as above, the following are equivalent.*

(i) *The constrained variational principle*

$$\delta \int_{t_1}^{t_2} \ell(u, a) \, dt = 0 \qquad (25)$$

holds on $\mathfrak{X}^s \times V^*$, using variations δu and δa of the form

Stochastic Geometric Mechanics with Diffeomorphisms

$$\delta u = dv + [d\chi_t, v], \qquad \delta a = -\mathcal{L}_v a, \tag{26}$$

where $v(t) \in \mathfrak{X}^s$ is arbitrary and vanishes at the endpoints in time for arbitrary times t_1, t_2.

(ii) The stochastic Euler-Poincaré equations hold on $\mathfrak{X}^s \times V^*$

$$d\frac{\delta \ell}{\delta u} + \mathcal{L}_{d\chi_t} \frac{\delta \ell}{\delta u} = \frac{\delta \ell}{\delta a} \diamond a\, dt, \tag{27}$$

and the advection equation

$$da + \mathcal{L}_{d\chi_t} a = 0. \tag{28}$$

Proof Using integration by parts and the endpoint conditions $v(t_1) = 0 = v(t_2)$, the variation can be computed to be

$$\begin{aligned}
\delta \int_{t_1}^{t_2} \ell(u, a)\, dt &= \int_{t_1}^{t_2} \left\langle \frac{\delta \ell}{\delta u}, \delta u \right\rangle + \left\langle \frac{\delta \ell}{\delta a}, \delta a \right\rangle dt \\
&= \int_{t_1}^{t_2} \left\langle \frac{\delta \ell}{\delta u}, dv + [d\chi_t, v] \right\rangle + \left\langle \frac{\delta \ell}{\delta a}\, dt, -\mathcal{L}_v a \right\rangle \\
&= \int_{t_1}^{t_2} \left\langle -d\frac{\delta \ell}{\delta u} - \mathcal{L}_{d\chi_t} \frac{\delta \ell}{\delta u} + \frac{\delta \ell}{\delta a} \diamond a\, dt, v \right\rangle \\
&= 0.
\end{aligned} \tag{29}$$

Since the vector field v is arbitrary, one obtains the stochastic Euler-Poincaré equations. Finally, the advection equation (28) follows by applying the KIW formula to $a(t) = g_{t*}a_0$. □

Remark 5 The stochastic Euler-Poincaré theorem is equivalent to the version presented in [16], which uses stochastic Clebsch constraints. In [16] one can also find an investigation the Itô formulation of the stochastic Euler-Poincaré equation.

Stochastic Lie-Poisson formulation. The stochastic Euler-Poincaré equations have an equivalent stochastic Lie-Poisson formulation. To obtain the Lie-Poisson formulation, one must Legendre transform the reduced Lagrangian. The Legendre transformation in the presence of stochasticity becomes itself stochastic in the following way

$$m := \frac{\delta \ell}{\delta u}, \qquad \hbar(m, a)\, dt + \sum_{i=1}^{N} \langle m, \xi_i \rangle \circ dW_t^i = \langle m, d\chi_t \rangle - \ell(u, a)\, dt. \tag{30}$$

The stochasticity enters the Legendre transformation because the momentum map m is coupled to the stochastic vector field $d\chi_t$. The left hand side of the transformation determines the Hamiltonian, which is a semimartingale. The underlying semidirect product group structure has not changed, it is still the H^s diffeomorphisms with a vector space, but the Hamiltonian has become a semimartingale. This implies that in the stochastic case the energy is not conserved, because Hamiltonian depends on time explicitly. Note that (30) emphasises that the Lagrangian does not feature stochasticity in this framework. Instead, the Lagrangian represents the physics in the problem, which does not change. The stochasticity is supposed to account for the difference between observed data and deterministic modelling. The stochastic Lie-Poisson equations are given by

$$\mathsf{d}(m, a) = -\mathrm{ad}^*_{(\frac{\delta h}{\delta m}, \frac{\delta h}{\delta a})}(m, a)\, dt - \sum_{i=1}^{N} \mathrm{ad}^*_{(\xi_i, 0)}(m, a) \circ dW^i_t, \qquad (31)$$

where ad* is given in (13). Since both the drift and the diffusion part use the same operator in (31), the stochastic Lie-Poisson equations preserve the same family of Casimirs (or integral conserved quantities) as the deterministic Lie-Poisson equations. The stochastic Euler-Poincaré theorem has a stochastic Kelvin-Noether circulation theorem as a corollary.

Let the manifold M be a submanifold of $\mathbb{R}^n$ with coordinates X. Then the volume form can be expressed with respect to a density. That is, $\mu(d^n X) = \rho_0(X) d^n X$. By pushing forward ρ_0 along the stochastic flow g_t, one obtains ρ. Let $\mathfrak{C}^s$ be the space of loops $\gamma : S^1 \to \mathfrak{D}^s$, which is acted upon from the left by $\mathfrak{D}^s$. Given an element $m \in \mathfrak{X}^s$, one can obtain a 1-form by formally dividing m by the density ρ.

The circulation map $\mathcal{K} : \mathfrak{C}^s \times V^* \to \mathfrak{X}^{s**}$ is defined by

$$\langle \mathcal{K}(\gamma, a), m \rangle = \oint_\gamma \frac{m}{\rho}. \qquad (32)$$

Given a Lagrangian $\ell : \mathfrak{X}^s \times V^* \to \mathbb{R}$, the *Kelvin-Noether quantity* is defined by

$$I(\gamma, u, a) := \oint_\gamma \frac{1}{\rho} \frac{\delta \ell}{\delta u}. \qquad (33)$$

One can now formulate the following stochastic Kelvin-Noether circulation theorem.

Theorem 6 (Stochastic Kelvin-Noether) *Let $u_t = u(t)$ satisfy the stochastic Euler-Poincaré equation (27) and $a_t = a(t)$ the stochastic advection equation (28). Let g_t be the flow associated to the vector field $\mathsf{d}\chi_t$. That is, $\mathsf{d}\chi_t = \mathsf{d}g_t \circ g_t^{-1} = u_t\, dt + \sum_{i=1}^{N} \xi_i \circ dW^i_t$. Let $\gamma_0 \in \mathfrak{C}^s$ be a loop. Denote by $\gamma_t = g_t \circ \gamma_0$ and define the Kelvin-Noether quantity $I(t) := I(\gamma_t, u_t, a_t)$. Then*

$$\mathsf{d}I(t) = \oint_{\gamma_t} \frac{1}{\rho} \frac{\delta \ell}{\delta a} \diamond a\, dt. \qquad (34)$$

Proof The statement of the stochastic Kelvin-Noether circulation theorem involves a loop that is moving with the stochastic flow. One can transform to stationary coordinates by pulling back the flow to the initial condition. This pull-back yields

$$I(t) = \oint_{\gamma_t} \frac{1}{\rho} \frac{\delta \ell}{\delta u} = \oint_{\gamma_0} g_t^* \left(\frac{1}{\rho} \frac{\delta \ell}{\delta u} \right) = \oint_{\gamma_0} \frac{1}{\rho_0} g_t^* \left(\frac{\delta \ell}{\delta u} \right). \tag{35}$$

An application of the Kunita-Itô-Wentzell formula (17) leads to

$$dI(t) = \oint_{\gamma_0} \frac{1}{\rho_0} g_t^* \left(d\frac{\delta \ell}{\delta u} + \mathcal{L}_{d\chi_t} \frac{\delta \ell}{\delta u} \right) = \oint_{\gamma_0} \frac{1}{\rho_0} g_t^* \left(\frac{\delta \ell}{\delta a} \diamond a \right) dt, \tag{36}$$

since u satisfies the stochastic Euler-Poincaré theorem. Transforming back to the moving coordinates by pushing forward with g_t yields the final result.

Thus, Theorem 6 explains how particle relabelling symmetry gives rise to the Kelvin-Noether circulation theorem via Noether's theorem. When the only advected quantity present is the mass density, the loop integral of the diamond terms vanishes. This means that circulation is conserved according to Noether's theorem for an incompressible fluid, or for a barotropically compressible fluid. The presence of other advected quantities breaks the symmetry further and introduces the *diamond terms* which generate circulation, as one can see in the Kelvin-Noether circulation theorem in equation (34). Consequently, the symmetry breaking due to additional order parameters can provide additional mechanisms for the generation of Kelvin-Noether circulation in ideal fluid dynamics.

Outlook. Stochastic geometric mechanics is an active field of mathematics which has recently established its utility for a broad range of applications in science. Basically, everything that can be done with Hamilton's principle for deterministic geometric mechanics can also be made stochastic in the sense of Stratonovich. This is possible because the variational calculus in Hamilton's principle requires only the product rule and chain rule from ordinary calculus. The happy emergence of the new science of stochastic geometric mechanics was celebrated with the publication of the book [2]. This book showcases some of the recent developments in stochastic geometric mechanics. Another collection of recent developments can be found in [6]. An ongoing development is in the direction of *rough geometric mechanics*, initiated with a rough version of the Euler-Poincaré theorem in [11]. Remarkably, variational principles which are driven by geometric rough paths again only require the product rule and the chain rule. Other directions involve the inclusion of jump processes, fractional derivatives and non-Markovian processes in geometric mechanics. For example, recent work by [3] shows that SDEs driven by semimartingales with jumps have weak symmetries and a corresponding extension of the reduction and reconstruction technique is discussed.

Acknowledgements We are enormously grateful for many encouraging discussions over the years with T.S. Ratiu, F. Gay-Balmaz, C. Tronci, S. Albeverio, A.B. Cruzeiro, F. Flandoli, and also with our friends in project STUOD (stochastic transport in upper ocean dynamics) and in the geometric mechanics research group at Imperial College London. The work of DDH was partially supported by European Research Council (ERC) Synergy grant STUOD - DLV-856408. EL was supported by EPSRC grant [grant number EP/L016613/1] and is grateful for the warm hospitality at the Imperial College London EPSRC Centre for Doctoral Training in the Mathematics of Planet Earth during the course of this work.

References

1. Abraham, R., Marsden, J.E.: Foundations of Mechanics, vol. 36. Benjamin/Cummings Publishing Company Reading, MA (1978)
2. Albeverio, S., Cruzeiro, A.B., Holm, D.: Stochastic Geometric Mechanics: CIB, Lausanne, Switzerland, Jan-June 2015, vol. 202. Springer (2017)
3. Albeverio, S., De Vecchi, F.C., Morando, P., Ugolini, S., et al.: Weak symmetries of stochastic differential equations driven by semimartingales with jumps. Electron. J. Probab. **25** (2020)
4. Bethencourt de Leon, A., Holm, D.D., Luesink, E., Takao, S.: Implications of Kunita–Itô–Wentzell formula for k-forms in stochastic fluid dynamics. J. Nonlinear Sci. 1–34 (2020)
5. Bismut, J.-M.: Mécanique aléatoire. In: Ecole d'Eté de Probabilités de Saint-Flour X-1980, pp. 1–100. Springer (1982)
6. Castrillón López, M., Martín de Diego, D., Ratiu, T.S., Tronci, C.: Journal of Geometric Mechanics: Special issue dedicated to Darryl D. Holm on the occasion of his 70th birthday, vol. 11 (2019)
7. Cotter, C., Crisan, D., Holm, D.D., Pan, W., Shevchenko, I: Modelling uncertainty using circulation-preserving stochastic transport noise in a 2-layer quasi-geostrophic model (2018). arXiv:1802.05711
8. Cotter, C., Crisan, D., Holm, D.D., Pan, W., Shevchenko, I.: Numerically modeling stochastic Lie transport in fluid dynamics. Multiscale Model. Simul. **17**(1), 192–232 (2019)
9. Cotter, C., Crisan, D., Holm, D.D., Pan, W., Shevchenko, I.: A particle filter for Stochastic Advection by Lie Transport (SALT): a case study for the damped and forced incompressible 2D Euler equation (2019). arXiv:1907.11884
10. Cotter, C.J., Gottwald, G.A., Holm, D.D.: Stochastic partial differential fluid equations as a diffusive limit of deterministic Lagrangian multi-time dynamics. Proc. R. Soc. A: Math. Phys. Eng. Sci. **473**(2205), 20170388 (2017)
11. Crisan, D., Holm, D.D., Leahy, J.-M., Nilssen, T.: Variational principles for fluid dynamics on rough paths (2020). arXiv:2004.07829
12. Cruzeiro, A.B., Holm, D.D., Ratiu, T.S.: Momentum maps and stochastic Clebsch action principles. Commun. Math. Phys. **357**(2), 873–912 (2018)
13. Ebin, D.G., Marsden, J.: Groups of diffeomorphisms and the motion of an incompressible fluid. Ann. Math. **92**(1), 102–163 (1970)
14. Ebin, D.G.: On the space of Riemannian metrics. Ph.D. thesis, Massachusetts Institute of Technology (1967)
15. Gelfand, I.M., Silverman, R.A., et al.: Calculus of variations. Courier Corporation (2000)
16. Holm, D.D.: Variational principles for stochastic fluid dynamics. Proc. R. Soc. A: Math. Phys. Eng. Sci. **471**(2176), 20140963 (2015)
17. Holm, D.D., Marsden, J.E., Ratiu, T.S.: The Euler–Poincaré equations and semidirect products with applications to continuum theories. Adv. Math. **137**(1), 1–81 (1998)
18. Kirillov, A.A.: Unitary representations of nilpotent Lie groups. RuMaS **17**(4), 53–104 (1962)
19. Kirillov, A.: Merits and demerits of the orbit method. Bull. Am. Math. Soc. **36**(4), 433–488 (1999)

20. Kostant, B.: Quantization and unitary representations. In: Lectures in modern analysis and applications III, pp. 87–208. Springer (1970)
21. Lázaro-Camı, J., Ortega, J.: Stochastic Hamiltonian dynamical systems. Rep. Math. Phys. (1) (2008)
22. Marsden, J.E., Raţiu, T., Weinstein, A.: Semidirect products and reduction in mechanics. Trans. Am. Math. Soc. **281**(1), 147–177 (1984)
23. Noether, E.: Invariante Variationsprobleme. Nachrichten von der Gesellschaft der Wissenschaften zu göttingen, 235-257. E. Noether: Gesammelte Abhandlungen, ed. N. Jacobson, pp. 248–270 (1918)
24. Noether, E.: Invariant variation problems. Transp. Theory Stat. Phys. **1**(3), 186–207 (1971)
25. Poincaré, H.: Sur une forme nouvelle des équations de la mécanique. CR Acad. Sci. **132**, 369–371 (1901)
26. Smale, S.: Topology and mechanics. i. Inventiones Mathematicae **10**(4), 305–331 (1970)
27. Smale, S.: Topology and mechanics. ii. Inventiones Mathematicae **11**(1), 45–64 (1970)
28. Souriau, J.-M.: Structure des systèmes dynamiques. Dunod, Paris (1970)

McKean Feynman-Kac Probabilistic Representations of Non-linear Partial Differential Equations

Lucas Izydorczyk, Nadia Oudjane, and Francesco Russo

Abstract This paper presents a partial state of the art about the topic of representation of generalized Fokker-Planck Partial Differential Equations (PDEs) by solutions of McKean Feynman-Kac Equations (MFKEs) that generalize the notion of McKean Stochastic Differential Equations (MSDEs). While MSDEs can be related to non-linear Fokker-Planck PDEs, MFKEs can be related to non-conservative non-linear PDEs. Motivations come from modeling issues but also from numerical approximation issues in computing the solution of a PDE, arising for instance in the context of stochastic control. MFKEs also appear naturally in representing final value problems related to backward Fokker-Planck equations.

Keywords Backward diffusion · McKean stochastic differential equation · Probabilistic representation of PDEs · Time reversed diffusion · HJB equation · Feynman-Kac measures

2010 AMS classification: 60H10 · 60H30 · 60J60 · 65C05 · 65C35 · 35K58

L. Izydorczyk · F. Russo (✉)
ENSTA-Paris, Institut Polytechnique de Paris, Unité de Mathématiques Appliquées (UMA), Palaiseau, France
e-mail: francesco.russo@ensta-paris.fr

L. Izydorczyk
e-mail: lucas.izydorczyk@ensta-paris.fr

N. Oudjane
EDF R&D, and FiME (Laboratoire de Finance des Marchés de l'Energie, (Dauphine, CREST, EDF R&D), Palaiseau, France
e-mail: nadia.oudjane@edf.fr

© Springer Nature Switzerland AG 2021
S. Ugolini et al. (eds.), *Geometry and Invariance in Stochastic Dynamics*, Springer Proceedings in Mathematics & Statistics 378,
https://doi.org/10.1007/978-3-030-87432-2_10

1 Introduction and Motivations

1.1 General Considerations

The idea of the present article is to focus on models which have a double macroscopic-microscopic face in the form of *perturbation* of a so called Fokker-Planck type equation that we call *generalized* Fokker-Planck equation. Our ambition is driven by two main reasons.

1. A *modeling* reason: the idea is to observe both from a macroscopic-microscopic point of view phenomena arising from physics, biology, chemistry or complex systems.
2. A *numerical simulation* reason: to provide Monte-Carlo suitable algorithms to approach PDEs.

The *target macroscopic* Fokker-Planck equation is

$$\begin{cases} \partial_t u = \frac{1}{2} \sum_{i,j=1}^{d} \partial_{ij}^2 \left((\sigma \sigma^\top)_{i,j}(t, x, u) u \right) - div \left(b(t, x, u, \nabla u) u \right) \\ \qquad + \Lambda(t, x, u, \nabla u) u , \quad \text{for } t \in]0, T] , \\ u(0, \cdot) = \mathbf{u_0}, \end{cases} \quad (1.1)$$

where $\mathbf{u_0}$ is a Borel probability measure $\sigma : [0, T] \times \mathbb{R}^d \times \mathbb{R} \to M_{d,p}(\mathbb{R})$, $b : [0, T] \times \mathbb{R}^d \times \mathbb{R} \to \mathbb{R}^d$, $\Lambda : [0, T] \times \mathbb{R}^d \times \mathbb{R} \times \mathbb{R}^d \to \mathbb{R}$ and ∇ denotes the gradient operator. The initial condition in (1.1) means that for every continuous bounded real function φ we have $\int \varphi(x) u(t, x) dx \to \int \varphi(x) \mathbf{u_0}(dx)$ when $t \to 0$. When $\mathbf{u_0}$ admits a density, we denote it by u_0. The unknown function $u :]0, T] \times \mathbb{R}^d \to \mathbb{R}$ is supposed to run in $L^1(\mathbb{R}^d)$ considered as a subset of the space of finite Radon measures $\mathcal{M}(\mathbb{R}^d)$. The idea consists in finding a probabilistic representation via the solution of a *Stochastic Differential Equation (SDE)* whose coefficients do not depend only on time and the position of the *particle* but also on its probability law. The *target microscopic equation* we have in mind is

$$\begin{cases} Y_t = Y_0 + \int_0^t \sigma\left(s, Y_s, u(s, Y_s)\right) dW_s + \int_0^t b\left(s, Y_s, u(s, Y_s)\right) ds \\ Y_0 \sim \mathbf{u_0} \\ \int \varphi(x) u(t, x) dx = \mathbb{E}\left[\varphi(Y_t) \exp\left\{ \int_0^t \Lambda(s, Y_s, u(s, Y_s), \nabla u(s, Y_s)) ds \right\} \right], \quad \text{for } t \in]0, T], \end{cases} \quad (1.2)$$

for any continuous bounded real valued test function φ. Sometimes we denominate the third line equation of (1.2) the *linking equation*. When $\Lambda = 0$, in Eq. (1.2), the linking equation simply says that $u(t, \cdot)$ coincides with the density of the marginal distribution $\mathcal{L}(Y_t)$. In this specific case, Eq. (1.2) reduces to a *McKean Stochastic Differential Equation (MSDE)*, which is in general an SDE whose coefficients, at time t, depend, not only on (t, Y_t), but also on the marginal law $\mathcal{L}(Y_t)$. With more general functions Λ, the role of the linking equation is more intricate since the whole history of the process $(Y_s)_{0 \le s \le t}$ is involved. This fairly general type of equations will be called *McKean Feynman-Kac Equation (MFKE)* to emphasize the fact that $u(t, x)dx$ now corresponds to a non-conservative Feynman-Kac measure.

An interesting feature of MSDEs (which means $\Lambda = 0$) is that the law of the process Y can often be characterized as the limiting empirical distribution of a large number of interacting particles, whose dynamics are described by a coupled system of classical SDEs. When the number of particles grows to infinity, the particles behave closely to a system of independent copies of Y. This constitutes the so called *propagation of chaos* phenomenon, already observed in the literature when the drift and diffusion coefficients are Lipschitz dependent on the solution marginal law, with respect to the Wasserstein metric, see e.g. [42, 52, 53, 55, 64]. Propagation of chaos is a common phenomenon arising in many physical contexts, see for instance [2] concerning Nelson stochastic mechanics.

When $\Lambda = 0$, Eq. (1.1) is a *non-linear* Fokker-Planck equation, it is conservative and it is known that, under mild assumptions, it describes the dynamics of the marginal probability densities, $u(t, \cdot)$, of the process Y. This correspondence between PDE (1.1) with MSDE (1.2) and interacting particles has extensive interesting applications. In physics, biology or economics, it is a way to relate a microscopic model involving interacting particles to a macroscopic model involving the dynamics of the underlying density. Numerically, this correspondence motivates Monte-Carlo approximation schemes for PDEs. In particular, [20] has contributed to develop stochastic particle methods in the spirit of McKean to provide original numerical schemes approaching a PDE related to Burgers equation providing also the rate of convergence.

Below we list some situations of particular interest where such correspondence holds.

1.2 Some Motivating Examples

Burgers Equation
We fix $d = p = 1$ and let $\nu > 0$ and u_0 be a probability density on $\mathbb{R}$. We consider two equivalent specific cases of (1.1). The first $\sigma \equiv \nu$, $b \equiv 0$, $\Lambda(t, x, u, z) = z$. The second $\sigma \equiv \nu$, $b(t, x, u) = \frac{u}{2}$, $\Lambda = 0$. Both instantiations correspond to the *viscid Burgers equation* in dimension $d = 1$, given by

$$\begin{cases} \partial_t u = \frac{\nu^2}{2}\partial_{xx}u - u\partial_x u, & (t, x) \in [0, T] \times \mathbb{R}, \\ u(0, \cdot) = \mathbf{u_0} . \end{cases} \quad (1.3)$$

Generalized Burgers-Huxley Equation
We fix $d = p = 1$ and let $\nu > 0$ and u_0 be a probability density on $\mathbb{R}$. We consider the particular cases of (1.1) where $\sigma \equiv \nu$, $b(t, x, u) = \alpha \frac{u^n}{n+1}$, $\Lambda(t, x, u) = \beta(1 - u^n)(u^n - \gamma)$, with fixed reals α, β, γ and a non-negative integer n. This instantiation

corresponds to a natural extension of Burgers equation called *Generalized Burgers-Huxley equation* or *Burgers-Fisher equation* which is of great importance to represent non-linear phenomena in various fields such as biology [3, 56], physiology [43] and physics [69]. These equations have the particular interest to describe the interaction between the reaction mechanisms, convection effect, and diffusion transport. Those are non-linear and non-conservative PDEs of the form

$$\begin{cases} \partial_t u = \frac{\nu}{2}\partial_{xx} u - \alpha u^n \partial_x u + \beta u(1 - u^n)(u^n - \gamma), & (t, x) \in [0, T] \times \mathbb{R}, \\ u(0, \cdot) = \mathbf{u}_0. \end{cases} \quad (1.4)$$

Fokker-Planck Equation with Terminal Condition

The present example does not properly integrate the framework of (1.1). In terms of application, we are interested by inverse problems that can be formulated by a PDE with terminal condition

$$\begin{cases} \partial_t u = \frac{1}{2}\sum_{i,j=1}^{d} \partial_{ij}^2 \left((\sigma\sigma^t)_{i,j}(t, x)u\right) - div\,(b(t, x)u) \\ \qquad + \Lambda(t, x)u, \quad \text{for } t \in]0, T[\,, \\ u(T, \cdot) = \mathbf{u}_T, \end{cases} \quad (1.5)$$

where $\mathbf{u}_T$ is a prescribed probability measure. Solving that equation by analytical means constitutes a delicate task. A probabilistic representation may help for studying well-posedness or providing numerical schemes.

Backward simulation of diffusions is a subject of active research in various domains of physical sciences and engineering, as heat conduction [13], material science [61] or hydrology [4]. In particular, *hydraulic inversion* is interested in inverting a diffusion phenomenon representing the concentration of a pollutant to identify the pollution source location when the final concentration profile is observed. The problem is in general ill-posed because either the solution is not unique or the solution is not stable. For this type of problem, the existence is ensured by the fact that the observed contaminant has necessarily originated from some place at a given time (as soon as the model is correct). To correct the lack of well-posedness two regularization procedures have been proposed in the literature: the first one relies on the notion of quasi-solution, introduced by Tikhonov [66], the second one on the method of quasi-reversibility, introduced by Lattes and Lions [45]. Besides well-posedness, a second crucial issue consists in providing a numerical approximating scheme to the backward diffusion equation. A probabilistic representation of (1.5) via the time-reversal of a diffusion could show those issues under a new light.

The Stochastic Fokker-Planck with Multiplicative Noise

We fix $p = d$, $\sigma(t, x, u) = \Phi(u)I_{d_d}$, where $\Phi : \mathbb{R} \to \mathbb{R}$ and $b = \Lambda \equiv 0$. Typical examples are the case of classical porous media type equation (resp. fast diffusion equation), when $\Phi(u) = u^q$, $1 \leq q$ (resp. $0 < q < 1$). The (singular) case $\Phi(u) = \gamma H(u - e_c)$, H being the Heaviside function and e_c a given threshold in $\mathbb{R}$, appears in the science of complex systems, more precisely in the so called *self-organized criticality*, see e.g. [5, 6, 24].

$$\begin{cases} \partial_t u = \frac{\gamma}{2}\Delta(H(u-e_c)u) \\ u(0,\cdot) = \mathbf{u_0}. \end{cases} \quad (1.6)$$

The phenomenon of *self-organized criticality* often is described in two scale phases: a fast dynamics (of *avalanch type*) described by the PDE (1.6) and a slower motion of *sand storming* modeled by the addition of a supplementary stochastic noise $\Lambda(t, x; \omega)$. In that case the *target macroscopic* equation is

$$\begin{cases} \partial_t u = \frac{\gamma}{2}\Delta(H(u-e_c)u) + \Lambda(t,x;\omega)u \\ u(0,\cdot) = \mathbf{u_0}, \end{cases} \quad (1.7)$$

where $\Lambda(t, x; \omega)$ is a quenched realization of a space-time colored (ideally white) noise. The SPDE will be represented by a MSDE in random environment, see Sect. 6.

1.3 Structure of the Paper

In the rest of the paper, to simplify notations, most of the results are stated in the one-dimensional setting. The generalization to the multi-dimensional case is straightforward.

The paper is organized as follows. Next section presents a brief review of basic situations where Fokker-Planck equations can be represented by MSDEs which in turn can be represented by interacting particles systems. Section 3, considers the case of generalized Fokker-Planck equations in the sense of (1.1) with a non-zero term Λ allowing to take into account non-conservative PDEs including a large class of semilinear PDEs. Section 5 highlights the correspondence between MFKEs and MSDEs with jumps which paves the way to a great variety of numerical approximations schemes for non-linear PDEs. Section 4 is devoted to a particular inverse problem which consists in modeling backwardly in time the evolution of a Fokker-Planck equation with a given terminal condition. This problem can be related to a time-reversed SDE which in turn can be represented by a MSDE. In Sect. 6 we analyze the well-posedness of generalized Fokker-Planck equation where the term Λ in (1.1) may involve an exogenous noise resulting in a stochastic non-linear PDE. Finally, in Sect. 7, we consider a stochastic control problem for which the associated Hamilton-Jacobi-Bellman equation can be represented by a MFKE.

2 McKean Representations of Non-linear Fokker-Planck Equations

In this section, we recall some standard situations where a Fokker-Planck PDE can be represented by an SDE which in turn can be approached by an interacting particles system.

2.1 Probabilistic Representation of Linear Fokker-Planck Equations

Suppose there exists a solution $(Y_t)_{t \in [0,T]}$ (in law) to the SDE

$$\begin{cases} Y_t = Y_0 + \int_0^t \sigma(s, Y_s) dW_s + \int_0^t b(s, Y_s) ds, t \in [0, T], \\ Y_0 \sim \mathbf{u_0}, \end{cases} \quad (2.1)$$

where W is a real valued Brownian motion on $[0, T]$ and $\mathbf{u_0}$ is a probability measure on $\mathbb{R}$. A direct application of Itô formula shows that the marginal probability laws $(\mu(t, \cdot) := \mathcal{L}(Y_t))_{t \in [0,T]}$ generate a distributional solution of the linear Fokker-Planck PDE

$$\begin{cases} \partial_t \mu = \frac{1}{2} \partial_{xx}^2 (\sigma^2(t, x)\mu) - \partial_x (b(t, x)\mu) \\ \mu(0, dx) = \mathbf{u_0}(dx). \end{cases} \quad (2.2)$$

This naturally suggests a Monte Carlo algorithm to approximate the above linear PDE, consisting in simulating N i.i.d. particles $(\xi^i)_{i=1,\cdots N}$ with N i.i.d. Brownian motions $(W^i)_{i=1,\cdots N}$ i.e.

$$\begin{cases} \xi_t^i = \xi_0^i + \int_0^t \sigma(s, \xi_s^i) dW_s^i + \int_0^t b(s, \xi_s^i) ds \\ \xi_0^i \text{ i.i.d.} \sim \mathbf{u_0} \\ \mu_t^N = \frac{1}{N} \sum_{j=1}^N \delta_{\xi_t^j}. \end{cases} \quad (2.3)$$

Then the law of large numbers provides the convergence of the empirical approximation $\mu_t^N \xrightarrow[N \to \infty]{} \mu(t, \cdot)$, the solution of the Fokker-Planck equation (2.2).

2.2 McKean Probabilistic Representation of Non-linear Fokker-Planck Equation

We consider the non-linear SDE in the sense of McKean (MSDE)

$$\begin{cases} Y_t = Y_0 + \int_0^t \sigma\Big(s, Y_s, (K*\mu)(s, Y_s)\Big) dW_s + \int_0^t b\Big(s, Y_s, (K*\mu)(s, Y_s)\Big) ds \\ Y_0 \sim \mathbf{u_0} \\ \mu(t, \cdot) \text{ is the probability law of } Y_t, t \in [0, T], \end{cases} \quad (2.4)$$

whose solution is a couple (Y, μ). Here σ, b are Lipschitz, $K : \mathbb{R} \times \mathbb{R} \to \mathbb{R}$ denotes a Lipschitz continuous convolution kernel such that $(K*\mu)(t, y) := \int K(y, z)\mu(t, dz)$ for any $y \in \mathbb{R}$. We emphasize that this type of regularized dependence of the drift and diffusion coefficients on μ is essentially different (and in general easier to handle) from a pointwise dependence where the coefficients b or σ may depend on the value of the marginal density at the current particle position $\frac{d\mu}{dx}(s, Y_s)$. This regularized or non-local dependence on the time-marginals $\mu(t, \cdot)$ is a particular case of the framework when the diffusion and drift coefficients are Lipschitz with respect to $\mu(t, \cdot)$ according to the the Wasserstein metric.

Again, by Itô formula, given a solution (Y, μ) of (2.4), μ solves the non-local non-linear PDE

$$\begin{cases} \partial_t \mu = \frac{1}{2}\partial_{xx}^2 \Big(\sigma^2(t, x, K*\mu)\mu\Big) - \partial_x \Big(b(t, x, K*\mu)\mu\Big) \\ \mu(0, dx) = \mathbf{u_0}(dx), \end{cases} \quad (2.5)$$

in the sense of distributions. In this setting, the well-posedness of (2.4) relies on a fixed point argument in the space of trajectories under the Wasserstein metric, see e.g. [64], at least in the case when the diffusion term does not depend on the law. We will denominate this situation as the *traditional* setting.

Deriving a Monte-Carlo approximation scheme from this probabilistic representation already becomes more tricky since it can no more rely on independent particles but should involve an interacting particles system as initially proposed in [42, 64]. Consider N interacting particles $(\xi^{i,N})_{i=1,\cdots N}$ with N i.i.d. Brownian motions (W^i), i.e.

$$\begin{cases} \xi_t^{i,N} = \xi_0^{i,N} + \int_0^t \sigma\Big(s, \xi_s^{i,N}, (K*\mu_s^N)(\xi_s^{i,N})\Big) dW_s^i + \int_0^t b\Big(s, \xi_s^{i,N}, (K*\mu_s^N)(\xi_s^{i,N})\Big) ds \\ \xi_0^{i,N} \text{ i.i.d.} \sim \mathbf{u_0} \\ \mu_t^N = \frac{1}{N}\sum_{j=1}^N \delta_{\xi_t^{j,N}}, \end{cases} \quad (2.6)$$

with $(K*\mu_t^N)(y) = \frac{1}{N}\sum_{j=1}^N K(y, \xi_t^{j,N})$. The above system defines a so-called *weakly interacting* particles system, as pointed out in [57]. This terminology underlines the fact that any particle interacts with the rest of the population with a vanishing impact of order $1/N$. In this setting, at least when the diffusion coefficient does not depend on the law, [64] proves the so called *chaos propagation* which means that $(\xi_t^{i,N})_{i=1,\cdots N}$ asymptotically behaves as an i.i.d. sample according to $\mu(t,\cdot)$ as the number of particles N grows to infinity, where μ is the solution of the regularized non-linear PDE (2.5). This in particular implies the convergence of the empirical measures $\mu_t^N \xrightarrow[N\to\infty]{} \mu(t,\cdot)$ with the rate $C/\sqrt{N}$ inherited from the law of large numbers.

As already announced, the case where the coefficients depend pointwise on the density law $u(t,\cdot)$ of $\mu(t,\cdot), t > 0$, is far more singular. Indeed the dependence of the coefficients on the law of Y is no more continuous with respect to the Wasserstein metric. In this context, well-posedness results rely generally on analytical methods. One important contribution in this direction is reported in [40], where strong existence and pathwise uniqueness are established when the diffusion coefficient σ and the drift b exhibit pointwise dependence on u but are assumed to satisfy strong smoothness assumptions together with the initial condition. In this case, the solution u is a classical solution of the PDE

$$\begin{cases} \partial_t u = \frac{1}{2}\partial_{xx}^2\left(\sigma^2(t,x,u(t,x))u\right) - \partial_x\left(b(t,x,u(t,x))u\right) \\ u(0,x) = \mathbf{u_0}(dx), \end{cases} \quad (2.7)$$

which is formally derived from (2.5) setting $K(x,y) = \delta_0(x-y)$. Let us fix K^ε being a mollifier (depending on a window-width parameter ε), such that $K^\varepsilon(x,y) = \frac{1}{\varepsilon^d}\phi(\frac{x-y}{\varepsilon}) \xrightarrow[\varepsilon\to 0]{} \delta_0(x-dy)$. As in (2.6), we consider the N interacting particles $(\xi^{i,N})_{i=1,\cdots N}$ solving

$$\begin{cases} \xi_t^{i,N} = \xi_0^{i,N} + \int_0^t \sigma\left(s, \xi_s^{i,N}, u_s^{N,\varepsilon}(\xi_s^{i,N})\right) dW_s^i + \int_0^t b\left(s, \xi_s^{i,N}, u_s^{N,\varepsilon}(\xi_s^{i,N})\right) ds \\ \xi_0^{i,N} \text{ i.i.d.} \sim \mathbf{u_0} \\ u_t^{N,\varepsilon} = \frac{1}{N}\sum_{j=1}^N K^\varepsilon(\cdot, \xi_t^{j,N}). \end{cases}$$

(2.8)

Under the smooth assumptions on b, σ, u_0 mentioned before and non-degeneracy of σ, [40] proved the convergence of the regularized particle approximation $u_t^{N,\varepsilon}$ to the solution u of the pointwise non-linear PDE (2.7) as soon as $\varepsilon(N) \xrightarrow[N\to\infty]{} 0$ slowly enough. According to [57], the system (2.8) defines a so-called *moderately interacting* particle system with $u_t^{N,\varepsilon}(x) = \frac{1}{N\varepsilon^d}\sum_{j=1}^N \phi(\frac{x-\xi_t^{j,N}}{\varepsilon})$. Indeed as the window width of the kernel, ε, goes to zero, the number of particles that significantly impact a single one is of order $N\varepsilon^d$ with a strength of interaction of order $\frac{1}{N\varepsilon^d}$. In contrast, when ε is fixed, we recover the weakly interacting situation in which case the strength of interaction of each particle is of order $\frac{1}{N}$ which is smaller than $\frac{1}{N\varepsilon^d}$. In this case of moderate interaction, the propagation of chaos occurs with a slower rate than $C/\sqrt{N}$ and depends exponentially on the space dimension. Reference [40] constitutes an extension of the weak propagation of chaos of moderately interacting particles proved in [57] for the limited case of identity diffusion matrix.

The peculiar case where the drift vanishes and the diffusion coefficient $\sigma(u(t, Y_t))$ has a pointwise dependence on the law density $u(t, \cdot)$ of Y_t has been more particularly studied in [16] for classical porous media type equations and [8, 9, 14, 15, 18] who obtain well-posedness results for measurable and possibly singular functions σ. In that case the solution u of the associated PDE (1.1), is understood in the sense of distributions.

3 McKean Feynman-Kac Representations for Non-conservative and Non-linear PDEs

The idea of generalizing MSDEs to MFKEs (1.2) was originally introduced in the sequence of papers [46–48], with an earlier contribution in [11], where $\Lambda(t, x, u, \nabla u) = \xi_t(x)$, ξ being the sample of a Gaussian noise random field, white in time and regular in space, see Sect. 6. The goal was to provide some probabilistic representation for non-conservative non-linear PDEs (1.1) by introducing some exponential weights defining Feynman-Kac measures instead of probability measures. An interesting aspect of this strategy is that it is potentially able to represent an extended class of second order non-linear PDEs. One particularity of MFKE equations is that the probabilistic representation involves the past of the process (via the exponential weights). In this context, it is worth to quote the recent paper [39] which proposes a probabilistic representation, which also includes a dependence on the past, in relation with Keller-Segel model with application to chemiotaxis.

It is important to consider carefully the two major features differentiating the MFKE (1.2) from the traditional setting of MSDEs. To recover the traditional setting one has to do the following.

1. First, one has to put $\Lambda = 0$ in the third line equation of (1.2) Then $u(t, \cdot)$ is explicitly given by the third line equation of (1.2) and reduces to the density of the marginal distribution, $\mathcal{L}(Y_t)$. When $\Lambda \neq 0$, the relation between $u(t, \cdot)$ and the process Y is more complex. Indeed, not only does Λ embed an additional non-linearity with respect to u, but it also involves the whole past trajectory $(Y_s)_{0 \leq s \leq t}$ of the process Y.
2. Secondly, one has to replace the pointwise dependence $b(s, Y_s, u(s, Y_s))$ in Eq. (1.2) with a mollified dependence $b(s, Y_s, \int_{\mathbb{R}^d} K(Y_s - y)u(s, y)dy)$, where the dependence with respect to $u(s, \cdot)$ is Wasserstein continuous. Here $K : \mathbb{R} \to \mathbb{R}$ is a convolution kernel.

One interesting aspect of probabilistic representation (1.2) is that it naturally yields numerical approximation schemes involving weighted interacting particle systems.

More precisely, we consider N interacting particles $(\xi^{i,N})_{i=1,\cdots N}$ with N i.i.d. Brownian motions $(W^i)_{i=1,\cdots N}$, i.e.

$$\begin{cases} \xi_t^{i,N} = \xi_0^{i,N} + \int_0^t \sigma\left(s, \xi_s^{i,N}, u_s^{N,\varepsilon}(\xi_s^{i,N})\right) dW_s^i + \int_0^t b\left(s, \xi_s^{i,N}, u_s^{N,\varepsilon}(\xi_s^{i,N})\right) ds \\ \xi_0^{i,N} \text{ i.i.d.} \sim \mathbf{u_0} \\ u_t^{N,\varepsilon}(\xi_t^i) = \sum_{j=1}^N \omega_t^{j,N} K^\varepsilon(\xi_t^{i,N} - \xi_t^{j,N}), \end{cases} \quad (3.1)$$

where the mollifier K^ε is such that $K^\varepsilon(x) = \frac{1}{\varepsilon^d}\phi(\frac{x}{\varepsilon}) \xrightarrow[\varepsilon \to 0]{} \delta_0$ and the weights $\omega_t^{j,N}$ for $j = 1, \cdots, N$ verify

$$\omega_t^{j,N} := \exp\left\{\int_0^t \Lambda\left(r, \xi_r^{j,N}, u_r^{\varepsilon,N}(\xi_r^{j,N}), \nabla u_r^{\varepsilon,N}(\xi_r^{j,N})\right) dr\right\}$$
$$= \omega_s^{j,N} \exp\left\{\int_s^t \Lambda\left(r, \xi_r^{j,N}, u_r^{\varepsilon,N}(\xi_r^{j,N}), \nabla u_r^{\varepsilon,N}(\xi_r^{j,N})\right) dr\right\}.$$

References [48, 49] consider the case of pointwise semilinear PDEs of the form

$$\begin{cases} \partial_t u = \frac{1}{2}\partial_{xx}^2(\sigma^2(t,x)u) - \partial_x b(t,x)u) + \Lambda(t,x,u,\nabla u)u \\ u(0,x) = u_0(x), \end{cases} \quad (3.2)$$

for which the target probabilistic representation is

$$\begin{cases} Y_t = Y_0 + \int_0^t \sigma\left(s, Y_s\right) dW_s + \int_0^t b\left(s, Y_s\right) ds \\ Y_0 \sim \mathbf{u_0} \\ \int \varphi(x) u_t(x) dx := \mathbb{E}\left[\varphi(Y_t) \exp\left\{\int_0^t \Lambda(s, Y_s, u_s(Y_s), \nabla u_s(Y_s)) ds\right\}\right]. \end{cases} \quad (3.3)$$

We set

$$L_t f := \frac{1}{2}\sigma^2(t,x)f''(x) + b(t,x)f'(x), t \in]0, T[, \quad \text{for any } f \in C^2(\mathbb{R}). \quad (3.4)$$

Let us consider the family of Markov transition functions $P(s, x_0, t, \cdot)$ associated with (L_t), see [49]. We recall that if X is a process solving the first line of (3.1) with $X_s \equiv x_0 \in \mathbb{R}$, then $\int_\mathbb{R} P(s, x_0, t, x) f(x) dx = \mathbb{E}(f(X_t)), t \geq s$, for every bounded Borel function $f : \mathbb{R} \to \mathbb{R}$. $u : [0, T] \times \mathbb{R} \to \mathbb{R}$ will be called **mild solution** of (3.2) (related to (L_t)) if for all $\varphi \in C_0^\infty(\mathbb{R}), t \in [0, T]$,

$$\int_{\mathbb{R}^d} \varphi(x) u(t,x) dx = \int_{\mathbb{R}^d} \varphi(x) \int_{\mathbb{R}^d} \mathbf{u_0}(dx_0) P(0, x_0, t, dx)$$
$$+ \int_{[0,t] \times \mathbb{R}^d} \left(\int_{\mathbb{R}^d} \varphi(x) P(s, x_0, t, dx)\right) \Lambda(s, x_0, u(s, x_0), \nabla u(s, x_0)) u(s, x_0) dx_0 ds.$$

The following theorem states conditions ensuring equivalence between (3.3) and (3.2) together with the convergence of the related particle approximation (3.1).

Theorem 3.1 *We suppose that σ and b are Lipschitz with linear growth and Λ is bounded.*

1. *Let $u : [0, T] \times \mathbb{R} \to \mathbb{R} \in L^1([0, T]; W^{1,1}(\mathbb{R}^d))$. u is a mild solution of PDE (3.2) if and only if u verifies (3.3).*
2. *Suppose that $\sigma \geq c > 0$ and Λ is uniformly Lipschitz w.r.t. to u and ∇u. There is a unique mild solution in $L^1([0, T]; W^{1,1}(\mathbb{R}) \cap L^\infty([0, T] \times \mathbb{R})$ of (3.2), therefore also of (3.3).*
3. *Under the same assumption of item 2., the particle approximation $u^{N,\varepsilon}$ (3.1) converges in $L^1([0, T]; W^{1,1}(\mathbb{R})$ to the solution of (3.2) as $N \to \infty$ and $\varepsilon(N) \to 0$ slowly enough.*

Item 1. was the object of Theorem 3.5 in [49]. Item 2. (resp. item 3.) was treated in Theorem 3.6 (resp. Corollary 3.5) in [49].

Remark 3.2 The error induced by the discrete time approximation of the particle system was evaluated in [48].

Reference [50] considers the case where b is replaced by $b + b_1$ where b is only supposed bounded Borel, without regularity assumption on the space variable. In particular they treat the pointwise semilinear PDEs of the form

$$\begin{cases} \partial_t u = \frac{1}{2}\partial_{xx}^2(\sigma^2(t, x)u) - \partial_x\Big(\big(b(t, x) + b_1(t, x, u)\big)u\Big) + \Lambda(t, x, u)u \\ u(0, x) = u_0(x), \end{cases} \quad (3.5)$$

for which the target probabilistic representation is

$$\begin{cases} Y_t = Y_0 + \int_0^t \sigma(s, Y_s)dW_s + \int_0^t [b(s, Y_s) + b_1\big(s, Y_s, u(s, Y_s)\big)]ds \\ Y_0 \sim u_0 \\ \int \varphi(x)u_t(x)dx := \mathbb{E}\Big[\varphi(Y_t) \exp\Big\{\int_0^t \Lambda\big(s, Y_s, u_s(Y_s)\big)ds\Big\}\Big]. \end{cases} \quad (3.6)$$

The following theorem states conditions ensuring equivalence between (3.6) and (3.5) together with well-posedness conditions for both equations.

Theorem 3.3 *We formulate the following assumptions.*

1. *The PDE in the sense of distributions $\partial_t u = L_t^* u_t$ admits as unique solution $u \equiv 0$, where L_t was defined in (3.4).*
2. *b is bounded measurable and σ is continuous $\sigma \geq c > 0$ for some constant $c > 0$.*
3. *$b_1, \Lambda : [0, T] \times \mathbb{R} \times \mathbb{R} \to \mathbb{R}$ is uniformly bounded, Lipschitz with respect to the third argument.*
4. *The family of Markov transition functions associated with (L_t), are of the form $P(s, x_0, t, dx) = p(s, x_0, t, x)dx$, i.e. they admit measurable densities p.*

5. The first order partial derivatives of the map $x_0 \mapsto p(s, x_0, t, x)$ exist in the distributional sense.
6. For almost all $0 \leq s < t \leq T$ and $x_0, x \in \mathbb{R}$ there are constants $C_u, c_u > 0$ such that

$$p(s, x_0, t, x) \leq C_u q(s, x_0, t, x) \quad \text{and} \quad |\partial_{x_0} p(s, x_0, t, x)| \leq C_u \frac{1}{\sqrt{t-s}} q(s, x_0, t, x),$$
(3.7)

where $q(s, x_0, t, x) := \left(\frac{c_u(t-s)}{\pi}\right)^{\frac{1}{2}} e^{-c_u \frac{|x-x_0|^2}{t-s}}$ is a Gaussian probability density.

The following results hold.

1. Let $u \in (L^1 \cap L^\infty)([0, T] \times \mathbb{R})$. u is a solution of PDE (3.5) in the sense of distributions if and only if u verifies (3.6) for a solution Y in the sense of probability laws.
2. There is a unique solution $u \in (L^1 \cap L^\infty)([0, T] \times \mathbb{R})$ in the sense of distributions of PDE (3.5) (and therefore of (3.6)).

The result 1. (resp. result 2.) was the object of Theorem 12. (resp. Proposition 16., Theorems 13., 22.) of [50].

Remark 3.4 Under more restrictive assumptions on b, item 3. of Theorem 13. in [50] states the well-posedness of (3.6)) in the sense of strong existence and pathwise uniqueness.

References [46, 47] studied a mollified version of (1.1), whose probabilistic representation falls into the Wasserstein continuous traditional setting mentioned above. Following the spirit of [64], a fixed point argument was carried out in the general case in [47] to prove well-posedness of

$$\begin{cases} Y_t = Y_0 + \int_0^t \sigma\left(s, Y_s, K * u_s(Y_s)\right) dW_s + \int_0^t b\left(s, Y_s, K * u_s(Y_s)\right) ds \\ Y_0 \sim u_0 \\ (K * u_t)(x) := \mathbb{E}\left[K(x - Y_t) \exp\left\{\int_0^t \Lambda(s, Y_s, K * u_s(Y_s)) ds\right\}\right], \end{cases}$$
(3.8)

where $K : \mathbb{R} \to \mathbb{R}$ is a mollified kernel. We remark that if (Y, u) is a solution of (3.8), then u is a solution (in the sense of distribution) of

$$\begin{cases} \partial_t u = \frac{1}{2} \partial_{xx}^2 (\sigma^2(t, x, K * u)u) - \partial_x (b(t, x, K * u)u) + \Lambda(t, x, K * u)u \\ u(0, x) = u_0(x). \end{cases}$$
(3.9)

Remark 3.5 1. Existence and uniqueness results (in the strong sense and in the sense of probability laws) for the MFKE (3.8) are established under various technical assumptions, see [46].
2. Chaos propagation for the interacting particle system (3.1) providing an approximation to the regularized PDE (3.9), as $N \to \infty$ (for fixed K), [47].

4 McKean Representation of a Fokker-Planck Equation with Terminal Condition

Let us consider the PDE with terminal condition (1.5) and $\Lambda = 0$.

$$\begin{cases} \partial_t u = \frac{1}{2} \sum_{i,j=1}^{d} \partial^2_{ij} \left((\sigma \sigma^t)_{i,j}(t,x) u\right) - div\, (b(t,x) u) \\ u(T, dx) = \mathbf{u}_T(dx), \end{cases} \quad (4.1)$$

where $\mathbf{u}_T$ is a given Borel probability measure. In the present section we assume the following.

Assumption 1 Suppose that (4.1) admits uniqueness, i.e. that there is at most one solution of (4.1).

Remark 4.1 Different classes of sufficient conditions for the validity of Assumption 1 are provided in [38]. In particular one significant result is Theorem 4.14 of [38] which states that previous Assumption 1 holds if σ, b are time-homogeneous and the following holds.

Assumption 2 1. $\Sigma = \sigma \sigma^*$ is strictly non-degenerate.
2. The functions σ is Lipschitz in space.
3. The functions $\sigma, b, (\nabla_x b_i)_{i \in [\![1,d]\!]}, (\nabla_x \Sigma_{ij})_{i,j \in [\![1,d]\!]}$ are continuous bounded and $\nabla_x^2 \Sigma$ is Hölder continuous with exponent $\alpha \in]0, 1[$.

Generalizations to the time-inhomogeneous setup are also available.

A natural representation of (4.1) is the following MSDE, where β is a Brownian motion.

$$\begin{cases} Y_t = \xi - \int_0^t \tilde{b}(s, Y_s; v_s)\, ds + \int_0^t \sigma(T-s, Y_s)\, d\beta_s, t \in [0, T] \\ \int_{\mathbb{R}^d} v_t(x) \varphi(x) dx = \mathbb{E}(\varphi(Y_t)), t \in [0, T] \\ \xi \sim \mathbf{u}_T, \end{cases} \quad (4.2)$$

where $\tilde{b}(s, y; v_s) = (\tilde{b}^1(s, y; v_s), \ldots, \tilde{b}^d(s, y; v_s))$ is defined as

$$\tilde{b}(s, y; v_s) := \left[\frac{div_y \left(\sigma \sigma^t_{j\cdot}(T-s, y) v_s(y)\right)}{v_s(y)} \right]_{j \in [\![1,d]\!]} - b(T-s, y). \quad (4.3)$$

For $d = 1$ previous expression gives

$$\tilde{b}(s, y; v_s) := \frac{\left(\sigma^2(T-s, \cdot) v_s\right)'}{v_s}(y) - b(T-s, y). \quad (4.4)$$

Remark 4.2 Equation (4.2) is in particular fulfilled if Y is the time reversal process $\hat{X}_t := X_{T-t}$ of a diffusion satisfying the SDE

$$\begin{cases} X_t = X_0 + \int_0^t b(s, X_s)ds + \sigma(s, X_s)dW_s, t \in [0, T] \\ X_0 \sim \mathbf{u_0} \in \mathcal{P}(\mathbb{R}). \end{cases} \quad (4.5)$$

This happens under locally Lipschitz conditions on σ and b and minimal regularity conditions on the law density p_t of X_t. Indeed in [35], the authors prove that

$$\hat{X}_t = X_T + \int_0^t \tilde{b}\left(s, \hat{X}_s; p_{T-s}\right) ds + \int_0^t \sigma\left(T - s, \hat{X}_s\right) d\beta_s, \ t \in [0, T], \quad (4.6)$$

where $\tilde{b}$ is defined in (4.3) and p_t is the density of X_t. We emphasize that the main difference between (4.2) and (4.6) is that in the first equation the solution is a couple (Y, v), in the second one, a solution is just Y, p being exogenously defined by (4.5).

We observe now that a solution (Y, v) of (4.2) provides a solution u of (4.1). This justifies indeed the terminology of probabilistic representation.

Proposition 4.3 1. Let (Y, v) be a solution of (4.2). Then $u(t, \cdot) := v(T - t, \cdot), t \in [0, T])$, is a solution of (4.1) with terminal value $\mathbf{u_T}$.
2. If (4.1) admits at most one solution, then there is at most one v such that (Y, v) solves (4.2).

Proof In order to avoid technicalities which complicate the task of the reader we express the proof for $d = 1$. We prove 1. since 2. is an immediate consequence of 1.

Let $\phi \in \mathcal{C}^\infty(\mathbb{R})$ with compact support and $t \in [0, T]$. Itô formula gives

$$\mathbb{E}\left[\phi\left(Y_{T-t}\right)\right] - \int_{\mathbb{R}^d} \phi(y) \mathbf{u_T}(dy) = \int_0^{T-t} \mathbb{E}\left[\tilde{b}(s, Y_s; v_s) \phi'(Y_s) + \frac{1}{2}\left(\sigma^2(T - s, Y_s) \phi''(Y_s)\right)\right] ds.$$

Fixing $s \in [0, T]$, we have

$$\mathbb{E}\left[\tilde{b}(s, Y_s; v_s) \phi'(Y_s)\right] = \int_{\mathbb{R}} (\sigma^2(T - s, \cdot) v_s)'(y) \phi'(y) dy - \int_{\mathbb{R}} b(T - s, y) \phi'(y) v_s(y) dy$$

$$= -\int_{\mathbb{R}} (\sigma^2)(T - s, y) \phi''(y) v_s(y) dy - \int_{\mathbb{R}} b(T - s, y), \phi'(y) v_s(y) dy.$$

Hence, we have the identity

$$\mathbb{E}\left[\phi\left(Y_{T-t}\right)\right] = \int_{\mathbb{R}} \phi(y) \mathbf{u_T}(dy) - \int_0^{T-t} \int_{\mathbb{R}} L_{T-s} \phi(y) v_s(y) dy ds.$$

Applying the change of variable $t \mapsto T - t$, we finally obtain the identity

$$\int_{\mathbb{R}} \phi(y) v_{T-t}(y) \, dy = \int_{\mathbb{R}} \phi(y) \mathbf{u_T}(dy) - \int_t^T \int_{\mathbb{R}^d} L_s \phi(y) v_{T-s}(y) \, dyds.$$

This means that $t \mapsto u_t$ is a solution of (1.2) with terminal value $\mathbf{u_T}$. □

Remark 4.4 Precise discussions on existence and uniqueness of (4.2) are provided in [38]. In particular we have the following.

1. There is at most one solution (in law) (Y, v) of (4.2) such that v is locally bounded in $[0, T[\times \mathbb{R}^d$.
2. There is at most one strong solution (Y, v) of (4.2) such that v is locally Lipschitz in $[0, T[\times \mathbb{R}^d$.

Item 1. is a consequence of Theorem 10.1.3 of [63]. Item 2. is a consequence of usual pathwise uniqueness arguments for SDEs.

5 Probabilistic Representation with Jumps for Non-conservative PDEs

In this section, we outline the link between non-conservative PDEs and non-linear jump diffusions. This kind of representation was emphasized in [25, 28] to design interacting jump particles systems to approximate time-dependent Feynman-Kac measures. For simplicity, we present this correspondence in the simple case of the non-conservative linear PDE (1.1) when the coefficients do not depend on the solution, see (1.1). However, the same ideas could be extended to the non-linear case where the coefficients σ, b, Λ may depend on the PDE solution.

Let us consider the SDE

$$\begin{cases} dX_t = b(t, X_t)dt + \sigma(t, X_t)dW_t \\ X_0 \sim \mathbf{u_0}, \end{cases} \quad (5.1)$$

where W is a one-dimensional Brownian motion. Assume that (5.1) admits a (weak) solution. Let Λ be a bounded and negative function defined on $[0, T] \times \mathbb{R}$. For any $t \in [0, T]$, we define the measure, $\gamma(t, \cdot)$ such that for any real-valued Borel measurable test function φ

$$\int \gamma(t, dx)\varphi(x) = \mathbb{E}\left[\varphi(X_t) \exp\left(\int_0^t \Lambda(s, X_s)ds\right)\right]. \quad (5.2)$$

We recall that by Sect. 3 we know that γ is a solution (in the distributional sense) of the linear and non-conservative PDE

$$\begin{cases} \partial_t \gamma = \frac{1}{2}\partial_{xx}^2(\sigma^2(t,x)\gamma) - \partial_x(b(t,x)\gamma) + \Lambda(t,x)\gamma \\ \gamma(0, \cdot) = \mathbf{u_0}. \end{cases} \quad (5.3)$$

Remark 5.1 If uniqueness of distributional solutions of (5.3) holds, then γ defined by (5.1), (5.2) is the unique solution of (5.3).

Let $\gamma(t,\cdot)$ be a solution of (5.2) which for each t is a positive measure. We introduce the family of probability measures $(\eta(t,\cdot))_{t\in[0,T]}$, obtained by normalizing $\gamma(t,\cdot)$, such that for any real valued bounded and measurable test function φ we have

$$\int \eta(t,dx)\varphi(x) := \frac{\int \gamma(t,dx)\varphi(x)}{\int \gamma(t,dx)}. \tag{5.4}$$

By simple differentiation of the above ratio and using the fact that γ satisfies (5.3), we obtain that η is a solution in the distributional sense of the integro-differential PDE

$$\begin{cases} \partial_t \eta = \frac{1}{2}\partial_{xx}(\sigma^2(t,x)\eta) - \partial_x(b(t,x)\eta) + \left(\Lambda(t,x) - \int \eta(t,dx)\Lambda(t,x)\right)\eta \\ \eta_0 = \mathbf{u}_0. \end{cases} \tag{5.5}$$

Besides one can express $\gamma(t,\cdot)$ as a function of $(\eta(s,\cdot))_{s\in[0,t]}$. Indeed, since γ solves the linear PDE (5.3) then in particular approaching the constant test function 1, yields

$$\partial_t \int \gamma(t,dx) = \int \gamma(t,dx)\Lambda(t,x) = \int \gamma(t,dx)\int \eta(t,dx)\Lambda(t,x),$$

which gives $\int \gamma(t,dx) = \exp\left(\int_0^t \int \eta(s,dx)\Lambda(s,x)ds\right)$. Then by definition (5.4) of η,

$$\gamma(t,\cdot) = \left(\int \gamma(t,dx)\right)\eta(t,\cdot) = \exp\left(\int_0^t \int \eta(s,dx)\Lambda(s,x)ds\right)\eta(t,\cdot). \tag{5.6}$$

We already know that for any solution γ of (5.3) one can build a solution η to (5.5) according to relation (5.4). Conversely, for any solution η of (5.5), by similar manipulations one can build a solution γ of (5.3) according to (5.6). Hence well-posedness of (5.3) is equivalent to well-posedness of (5.5).

We propose now an alternative probabilistic representation to (5.1) and (5.2) of (5.3). Let us introduce the non-linear jump diffusion Y (if it exists), which evolves between two jumps according to the diffusion dynamics (5.1) and jumps at exponential times with intensity $-\Lambda_t(Y_t) \geq 0$ to a new point independent of the current position and distributed according to the current law, $\mathcal{L}(Y_t)$. More specifically, we consider a process Y solution of the following non-linear (in the sense of McKean) SDE with jumps

$$\begin{cases} dY_t = b(t,Y_{t-})dt + \sigma(t,Y_{t-})dW_t + \int_{\mathbb{R}} x\mathbf{1}_{|x|>1} J_t(\mu_{t-},Y_{t-},dx)dt + \int_{\mathbb{R}} x\mathbf{1}_{|x|\leq 1}(J_t - \tilde{J}_t)(\mu_{t-},Y_{t-},dx)dt \\ Y_0 \sim \mathbf{u}_0 \\ \mu_{t-} = \mathcal{L}(Y_{t-}). \end{cases}$$

$$\tag{5.7}$$

where J denotes the jump measure and $\bar{J}$ is the associated predictable compensator such that for any probability measure ν on $\mathbb{R}$

$$\bar{J}_t(\nu, y, d(y'-y)) = -\Lambda_t(y)\nu(dy'), \quad \text{for any } y, y' \in \mathbb{R}.$$

Note that well-posedness analysis of the above equation constitutes a difficult task. In particular, [41] analyzes well-posedness and particle approximations of some types of non-linear jump diffusions. However, contrarily to (5.7), the nonlinearity considered in [41] is concentrated on the diffusion matrix (assumed to be Lipschitz in the time-marginals of the process w.r.t. Wasserstein metric) and does not involve the jump measure which is assumed to be given.

Assume that MSKE (5.7) admits a weak solution. By application of Itô formula, we observe that the marginals of Y are distributional solutions of (5.5). Indeed, for any real valued test function in $C_0^\infty(\mathbb{R})$

$$\mathbb{E}[\varphi(Y_t)] = \mathbb{E}[\varphi(Y_0)]$$
$$+ \int_0^t \mathbb{E}\left[b(s, Y_{s-})\varphi'(Y_{s-}) + \frac{1}{2}\sigma^2(s, Y_{s-})\varphi''(Y_{s-})\right] ds$$
$$+ \int_0^t \mathbb{E}\left[\int \varphi(Y_{s-} + x)\bar{J}_s(\mu_{s-}, Y_{s-}, dx)\right] ds$$
$$- \int_0^t \mathbb{E}\left[\varphi(Y_{s-})\bar{J}_s(\mu_{s-}, Y_{s-}, \mathbb{R})\right] ds. \tag{5.8}$$

Conclusion 5.2 *Suppose that (5.3) admits a unique distributional solution γ; let η defined by (5.4). Suppose the existence of a (weak) solution X (resp. Y) of (5.1) (resp. (5.7)).*

1. *η is the unique solution (in the sense of distributions) of (5.5). Moreover $\int_\mathbb{R} \varphi(x)\eta(t, dx) = \mathbb{E}[\varphi(Y_t)], t \geq 0$.*
2. *We obtain the following identities for γ and η:*

$$\int \gamma(t, dx)\varphi(x) = \mathbb{E}\left[\varphi(X_t) \exp\left(\int_0^t \Lambda(s, X_s)ds\right)\right]$$
$$= \exp\left(\int_0^t \int \eta(s, dx)\Lambda(s, x)ds\right) \eta_t(\varphi)$$
$$= \exp\left(\int_0^t \mathbb{E}[\Lambda(s, Y_s)]ds\right) \mathbb{E}[\varphi(Y_t)]. \tag{5.9}$$

Using the above identities allows to design discrete time interacting particles systems with geometric interacting jump processes. In particular, in [27] the authors provide non asymptotic bias and variance theorems w.r.t. the time step and the size of the system, allowing to numerically approximate the time-dependent family of Feynman-Kac measures γ.

6 McKean SDEs in Random Environment

6.1 The (S)PDE and the Basic Idea

Let $(\Omega, \mathcal{F}, (\mathcal{F}_t), \mathbb{P})$ be a filtered probability space. We consider a progressively measurable random field $(\xi(t, x))$. We want to discuss probabilistic representations of

$$\begin{cases} \partial_t u = \frac{1}{2}\Delta(\beta(u)) + \partial_t \xi(t, x) u(t, x), & \text{with } \beta(u) = \sigma^2(u)u, \\ u(0, \cdot) = \mathbf{u_0}. \end{cases} \quad (6.1)$$

Suppose for a moment that ξ has random realizations which are smooth in time so that

$$\partial_t \xi(t, x) = \Lambda(t, x; \omega). \quad (6.2)$$

Under some regularity assumptions on Λ, (6.1) can be observed as a randomization of a particular case of the PDE (1.1). For each random realization $\omega \in \Omega$, the natural (double) probabilistic representation is

$$\begin{cases} Y_t = Y_0 + \int_0^t \sigma\left(u(s, Y_s)\right) dW_s \\ Y_0 \sim \mathbf{u_0} \\ \int \varphi(x) u(t, x) dx = \mathbb{E}^\omega \left[\varphi(Y_t) \exp\left\{ \int_0^t \Lambda(s, Y_s; \omega)) ds \right\} \right], & \text{for } t \in [0, T], \end{cases} \quad (6.3)$$

where $\mathbb{E}^\omega$ denotes the expectation with frozen ω. However the assumption (6.2) is not realistic and we are interested in $\partial_t \xi$ being a white noise in time. Let $N \in \mathbb{N}^*$. Let $B^1, \ldots, B^N$ be N independent $(\mathcal{F}_t)$-Brownian motions, $e^1, \ldots, e^N$ be functions in $C_b^2(\mathbb{R})$. In particular they are H^{-1}-multiplier, i.e. the maps $\varphi \to \varphi e^i$ are continuous in H^{-1}.

We define the random field $\xi(t, x) = \sum_{i=0}^N e^i(x) B_t^i$, where $B_t^0 \equiv t$ and we consider the SPDE (6.1) in the sense of distributions, i.e.

$$\int_\mathbb{R} \varphi(x) u(t, x) dx = \int_\mathbb{R} \varphi(x) \mathbf{u_0}(dx) + \frac{1}{2} \int_0^t \int_\mathbb{R} \varphi''(x) \sigma^2(u(s, x)) ds dx + \int_0^t \int_\mathbb{R} \varphi(x) u(s, x) \xi(ds, x) dx, \quad (6.4)$$

where the latter stochastic integral is intended in the Itô sense.

6.2 Well-Posedness of the SPDE

The theorem below contains results taken from [10, 62].

Theorem 6.1 *Suppose that β is Lipschitz.*

- *Suppose that $u_0 \in L^2(\mathbb{R})$. There is a solution to Eq. (6.1).*

- Assume further that β is non-degenerate, i.e. $\beta(r) \geq ar^2$, $r \in \mathbb{R}$, where $a > 0$. Then, there is a solution u to (6.1) for any probability $\mathbf{u_0}(dx)$ (even in $H^{-1}(\mathbb{R})$).
- There is at most one solution in the class of random fields u such that $\int_{[0,T] \times \mathbb{R}} u^2(t, x) dt dx < \infty$ a.s.

Remark 6.2 • Previous result extends to the case of an infinite number of modes e^i and for $d \geq 1$.
- We remark that the $\partial_t \xi(t, x)$ is a colored noise (in space). The case of space-time white noise seems very difficult to treat.

6.3 McKean Equation in Random Environment

Given a local martingale M, $\mathcal{E}(M)$ denotes the Doléans exponential of M i.e. $\exp(M_t - \frac{1}{2}[M]_t)$, $t \geq 0$. We say that a filtered probability space $(\Omega_0, \mathcal{G}, (\mathcal{G}_t), Q)$ is a **suitable enlarged space** of $(\Omega, \mathcal{F}, (\mathcal{F}_t), P)$, if the following holds.

1. There is a measurable space $(\Omega_1, \mathcal{H})$ with $\Omega_0 = \Omega \times \Omega_1$, $\mathcal{G} = \mathcal{F} \otimes \mathcal{H}$ and a random kernel $(\omega, H) \mapsto \mathbb{Q}^\omega(H)$ defined on $\Omega \times \mathcal{H} \to [0, 1]$ such that the probability $\mathbb{Q}$ on $(\Omega_0, \mathcal{G})$ is defined by $d\mathbb{Q}(\omega, \omega_1) = d\mathbb{P}(\omega)\mathbb{Q}^\omega(\omega_1)$.
2. The processes $B^1, \ldots, B^N$ are $(\mathcal{G}_t)$- Brownian motions where $\mathcal{G}_t = \mathcal{F}_t \vee \mathcal{H}$.

Definition 6.3 We say that the *non-linear doubly-stochastic diffusion*

$$\begin{cases} Y_t &= Y_0 + \int_0^t \Phi(u(s, Y_s)) dW_s, \\ \int \varphi(x) u(t, x) dx &= \mathbb{E}^{\mathbb{Q}^\omega} \left(\varphi(Y_t(\omega, \cdot)) \mathcal{E}_t \left(\int_0^\cdot \xi(ds, Y_s)(\omega, \cdot) \right) \right), \\ \xi - \text{Law}(Y_0) &= \mathbf{u_0}(dx), \end{cases} \quad (6.5)$$

admits **weak existence** on $(\Omega, \mathcal{F}, (\mathcal{F}_t), \mathbb{P})$ if there is a suitably enlarged probability space $(\Omega_0, \mathcal{G}, (\mathcal{G}_t), \mathbb{Q})$ an $(\mathcal{G}_t)$-Brownian motion W such that (6.5) is verified. The couple (Y, u) will be called **weak solution** of (6.5).

Remark 6.4 • We remark that the second line in (6.5) represents a sort of ξ-**marginal weighted law** of Y_t.
- Let (Y, u) be a solution to (6.5). Then u is a solution to (6.1).
- Such representation allows to show that $u(t, x) \geq 0$, $d\mathbb{P} dt dx$ a.e. and, at least if $e^0 = 0$, $\mathbb{E}\left(\int_\mathbb{R} u(t, x) dx \right) = 1$, so that the conservativity is maintained at the expectation level.

Definition 6.5 Let two measurable random fields $u^i : \Omega \times [0, T] \times \mathbb{R} \to \mathbb{R}$, $i = 1, 2$ on $(\Omega, \mathcal{F}, \mathbb{P}, (\mathcal{F}_t))$, and Y^i, on a suitable extended probability space $(\Omega_0^i, \mathcal{G}^i, (\mathcal{G}_t^i), \mathbb{Q}^i)$, $i = 1, 2$, such that (Y^i, u^i) are (weak) solutions of (6.5) on $(\Omega, \mathcal{F}, (\mathcal{F}_t), \mathbb{P})$. If we always have that $(Y^1, B^1, \ldots, B^N)$ and $(Y^2, B^1, \ldots, B^N)$ have the same law, then we say that (6.5) admits **weak uniqueness** (on $(\Omega, \mathcal{F}, (\mathcal{F}_t), \mathbb{P})$).

Theorem 6.6 *Under the assumption of Theorem 6.1 Eq. (6.5) admits (weak) existence and uniqueness on $(\Omega, \mathcal{F}, (\mathcal{F}_t), \mathbb{P})$.*

7 McKean Representation of Stochastic Control Problems

7.1 Stochastic Control Problems and Non-linear Partial Differential Equations

There are several connections between stochastic control and McKean type SDEs, see e.g. [1]. Here, we propose an original (and maybe unexpected) point of view. Let us briefly recall the link between stochastic control and non-linear PDEs given by the Hamilton-Jacobi-Bellman (HJB) equation. We refer for instance to [31, 60, 67] for more details. Consider a *state process* $(X_s^{t_0,x,\alpha})_{t_0 \leq s \leq T}$ on $[t_0, T] \times \mathbb{R}^d$ solution to the controlled SDE

$$\begin{cases} dX_s^{t_0,x,\alpha} = b\left(s, X_s^{t_0,x,\alpha}, \alpha(s, X_s^{t_0,x,\alpha})\right)ds + \sigma\left(s, X_s^{t_0,x,\alpha}, \alpha(s, X_s^{t_0,x,\alpha})\right)dW_s \\ X_{t_0}^{t_0,x,\alpha} = x, \end{cases} \quad (7.1)$$

where W denotes the Brownian motion on $[t_0, T] \times \mathbb{R}^d$, and $\alpha(s, X_s^{t_0,x,\alpha})$ represents *Markovian* control in the sense that the control at time t is supposed here to depend on t and on the current value of the state process:

$$\alpha \in \mathcal{A}_{t_0,T} := \left\{ \alpha : (t, x) \in [t_0, T] \times \mathbb{R}^d \mapsto \alpha(t, x) \in A \subset \mathbb{R}^k \right\}, \quad (7.2)$$

A being a subset of $\mathbb{R}^k$. For a given initial time and state $(t_0, x) \in [0, T] \times \mathbb{R}^d$, we are interested in maximizing, over the Markovian controls $\alpha \in \mathcal{A}_{t_0,T}$, the criteria

$$J(t_0, x, \alpha) := \mathbb{E}\left[g(X_T^{t_0,x,\alpha}) + \int_{t_0}^{T} f\left(s, X_s^{t_0,x,\alpha}, \alpha(s, X_s^{t_0,x,\alpha})\right)ds \right]. \quad (7.3)$$

In the above criteria, the function f is called the *running gain* whereas g is called the *terminal gain*.

Remark 7.1 At first glance, the set of control processes of the form $\alpha_t = \alpha(t, X_t)$ defined in (7.2) may appear too restrictive compared to a larger set of non-anticipative controls (α_t) which may depend on all the past history of the state process (X_t). However, in the framework of Markov control problems (for which the state process $(X_t^{t_0,x,a})$ is Markov, as soon as the control is fixed to a deterministic value $\alpha_t = a \in A$, for all $t \in [t_0, T]$), it is well-known that the optimal control process (α_t) lies in the set of Markovian controls verifying $\alpha_t = \alpha(t, X_t)$. Hence, considering controls of the particular form (7.2) is done here without loss of generality.

To tackle this finite horizon stochastic control problem, the usual approach consists in introducing the associated value (or *Bellman*) function $v : [t_0, T] \times \mathbb{R}^d \to \mathbb{R}$ representing the maximum gain one can expect, starting from time t at state x, i.e.

$$v(t, x) := \sup_{\alpha \in \mathcal{A}_{t,T}} J(t, x, \alpha), \quad \text{for } t \in [t_0, T]. \quad (7.4)$$

Note that the terminal condition is known, which fixes $v(T, x) = g(x)$, whereas the initial condition $v(t_0, x)$ corresponds to the solution of the original minimization problem. The value function is then proved to verify the *Dynamic Programming Principle (DPP)* which consists in the backward induction

$$v(t, x) = \sup_{\alpha \in \mathcal{A}_{t,\tau}} \mathbb{E}\Big[\int_t^\tau f(s, X_s^{t,x,\alpha}, \alpha(s, X_s^{t,x,\alpha}))ds + v(\tau, X_\tau^{t,x,\alpha}) \Big], \quad \text{for any stopping time } \tau \in]t, T]. \tag{7.5}$$

Under continuity assumptions on b, σ, f, g, using DPP together with Itô formula allows to characterize v as a viscosity solution of the HJB equation

$$\begin{cases} v(T, x) = g(x) \\ \partial_t v(t, x) + H(t, x, \nabla v(t, x), \nabla^2 v(t, x)) = 0, \end{cases} \tag{7.6}$$

where ∇ and ∇^2 denote the gradient and the Hessian operators and the so-called, Hamiltonian, H denotes the real valued function defined on $[0, T] \times \mathbb{R}^d \times \mathbb{R}^d \times \mathcal{S}^d$ ($\mathcal{S}^d$ denoting the set of symmetric matrices in $\mathbb{R}^{d \times d}$), such that

$$H(t, x, \delta, \gamma) := \sup_{a \in A} \Big\{ f(t, x, a) + b(t, x, a)^\top \delta(t, x) + \frac{1}{2} Tr[\sigma \sigma'(t, x, a) \gamma(t, x)] \Big\}. \tag{7.7}$$

Note that (7.6) is a non-linear PDE because of the nonlinearity in the Hamiltonian induced by the supremum operator. Besides, assuming that, for all $(t, x) \in [t_0, T] \times \mathbb{R}^d$, the supremum in (7.7) is attained at a unique maximizer, then the optimal control α^* is directly obtained as a function of the Bellman function and its derivatives, i.e.

$$\alpha^*(t, x) = \arg\max_{a \in A} \Big\{ f(t, x, a) + b(t, x, a)^\top \nabla v(t, x) + \frac{1}{2} Tr[\sigma \sigma^\top (t, x, a) \nabla^2 v(t, x)] \Big\}. \tag{7.8}$$

Except in some very concrete cases such as the Linear Quadratic Gaussian (LQG) setting (where the states dynamics involve an affine drift with Gaussian noise and the cost is quadratic both w.r.t. the control and the state), there is no explicit solution to stochastic control problems. To numerically approximate the solution of equation (7.6), several approaches have been proposed, mainly differing in the way the value function v is interpreted. Indeed, as pointed out, v can be viewed either as the solution to the control problem (7.4), or as a (viscosity) solution of the non-linear PDE (7.6).

1. When v is defined as the solution to the control problem (7.4), a natural approach consists in discretizing the time continuous control problem and apply the time discrete Dynamic Programming Principle [17]. Then the problem consists in maximizing over the controls, backwardly in time, the conditional expectation of the value function related to (7.5). The maximization at time step t_k can be done via a parametrization of the control $x \mapsto \alpha_{t_k}^\theta(x)$ via a parameter θ so that parametric optimization methods such as the stochastic gradient algorithm could be applied to maximize the expectation over θ. It remains to approximate the conditional expectations by numerical methods such as PDE, Fourier, Monte Carlo, Quanti-

zation or lattice methods... A great variety of numerical approximation schemes has been developed in the specific Bermudan option valuation test-bed [7, 22, 23, 26, 51, 68]. Alternatively, one can use Markov chain approximation method [44] which consists in a time-space discretization designed to obtain a proper Markov chain.

2. In the second approach we recall that v is viewed as the solution of (7.6). The problem amounts then to discretize a non-linear PDE. Then one can rely on numerical analysis methods (e.g. finite differences, or finite elements) and use monotone approximation schemes in the sense of Barles and Souganidis [12] to build converging approximation schemes, e.g. [19, 33]. This type of approach is in general limited to state space dimension lower than 4. To tackle higher dimensional problems, one approach consists in converting the PDE into a probabilistic setting in order to apply Monte Carlo types algorithms. To that end, various kinds of probabilistic representations of non-linear PDEs are available. Forward Backward Stochastic Differential Equations (FBSDE) were introduced in [59] as probabilistic representations of semi-linear PDEs. Then various types of numerical schemes for FBSDE have been developed. They mainly differ in the approach of evaluating conditional expectations: [21] (resp. [34], [29, 58]) use kernel (resp. regression, quantization) methods. Recently, important progresses have been done performing machine learning techniques, see e.g. [30, 37]. Branching processes [36, 54] can also provide probabilistic representations of semi-linear PDEs via Feynman-Kac formula. Non-linear SDEs in the sense of McKean [53] are another approach that constitutes the subject of the present paper.

3. Other approaches take advantage of both interpretations see for instance [32] and in [65].

7.2 McKean Type Representation in a Toy Control Problem Example

In order to illustrate the application of MFKEs to control problems, we consider a simple example corresponding to an inventory problem, for which the Hamiltonian maximization (7.7), (7.8) is explicit. The state $(X_t)_{t \in [t_0, T]}$ denotes the stock level evolving randomly with a control of the drift α:

$$\begin{cases} dX_t^{t_0, x, \alpha} = -\alpha(t, X_t^{t_0, x, \alpha}) dt + \sigma dW_t \\ J(t_0, x, \alpha) = \sup_{\alpha \in \mathcal{A}_{t_0, T}} \mathbb{E}\Big[g(X_T^{t_0, x, \alpha}) - \int_{t_0}^T \Big[(\alpha(t, X_t^{t_0, x, \alpha}) - D_t)^2 + h(X_t^{t_0, x, \alpha}) \Big] dt \Big]. \end{cases}$$

Bound constraints on the storage level are implicitly forced by the penalization h. A target terminal level is indicated by the terminal gain g, supposed here to be Lebesgue integrable. The objective is then to follow a deterministic target profile $(D_t)_{t \in [0, T]}$, on a given finite horizon $[t_0, T]$. When the admissible set in which the controls take their values $A = \mathbb{R}$, one can explicitly derive the optimal control as a function of the value function derivative

$$\alpha^*(t, \cdot) = D_t + \frac{1}{2}(\partial_x v)(t, \cdot),$$

which yields the following HJB equation

$$\partial_t v + \frac{1}{4}(\partial_x v)^2 + D_t \partial_x v + \frac{\sigma^2}{2}\partial_{xx} v - h = 0.$$

Reversing the time, (with $t_0 = 0$) gives $(t, x) \mapsto u(t, x) := v(T - t, x)$ solution of

$$\begin{cases} \partial_t u = \frac{1}{4}(\partial_x u)^2 + \frac{\sigma^2}{2}\partial_{xx} u + D_t \partial_x u - h, \\ u(0, x) = g(x). \end{cases} \quad (7.9)$$

We recover the framework of (1.1), with $\Lambda(t, x, y, z) = \frac{1}{4}\frac{|z|^2}{y} - \frac{h(x)}{y}$ and $b(t, x, y) = -D_t$. Consequently the Bellman function v can be represented via

$$\begin{cases} Y_t = Y_0 + \sigma W_t - \int_0^t D_s ds \\ Y_0 \sim \frac{g(x)dx}{\int_{\mathbb{R}} g(y)dy} \\ \int \varphi(x)v(t, x)dx = \left(\int_{\mathbb{R}} g(y)dy\right) \mathbb{E}\left[\varphi(Y_{T-t}) \exp\left\{\int_t^T \Lambda(s, Y_{T-s}, v(s, Y_{T-s}), \nabla v(s, Y_{T-s}))ds\right\}\right], \\ \text{for } t \in [0, T]. \end{cases}$$

$$(7.10)$$

Acknowledgements The authors are grateful to the Referee for her/his suggestions. The work was supported by a public grant as part of the *Investissement d'avenir project*, reference ANR-11-LABX-0056-LMH, LabEx LMH, in a joint call with Gaspard Monge Program for optimization, operations research and their interactions with data sciences.

References

1. Albeverio, S., De Vecchi, F.C., Romano, A., Ugolini, S.: Mean-field limit for a class of stochastic ergodic control problems (2020). arXiv:2003.06469. To appear in J. Control and Optimization (2022)
2. Albeverio, S., Ugolini, S.: A Doob h-transform of the Gross-Pitaevskii Hamiltonian. J. Stat. Phys. **161**(2), 486–508 (2015)
3. Aronson, D.G., Weinberger, H.F.: Multidimensional nonlinear diffusion arising in population genetics. Adv. Math. **30**, 33–76 (1978)
4. Bagtzoglou, A.C., Atmadja, J.: Marching-jury backward beam equation and quasi-reversibility methods for hydrologic inversion: application to contaminant plume spatial distribution recovery. Water Resour. Res. **39**(2) (2003)
5. Bak, P.: How nature works. Copernicus, New York (1996). The science of self-organized criticality
6. Bak, P., Tang, C., Wiesenfeld, K.: Self-organized criticality. Phys. Rev. A (3) **38**(1), 364–374 (1988)

7. Bally, V., Pagès, G., Printems, J.: A quantization tree method for pricing and hedging multidimensional American options. Math. Finance **15**(1), 119–168 (2005)
8. Barbu, V., Röckner, M.: Probabilistic representation for solutions to nonlinear Fokker-Planck equations. SIAM J. Math. Anal. **50**(4), 4246–4260 (2018)
9. Barbu, V., Röckner, M., Russo, F.: Probabilistic representation for solutions of an irregular porous media type equation: the irregular degenerate case. Probab. Theory Relat. Fields **151**(1–2), 1–43 (2011)
10. Barbu, V., Röckner, M., Russo, F.: Stochastic porous media equations in $\mathbb{R}^d$. J. Math. Pures Appl. (9) **103**(4), 1024–1052 (2015)
11. Barbu, V., Röckner, M., Russo, F.: Doubly probabilistic representation for the stochastic porous media type equation. Ann. Inst. Henri Poincaré Probab. Stat. **53**(4), 2043–2073 (2017)
12. Barles, G., Souganidis, P.E.: Convergence of approximation schemes for fully nonlinear second order equations. Asymptot. Anal. **4**(3), 271–283 (1991)
13. Beck, J.V., Blackwell, B., St. Clair Jr. Ch.R.: Inverse heat conduction: Ill-posed problems. James Beck (1985)
14. Belaribi, N., Cuvelier, F., Russo, F.: Probabilistic and deterministic algorithms for space multidimensional irregular porous media equation. SPDEs: Anal. Comput. **1**(1), 3–62 (2013)
15. Belaribi, N., Russo, F.: Uniqueness for Fokker-Planck equations with measurable coefficients and applications to the fast diffusion equation. Electron. J. Probab. **17**(84), 28 (2012)
16. Benachour, S., Chassaing, P., Roynette, B., Vallois, P.: Processus associés à l'équation des milieux poreux. Ann. Scuola Norm. Sup. Pisa Cl. Sci. (4) **23**(4), 793–832 (1996)
17. Bertsekas, D.P., Shreve, S.E.: Stochastic optimal control, vol. 139. Mathematics in Science and Engineering. Academic Press, Inc. [Harcourt Brace Jovanovich, Publishers], New York-London (1978). The discrete time case
18. Blanchard, P., Röckner, M., Russo, F.: Probabilistic representation for solutions of an irregular porous media type equation. Ann. Probab. **38**(5), 1870–1900 (2010)
19. Bonnans, J.F., Zidani, H.: Consistency of generalized finite difference schemes for the stochastic HJB equation. SIAM J. Numer. Anal. **41**(3), 1008–1021 (2003)
20. Bossy, M., Talay, D.: Convergence rate for the approximation of the limit law of weakly interacting particles: application to the Burgers equation. Ann. Appl. Probab. **6**(3), 818–861 (1996)
21. Bouchard, B., Touzi, N.: Discrete-time approximation and Monte Carlo simulation of backward stochastic differential equations. Stoch. Process. Appl. **111**, 175–206 (2004)
22. Bouchard, B., Warin, X.: Monte-Carlo valorisation of American options: facts and new algorithms to improve existing methods. In: Numerical Methods in Finance. Springer (2012)
23. Broadie, M., Glasserman, P., et al.: A stochastic mesh method for pricing high-dimensional American options. J. Comput. Finance **7**, 35–72 (2004)
24. Cafiero, R., Loreto, V., Pietronero, L., Vespignani, A., Zapperi, S.: Local rigidity and self-organized criticality for avalanches. Europhys. Lett. **29**, 111–116 (1995)
25. Del Moral, P.: Feynman-Kac formulae. Probability and its Applications (New York). Springer, New York (2004). Genealogical and interacting particle systems with applications
26. Del Moral, P., Hu, P., Oudjane, N., Rémillard, B.: On the robustness of the Snell envelope. SIAM J. Financ. Math. **2**(1), 587–626 (2011)
27. Del Moral, P., Jacob, P.E., Lee, A., Murray, L., Peters, G.W.: Feynman-Kac particle integration with geometric interacting jumps. Stoch. Anal. Appl. **31**(5), 830–871 (2013)
28. Del Moral, P., Miclo, L.: Branching and interacting particle systems approximations of Feynman-Kac formulae with applications to non-linear filtering. In: Séminaire de Probabilités, XXXIV. Lecture Notes in Math, vol. 1729, pp. 1–145. Springer, Berlin (2000)
29. Delarue, F., Menozzi, S.: An interpolated stochastic algorithm for quasi-linear PDEs. Math. Comp. **77**(261), 125–158 (electronic) (2008)
30. Hutzenthaler, W.E.M., Jentzen, A., Kruse, Th.: On multilevel Picard numerical approximations for high-dimensional nonlinear parabolic partial differential equations and high-dimensional nonlinear backward stochastic differential equations. J. Sci. Comput. **79**(3), 1534–1571 (2019)

31. Fabbri, G., Gozzi, F., Święch, A.: Stochastic optimal control in infinite dimension. In: Probability Theory and Stochastic Modelling, vol. 82. Springer, Cham (2017). Dynamic programming and HJB equations, With a contribution by Marco Fuhrman and Gianmario Tessitore
32. Fahim, A., Touzi, N., Warin, X.: A probabilistic numerical method for fully nonlinear parabolic PDEs. Ann. Appl. Probab. **21**(4), 1322–1364 (2011)
33. Forsyth, P.A., Vetzal, K.R.: Numerical methods for nonlinear pdes in finance. In: Handbook of Computational Finance, pp. 503–528. Springer (2012)
34. Gobet, E., Lemor, J.-P., Warin, X.: A regression-based Monte Carlo method to solve backward stochastic differential equations. Ann. Appl. Probab. **15**(3), 2172–2202 (2005)
35. Haussmann, U.G., Pardoux, É.: Time reversal of diffusions. Ann. Probab. **14**(4), 1188–1205 (1986)
36. Henry-Labordère, P., Oudjane, N., Tan, X., Touzi, N., Warin, X.: Branching diffusion representation of semilinear pdes and Monte Carlo approximations (2016). http://arxiv.org/pdf/1603.01727v1.pdf
37. Huré, C., Pham, H., Warin, X.: Deep backward schemes for high-dimensional nonlinear PDEs. Math. Comp. **89**(324), 1547–1579 (2020)
38. Izydorczyk, L., Oudjane, N., Russo, F., Tessitore, M.: Fokker-Planck equations with terminal condition and related McKean probabilistic representation (2020). Preprint hal-02902615
39. Jabir, J.-F., Talay, D., Tomašević, M.: Mean-field limit of a particle approximation of the one-dimensional parabolic-parabolic Keller-Segel model without smoothing. Electron. Commun. Probab. **23**(Paper No. 84), 14 (2018)
40. Jourdain, B., Méléard, S.: Propagation of chaos and fluctuations for a moderate model with smooth initial data. Ann. Inst. H. Poincaré Probab. Statist. **34**(6), 727–766 (1998)
41. Jourdain, B., Méléard, S., Woyczynski, W.A.: Nonlinear SDEs driven by Lévy processes and related PDEs. ALEA Lat. Am. J. Probab. Math. Stat. **4**, 1–29 (2008)
42. Kac, M.: Probability and related topics in physical sciences. In: With special lectures by G.E. Uhlenbeck, A.R. Hibbs, and B. van der Pol. Lectures in Applied Mathematics. Proceedings of the Summer Seminar, Boulder, Colo, vol. 1957. Interscience Publishers, London-New York (1959)
43. Keener, J.P., Sneyd, J.: Mathematical Physiology II: Systems Physiology. Springer, New York (2008)
44. Kushner, H.J., Dupuis, P.G.: Numerical methods for stochastic control problems in continuous time. Applications of Mathematics (New York), vol. 24. Springer, New York (1992)
45. Lattès, R., Lions, J.-L.: The method of quasi-reversibility. Applications to partial differential equations. Translated from the French edition and edited by Richard Bellman. Modern Analytic and Computational Methods in Science and Mathematics, No. 18. American Elsevier Publishing Co., Inc., New York (1969)
46. Le Cavil, A., Oudjane, N., Russo, F.: Particle system algorithm and chaos propagation related to a non-conservative McKean type stochastic differential equations. In: Stochastics and Partial Differential Equations: Analysis and Computation, pp. 1–37 (2016)
47. Le Cavil, A., Oudjane, N., Russo, F.: Probabilistic representation of a class of non-conservative nonlinear partial differential equations. ALEA Lat. Am. J. Probab. Math. Stat. **13**(2), 1189–1233 (2016)
48. Le Cavil, A., Oudjane, N., Russo, F.: Monte-Carlo algorithms for a forward Feynman-Kac-type representation for semilinear nonconservative partial differential equations. Monte Carlo Methods Appl. **24**(1), 55–70 (2018)
49. Le Cavil, A., Oudjane, N., Russo, F.: Forward Feynman-Kac type representation for semilinear nonconservative partial differential equations. In: Stochastics: an International Journal of Probability and Stochastic Processes, to appear. First version 2016, Preprint hal-01353757
50. Lieber, J., Oudjane, N., Russo, F.: On the well-posedness of a class of McKean Feynman-Kac equations. Markov Processes and Related Fields. To appear. arXiv:1810.10205
51. Longstaff, F.A., Schwartz, E.S.: Valuing American options by simulation: a simple least-squares approach. Rev. Fin. Stud. **14**, 113–147 (2001)

52. Jr. McKean, H.P.: A class of Markov processes associated with nonlinear parabolic equations. In: Proc. Nat. Acad. Sci. U.S.A., 1966), pp. 1907–1911 (1966)
53. Jr. McKean, H.P.: Propagation of chaos for a class of non-linear parabolic equations. In: Stochastic Differential Equations (Lecture Series in Differential Equations, Session 7, Catholic Univ., 1967), pp. 41–57. Air Force Office Sci. Res., Arlington, Va. (1967)
54. Jr. McKean, H.P.: Application of Brownian motion to the equation of Kolmogorov-Petrovskii-Piskunov. Comm. Pure Appl. Math. **28**(3), 323–331 (1975)
55. Méléard, S., Roelly-Coppoletta, S.: A propagation of chaos result for a system of particles with moderate interaction. Stoch. Process. Appl. **26**(2), 317–332 (1987)
56. Murray, J.D.: Mathematical biology. I, 3rd edn. Interdisciplinary Applied Mathematicsvol. 17. Springer, New York (2002). An introduction
57. Oelschläger, K.: A martingale approach to the law of large numbers for weakly interacting stochastic processes. Ann. Probab. **12**(2), 458–479 (1984)
58. Pagès, G., Sagna, A.: Improved error bounds for quantization based numerical schemes for BSDE and nonlinear filtering. Stoch. Process. Appl. **128**(3), 847–883 (2018)
59. Pardoux, É., Peng, S.: Backward stochastic differential equations and quasilinear parabolic partial differential equations. In: Stochastic Partial Differential Equations and Their Applications (Charlotte, NC, 1991). Lecture Notes in Control and Information Sciences, vol. 176 , pp. 200–217. Springer, Berlin (1992)
60. Pham, H.: Continuous-time stochastic control and optimization with financial applications. Stochastic Modelling and Applied Probability, vol. 61. Springer, Berlin (2009)
61. Renardy, M., Hrusa, W.J., Nohel, J.A.: Mathematical problems in viscoelasticity. In: Pitman Monographs and Surveys in Pure and Applied Mathematics, vol. 35. Longman Scientific & Technical, Harlow; Wiley, Inc., New York (1987)
62. Röckner, M., Russo, F.: Uniqueness for a class of stochastic Fokker-Planck and porous media equations. J. Evol. Equ. **17**(3), 1049–1062 (2017)
63. Stroock, D.W., Varadhan, S.R.S.: Multidimensional diffusion processes. Classics in Mathematics. Springer, Berlin (2006). Reprint of the 1997 edition
64. Sznitman, A.-S.: Topics in propagation of chaos. In: École d'Été de Probabilités de Saint-Flour XIX—1989. Lecture Notes in Math, , vol. 1464, pp. 165–251. Springer, Berlin (1991)
65. Tan, X.: Probabilistic numerical approximation for stochastic control problems. preprint (2012)
66. Tikhonov, A.N., Arsenin, V.Y.: Solutions of ill-posed problems. V. H. Winston & Sons, Washington, D.C.: Wiley, New York-Toronto, Ont.-London (1977). Translated from the Russian, Preface by translation editor Fritz John, Scripta Series in Mathematics
67. Touzi, N.: Optimal stochastic control, stochastic target problems, and backward SDE. Fields Institute Monographs, vol. 29. Springer, New York; Fields Institute for Research in Mathematical Sciences, Toronto, ON, (2013). With Chapter 13 by Agnès Tourin
68. Tsitsiklis, J.N., Van Roy, B.: Regression methods for pricing complex American-style options. IEEE Trans. Neural Netw. **12**(4), 694–703 (2001)
69. Wang, X.Y., Zhu, Z.S., Lu, Y.K.: Solitary wave solutions of the generalized Burgers-Huxley equation. J. Phys. A: Math. Gen. **23**, 271–274 (1990)

Bernstein Processes, Isovectors and Mechanics

Paul Lescot and Laurène Valade

Abstract We investigate the symmetries of a class of diffusions processes ("Bernstein's reciprocal" processes) introduced in the eighties for the solution of a problem stated by Schrödinger in 1931. Those diffusions satisfy two unusual properties. Although typically not time-homogeneous, they are time reversible. Also their infinitesimal coefficients are specific functions of positive solutions of time adjoint parabolic equations. The symmetries of these PDEs will therefore be transformed into symmetries of the diffusions and provide relations between them hard to guess otherwise. We shall use an algebraico-geometric method ("of isovectors") and mention applications in finance and mathematical physics. As can be expected Schrödinger's initial motivation was quantum mechanics.

Keywords Isovectors · Euclidian Quantum Mechanics · Bernstein Processes · Finance

1 Introduction

Since, according to K. Itô, his stochastic calculus should be regarded as a deformation of Leibniz-Newton calculus along Brownian-like trajectories, it is a natural hope that the study of the symmetries of stochastic differential equations should have the same impact as its deterministic counterpart (essentially due to S. Lie) for ODEs. A number of works in this direction have been published along the years. Recently we can mention, in particular, [1, 4] or [5].

The option chosen here is different and follows directly from the above Schrödinger's idea. Instead of look for a general theory valid for all kinds of stochastic differential equations, we focus on those which are solutions of Schrödinger's problem. What we shall lose in generality will be compensated by their strong analogy with quantum mechanics which was precisely at the origin of his idea. Or,

P. Lescot (✉) · L. Valade
University of Rouen, Mont-Saint-Aignan, France
e-mail: paul.lescot@univ-rouen.fr

equivalently, with the benefit of knowing the future of the early thirties, we can be inspired by Feynman's path integrals approach to quantum mechanics, involving purely symbolic diffusion processes. Indeed Bernstein's reciprocal processes share a lot of qualitative properties with those "diffusions". This perspective is described in detail in the review [22].

We shall start with a short summary of Schrödinger's problem and its solutions for the class of elementary mechanical systems (originally this author considered only the one dimensional free case). This will be sufficient to understand the origin of the specific class of processes and our transfer from symmetries of underlying PDEs to the ones of associated diffusions. In particular, we shall limit ourselves to one type of those PDEs, backward parabolic equations (well defined for final boundary conditions). They correspond, in our framework, to the usual description of diffusions in term of increasing filtrations. The adjoint description can be obtained without further computations.

Then we shall describe the method of isovectors. Inspired by Cartan's approach to partial differential equations in term of differential forms and their exterior differential calculus, it has been formulated by Harrison and Estabrook as a tool for the study of the invariance group of these equations [7]. This method is particularly suitable to analyse the symmetries of the class of Bernstein's reciprocal processes.

A number of examples will be given, motivated by Stochastic finance or the analogy with Quantum Physics which was the initial motivation of Schrödinger.

2 Summary

- A summary of Schrödinger's problem and its solution
- The method of isovectors
- Bernstein processes
- Parametrization of a one-factor affine model
- Generalized Brownian Bridges
- References

3 A Summary of Schrödinger's Problem and Its Solution

Consider a real valued diffusion process $Z(t)$ defined on a finite time interval, say $[-\frac{T}{2}, \frac{T}{2}]$. Let's suppose that, in addition to the data of its initial probability $\mu_{-\frac{T}{2}} = \delta_{-\frac{T}{2}}(dx)$ we are given also a final one $\mu_{-\frac{T}{2}} = \delta_{-\frac{T}{2}}(dz)$ distinct from the probability associated with the evolution of $\delta_{-\frac{T}{2}}$ under the diffusion equation. Such evolutions may be rare but they are possible and result manifestly from a conditioning. Schrödinger's problem is : what is the most probable evolution between those two probabilities? [22]. The answer was given by the same author: the optimal process Z is such that $P(Z(t) \in A) = \int_A \eta^*(t, q) \eta(t, q) dq$, for A a Borelian, where η^* and η

are respectively positive solutions of two adjoint parabolic partial differential equations:

$$\begin{cases} \frac{\partial \eta^*}{\partial t} = H_0 \eta^* \\ \eta^*(\frac{-T}{2}, x) = \eta^*_{\frac{-T}{2}}(x) \end{cases}$$

and

$$\begin{cases} -\frac{\partial \eta}{\partial t} = H_0 \eta \\ \eta(\frac{-T}{2}, z) = \eta_{\frac{-T}{2}}(z) \end{cases}$$

and $H_0 = \frac{-\kappa}{2}\Delta$, for κ a positive constant and $\eta^*_{\frac{-T}{2}}, \eta_{\frac{-T}{2}}$ are two unspecified positive functions.

A corresponding pair of variational principles inspired by stochastic Optimal Control theory provides the drift of the SDE solved by the optimal process: $dZ(t) = \kappa \frac{\nabla \eta}{\eta}(t, Z(t))dt + \kappa^{\frac{1}{2}} dW(t)$ for $W(t)$ a Wiener process and an initial distribution $\delta_{\frac{-T}{2}}$ and an associated filtration $\mathcal{P}_t$ increasing on the time interval. The stochastic variational principle corresponding to the adjoint PDE provides a backward SDE whose increment $d_*Z(t)$ points towards the past of $Z(t)$: $d_*Z(t) = -\kappa \frac{\nabla \eta_*}{\eta_*}(t, Z(t))dt + \kappa^{\frac{1}{2}} d_* W_*(t)$ for an associated decreasing filtration $\mathcal{F}_t$ from a final data $\delta_{\frac{T}{2}}$. W_* is defined in complete analogy with $W(t)$ and $\mathcal{P}_t$. For the purpose of the symmetry analysis it is sufficient to consider the more familiar forward SDE as no new computaions are needed.

Bernstein measures are constructed from the joint probability $M(dq, dZ)$ at the boundaries of the time interval $[\frac{-T}{2}, \frac{T}{2}]$. In the special Markovian case considered by Schrödinger, it is of the form $M(dq, dZ) = u^*_{\frac{-T}{2}}(q)h(q, T, Z)u_{\frac{T}{2}}(Z)dqdZ$ where $h(q, T, Z)$ is the integral kernel $e^{\frac{-1}{\kappa}TH}(q, Z)$, and we have allowed a more general Hamiltonian operator $H = H_0 + V(Z)$ for a given bounded below scalar potential V.

Then the two marginals of M, for given boundary probabilities $\delta_{\frac{-T}{2}}$ and $\delta_{\frac{T}{2}}$ provide a nonlinear integral system of equations for $u^*_{\frac{-T}{2}}$ and $u_{\frac{-T}{2}}$. A. Beurling proved in 1960 the existence and uniqueness of its positive solutions [2] under general conditions. In other words the Markovian Bernstein diffusions are well defined ([22] and references therein for more details).

In recent years Schrödinger's problem and its solution have been re-interpreted in a measure theoretic perspective and in the context of Mass Transportation theory ([11] and references therein). In particular, the reciprocal Bernstein diffusions solving Schrödinger's problem can, curiously, be regarded as regularization of the Monge-Kantorovich problem solved ten years after Schrödinger by Kantorovich.

Schrödinger's problem is at the origin of the program of Stochastic Deformation [22]. Its aim is the elaboration of the randomized mechanics whose existence was suggested, in particular by Bernstein and whose Feynman's Path integral theory can be regarded as a (very) informal description.

4 The Method of Isovectors

The method of isovectors was introduced in [4] in order to classify up to equivalence (systems of) partial differential equations appearing in mathematical physics.

Given a system of partial differential equations, one expresses it, adding if necessary some of the derivatives of the unknown function as auxiliary unknowns, as the vanishing of a family of first-order differential forms. An isovector is then defined as a vector field in all the variables preserving the differential ideal generated by the forms.

For the one–dimensional heat equation, it was determined (using a different language) by Bluman and Cole ([3]).

Olver's *prolongation method* ([19]) provides a somewhat different approach.

Let us now give some details. We shall consider an equation of the shape

$$\frac{\partial u}{\partial t} = G(t, q, u, \frac{\partial u}{\partial q}, ..., \frac{\partial^{n-1} u}{\partial q^{n-1}}) + \lambda \frac{\partial^n u}{\partial q^n}$$

for $n \geq 2$, $t \in J$ (an interval of $\mathbf{R}$), $q \in O$ (an open set in $\mathbf{R}^n$), u a function of t and q, λ a constant and $G : J \times O \times \mathbf{R}^n \to \mathbf{R}$ a C^∞ function of t, q u and its derivatives in space.

In order to study the symmetries of the equation, we shall temporarily consider $u, \frac{\partial u}{\partial t}, \frac{\partial u}{\partial q}, ..., \frac{\partial^{n-1} u}{\partial q^{n-1}}$ as *independent* variables.

We now consider a differential equation ($\mathcal{E}$) of order $n = 2$.

We note $u(t, q)$ a solution of this equation.

We define $B_1 = \frac{\partial u}{\partial q}$.

We consider u, $A = \frac{\partial u}{\partial t}$ and B_1 as independent variables.

Let $\alpha = du - A dt - \sum_{i=1}^{n} B_1^i dq_i$, a 1-differential form.

So we have $d\alpha = -dA dt - dB_1 dq$.

The equation ($\mathcal{E}$) is equivalent to the vanishing of the forms α, $d\alpha$ and of a certain number of differential forms of degrees at least 2 noted δ_j, $d\delta_j$ où $j \in \mathbb{N}$.

We pose $I = <\alpha, d\alpha, \delta_j, d\delta_j>$. It is a closed differential ideal of the algebra of the differential forms.

Lemma 4.1 *Each 2-form of I can be written $a\alpha + z d\alpha + \xi$ with z a 0-form, $\xi \in < \delta_j, d\delta_j >$ for $j \in \mathbb{N}$ and a a 1-form of the type $a_1 dt + a_2 dq + a_3 du + a_4 dA + a_5 dB_1$ for which we can suppose $a_3 = 0$.*

Proof If $a_3 \neq 0$, we replace $a\alpha$ by $a'\alpha$ where

$$a' = (a_1 + A a_3) dt + (a_2 + B_1 a_3) dq + a_4 d A a_5 d B_1$$

We note

$$N = N^t \frac{\partial}{\partial t} + N^q \frac{\partial}{\partial q} + N^u \frac{\partial}{\partial u} + N^A \frac{\partial}{\partial A} + N^{B_1} \frac{\partial}{\partial B_1}$$

Definition 4.2 We say that N is an isovector of $(\mathcal{E})$ if $\mathcal{L}_N(I) \subset I$.

To compute these isovectors, we use the method developed by Harrison Estabrook in [7] and already set up to find the symmetries of many equations for instance in ([12–17, 20, 21]).

We denote by $\mathcal{G}$ the set of isovectors.

Lemma 4.3 $\mathcal{G}$ *is a Lie algebra for the usual bracket of vector fields.*

Lemma 4.4 *Let $N \in \mathcal{G}$. There is a function $F = F_N(t, q, u, A, B_1)$ such that*

$$N^t = -F_A$$
$$N^q = -F_{B_1}$$
$$N^u = F + AN^t + B_1 N^q$$
$$N^A = AF_u + F_t$$
$$N^{B_1} = B_1 F_u + F_q$$

Remark 4.5 This lemma is present in ([7, p. 657]).

Proposition 4.6 *N^t depends only of t. We note $N^t = -f(t)$.*

For an equation of the following form:

$$(\mathcal{E}_{g,2}) : Q(t, q, u) \frac{\partial^2 u}{\partial q^2} + T(t, q) \frac{\partial u}{\partial t} + R(t, q, u, B_1) = 0$$

with $Q \neq 0$, we have complementary results:

Theorem 4.7 • *Under the previous hypothesis, considering $(\mathcal{E}_{g,2})$ we have:*

$$N^q = N^q(t, q)$$

We note $N^q = -w(t, q)$.
• *If we suppose also that $Q = Q(t, q)$ and R is linear in B_1 then N^u is affine in u. We pose $N^u = um(t, q) + l(t, q)$ where l and m are functions of t and q. We have also relations between the various auxiliary functions introduced:*
$-QT_t f + TQ_t f + TQf_t - QT_q w - 2TQw_q + TQ_q w = 0$ and

$QTB_1 w_t + QTum_t + QTl_t - QR_t f - QR_q w + QR_u um + QR_u l + QR_{B_1} B_1 m + QR_{B_1} B_1 w_q$
$+ QR_{B_1} um_q + QR_{B_1} l_q - QRw_q + Q^2 B_1 m_q + Q^2 B_1 w_{qq} + Q^2 um_{qq} + Q^2 l_{qq} + Q^2 B_1 m_q$
$= -RQ_t f - RQ_q w + QR(m + w_q)$

Proof The proof of this theorem will be detailed in the second author's Ph.D. thesis [21].

4.1 Examples

4.1.1 Black Scholes

We can apply the previous results to find the symmetries of the Black Scholes equation. It is the most famous equation in financial mathematics:

$$\frac{\partial C}{\partial t} + \frac{1}{2}\sigma^2 S^2 \frac{\partial^2 C}{\partial S^2} + rS\frac{\partial C}{\partial S} - rC = 0$$

where $C(t, S)$ denotes the price of a call option with maturity T and strike price K on an underlying asset satisfying $S_t = S$, σ is the volatility and r the risk free interest rate.

We pose $q = \ln(S)$ and $u(t, q) = C(t, \exp(q)) = C(t, S)$. The equation becomes:

$$\frac{\partial u}{\partial t} + \frac{1}{2}\sigma^2 \frac{\partial^2 u}{\partial q^2} + \left(r - \frac{\sigma^2}{2}\right)\frac{\partial u}{\partial q} - ru = 0$$

According to our notations we have $T = 1$, $Q = \frac{1}{2}\sigma^2$ and $R = (r - \frac{\sigma^2}{2})B_1 - ru$.
We note $\tilde{r} = r - \frac{\sigma^2}{2}$ and $\tilde{s} = r + \frac{\sigma^2}{2}$.
By Theorem 3.7 we obtain:

- $\frac{\sigma^2}{2} f_t - 2\frac{\sigma^2}{2} w_q = 0$ so
$$f_t = 2w_q$$

and
$$w_{qq} = 0$$

- We divide by Q to find:

$$B_1 w_t + um_t + l_t - r(um + l) + \left(r - \frac{\sigma^2}{2}\right)(B_1 m + B_1 w_q + um_q + l_q) - \left(r - \frac{\sigma^2}{2}\right) B_1 w_q + ruw_q$$
$$+ \frac{\sigma^2}{2}(B_1 m_q + um_{qq} + l_{qq} + B_1 m_q) = \left(r - \frac{\sigma^2}{2}\right) B_1(m + w_q) - ru(m + w_q)$$

Differentiating with respect to B_1 we obtain:

$$w_t = \left(r - \frac{\sigma^2}{2}\right)w_q - \sigma^2 m_q$$

As $w_{qq} = 0$, $w_{tqq} = \sigma^2 m_{qqq} = 0$ so

$$m_{qqq} = 0.$$

Differentiating with respect to u we have:

$$m_t = -\left(r - \frac{\sigma^2}{2}\right)m_q - 2rw_q - \frac{\sigma^2}{2}m_{qq}$$

In addition we have $m_{tqq} = 0$, so

$$m_{qq} = constant.$$

Taking $u = B_1 = 0$, we find that l is a solution of the equation.
Working with the different equations obtained above, we have the following relations:
$$f_{tt} = 2w_{qt} = -2\sigma^2 m_{qq} = 2C_1$$

$$f = C_1 t^2 + C_2 t + C_3$$

$$w_q = C_1 t + \frac{1}{2}C_2$$

So
$$m_q = -\frac{C_1}{\sigma^2}q + v(t)$$

$$m_{tt} = -\left(r - \frac{\sigma^2}{2}\right)v_t - 2C_1 r$$

Hence
$$m_{ttq} = 0$$

But by the expression of m_q we deduce that $v_t = constant$.
The expression of m_t gives us $m_{tq} = \left(r - \frac{\sigma^2}{2}\right)\frac{C_1}{\sigma^2}$, so

$$v_t = \left(r - \frac{\sigma^2}{2}\right)\frac{C_1}{\sigma^2}$$

$w_{tt} = \frac{1}{2}\left(r - \frac{\sigma^2}{2}\right) f_{tt} + \sigma^2 m_{tq} = \frac{1}{2}\left(r - \frac{\sigma^2}{2}\right)2C_1 - \sigma^2\left(r - \frac{\sigma^2}{2}\right)\left(\frac{C_1}{\sigma^2}\right)$ by the expression of m_t.
So
$$w_{tt} = 0$$

As well, $w_t = \frac{1}{2}(r - \frac{\sigma^2}{2})C_2 + (r - \frac{\sigma^2}{2})C_1 t + \sigma^2(\frac{C_1}{\sigma^2}q) - \sigma^2 v(t) = C_4$
Hence

$$v = \frac{(r-\frac{\sigma^2}{2})C_1}{\sigma^2}t + \frac{(r-\frac{\sigma^2}{2})}{2\sigma^2}C_2 - \frac{C_4}{\sigma^2}$$

So we have :

$$f = C_1 t^2 + C_2 t + C_3$$

$$w = C_1 tq + \frac{1}{2}C_2 q + C_4 t - C_5$$

$$m = \frac{\tilde{r}C_1}{\sigma^2}tq - \frac{C_1}{2\sigma^2}q^2 + \left(\frac{\tilde{r}C_2}{2\sigma^2} - \frac{C_4}{\sigma^2}\right)q - \frac{\tilde{s}^2 C_1}{2\sigma^2}t^2 + \left(-\frac{\tilde{s}^2 C_2}{2\sigma^2} + \frac{r^2 C_4}{\sigma^2} + \frac{C_1}{2}\right)t + C_6$$

Finally we find a basis $(N_i)_{1 \leq i \leq 6}$ of 6 isovectors, for which

$$\tilde{N}_1 = -t^2 \frac{\partial}{\partial t} - tq\frac{\partial}{\partial q} + \frac{\tilde{r}}{\sigma^2}tqu\frac{\partial}{\partial u} - \frac{1}{2\sigma^2}q^2 u\frac{\partial}{\partial u} - \frac{\tilde{s}^2}{2\sigma^2}ut^2\frac{\partial}{\partial u} + \frac{1}{2}ut\frac{\partial}{\partial u}$$

$$\tilde{N}_2 = -t\frac{\partial}{\partial t} + \frac{1}{2}q\frac{\partial}{\partial q} + \frac{\tilde{r}}{2\sigma^2}qu\frac{\partial}{\partial u} - \frac{\tilde{s}^2}{2\sigma^2}tu\frac{\partial}{\partial u}$$

$$\tilde{N}_3 = -\frac{\partial}{\partial t}$$

$$\tilde{N}_4 = -t\frac{\partial}{\partial q} - \frac{1}{\sigma^2}qu\frac{\partial}{\partial u} + \frac{r^2}{\sigma^2}tu\frac{\partial}{\partial u}$$

$$\tilde{N}_5 = \frac{\partial}{\partial q}$$

$$\tilde{N}_6 = u\frac{\partial}{\partial u}$$

4.1.2 Backward Heat Equation with Potential V

As observed before, we can limit ourselves to this equation, well defined for a final boundary condition, in the context of Schrödinger's problem.

The theorem can also be applied to determine the isovectors and so the symmetries of the backward heat equation with a potential $V(t, q)$. Although we assumed in the first paragraph that the potential V is time independant, the symmetry analysis does not require this hypothesis.

This equation is very important in the theory of Bernstein process, especially concerning their applications to the Euclidian Quantum Mechanics.

It is also present in the theory of affine interest rate models.

The results are already obtained by P. Lescot, H. Quintard and J.C. Zambrini for the equivalent Hamilton-Jacobi-Bellman equation ([14, 16, 17]).

We consider the equation:

$$(\mathcal{E}_\kappa^V) : \kappa \frac{\partial u}{\partial t} + \frac{\kappa^2}{2} \frac{\partial^2 u}{\partial q^2} - V(t,q)u = 0$$

Applying Theorem 3.7 we obtain:

$$f_t = 2w_q$$

So we can pose

$$w = \frac{1}{2} f_t q + a(t).$$

The second equation is:

$$\kappa(B_1 w_t + u m_t + l_t) + V_t u f + V_q u w - V u m - V l + V u w_q$$
$$+ \frac{\kappa^2}{2}(B_1 m_q + u m_{qq} + l_{qq} + B_1 m_q) + V u(m + w_q) = 0$$

For $u = B_1 = 0$, l is a solution of the equation.
Differentiating according to B_1 we have

$$w_t = -\kappa m_q$$

So we can pose

$$m = \frac{-1}{4K} f_{tt} q^2 - \frac{1}{K} a_t q + b(t).$$

Differentiating according to u we obtain

$$m_t = -\frac{1}{\kappa}(V_t f + V_q w + 2V w_q + \frac{\kappa^2}{2} m_{qq})$$

That is

$$\frac{-1}{4\kappa} f_{ttt} q^2 - \frac{1}{\kappa} a_{tt} q + b_t = \frac{-1}{\kappa}(V_t f + \frac{1}{2} V_q f_t q + V_q a + V f_t - \frac{\kappa}{4} f_{tt})$$

Particular Case $V = 0$
We obtain

$$\frac{-1}{4} f_{ttt} q^2 - a_{tt} q + b_t = \frac{\kappa}{4} f_{tt}$$

Differentiating twice with respect to q we have $f_{ttt} = 0$.
So
$$f = 2C_1 t^2 + C_2 t + C_3.$$

Differentiating with respect to q we have $a_{tt} = 0$.
Hence
$$w = 2C_1 t q + \frac{C_2}{2} q + C_4 t + C_5$$

Finally $b_t = \kappa C_1$ and
$$m = -\frac{C_1}{\kappa} q^2 - \frac{C_4}{\kappa} q + \kappa C_1 t + C_6$$

Hence we have a basis of isovectors for which :

$$\tilde{N}_1 = -2t^2 \frac{\partial}{\partial t} - 2tq \frac{\partial}{\partial q} + \kappa t u \frac{\partial}{\partial u}$$

$$\tilde{N}_2 = -t \frac{\partial}{\partial t} - \frac{1}{2} q \frac{\partial}{\partial q}$$

$$\tilde{N}_3 = \frac{\partial}{\partial t}$$

$$\tilde{N}_4 = -t \frac{\partial}{\partial q} - \frac{1}{\kappa} q u \frac{\partial}{\partial u}$$

$$\tilde{N}_5 = \frac{\partial}{\partial q}$$

$$\tilde{N}_6 = u \frac{\partial}{\partial u}$$

We find a basis of isovectors similar to that of ([16, 19]).

Particular Case $V(t, q) = \frac{C}{q^2} + D q^2$:
For this potential we obtain:

$$\frac{1}{4} f_{ttt} q^2 + a_{tt} q - \kappa b_t - 2D f_t q^2 + \frac{2Ca}{q^3} - 2Daq + \frac{\kappa}{4} f_{tt} = 0$$

Because f, a and b are independent of q, after multiplying by q^3, we can equal the terms in q^5, q^4, q^3 and the constant terms in q.

$$\begin{cases} f_{ttt} = 8Df_t \\ a_{tt} = 2Da \\ 2Ca = 0 \\ b_t = \frac{1}{4}f_{tt} \end{cases}$$

We find exactly the same system as in [14, 20].

5 Bernstein Processes

For simplicity, we shall only consider processes with values in **R**.

Many familiar stochastic processes can be viewed as Bernstein processes, solving Schrödinger's problem. Instead of a time interval centred in 0, we shall consider, here, an interval $[0, T]$, and denote by θ the constant $\sqrt{\kappa}$.

Let us mention a few classical examples:

(1) **The Brownian Motion** (see e.g. [16, p. 200])
   For $t_0 = 0, t_1 = T > 0, \mu_0 = \delta_0$ and $\mu_1 = \mathcal{N}(0, \theta^2 T)$, $z(t) = \theta w(t)$ is a Bernstein process for $V = 0$, $\eta(t, q) = 1$ and $\eta_*(t, q) = \dfrac{1}{\theta\sqrt{2\pi t}} e^{-\frac{q^2}{2\theta^2 t}}$.
   Here
   $$dz(t) = \theta dw(t).$$

(2) **The Brownian Bridge** (see [16, p. 201]).
   Here $t_0 = 0, t_1 = 1, V = 0, \mu_0 = \mu_1 = \delta_0$ and
   $$\eta(t, q) = \frac{1}{\sqrt{1-t}} e^{-\frac{q^2}{2\theta^2(1-t)}}.$$
   Then
   $$dz(t) = \theta dw(t) - \frac{z(t)}{1-t} dt.$$

(3) **The Ornstein–Uhlenbeck process** starting from 0
   Here $t_0 = 0, t_1 = T > 0, \mu_0 = \delta_0, \mu_1 = \mathbf{N}(0, \dfrac{\theta^2}{2\omega}(1 - e^{-2\omega T}))$, $V = \dfrac{\omega^2 q^2}{2}$ and
   $$\eta(t, q) = e^{\frac{\omega}{2\theta^2}(\theta^2 t - q^2)}.$$
   We have
   $$dz(t) = \theta dw(t) - \omega z(t) dt.$$

6 Parametrization of a One-Factor Affine Model in Stochastic Finance

A *one–factor affine interest rate model* is characterized by an instantaneous rate $r(t)$, satisfying a stochastic differential equation of the following type:

$$dr(t) = \sqrt{\alpha r(t) + \beta}\, dw(t) + (\phi - \lambda r(t))\, dt ,$$

under the risk–neutral probability Q ($\alpha = 0$ corresponds to the Vasicek model, and $\beta = 0$ corresponds to the Cox–Ingersoll–Ross model; see [8, 10]).

Assuming $\alpha \neq 0$, let us set

$$\theta = \frac{\alpha}{2},$$

$$\tilde{\phi} := \phi + \frac{\lambda\beta}{\alpha},$$

$$\delta := \frac{4\tilde{\phi}}{\alpha},$$

$$C := \frac{\alpha^2}{8}\left(\tilde{\phi} - \frac{\alpha}{4}\right)\left(\tilde{\phi} - \frac{3\alpha}{4}\right)$$

$$= \frac{\alpha^4}{128}(\delta - 1)(\delta - 3),$$

$$D := \frac{\lambda^2}{8},$$

and define the potential

$$V(t, q) = \frac{C}{q^2} + Dq^2 .$$

Theorem 6.1 (see [14, Theorem 5.4]) *Consider the process*

$$z(t) = \sqrt{\alpha r(t) + \beta}$$

and the stopping time

$$\tilde{T} = \inf\{t > 0 | \alpha r(t) + \beta = 0\}.$$

(1) *One has $\tilde{T} = +\infty$ a.s. for $\delta \geq 2$, and $\tilde{T} < +\infty$ a.s. for $\delta < 2$.* (2) *There exists a process $y(t)$ satisfying the stochastic differential equation*

$$\forall t > 0$$

$$dy(t) = \theta dw(t) + \tilde{B}(t, y(t))dt$$

relatively to the canonical increasing filtration of the Brownian w, where

$$\tilde{B} \equiv_{def} \theta^2 \frac{\frac{\partial \eta}{\partial q}}{\eta}.$$

for a certain solution η of $(\mathcal{E}_V)$

For each given $t > 0$, the law of $y(t)$ is $\eta(t, q)\eta_*(t, q)dq$, where η_* satisfies the dual equation $(\mathcal{E}'_V)$. One has

$$z(t) = y(t) \quad \forall t \in [0, \tilde{T}[$$

In particular, in case $\delta \geq 2$, z itself is a Bernstein process on any interval $[0, T_0]$ ($T_0 > 0$).

Proposition 6.2 *The isovector algebra associated with the affine model has dimension 6 if and only if $\delta \in \{1, 3\}$; in the opposite case, it has dimension 4.*

Let us analyze more closely the first case; the general case is considered in [14].
1) $\tilde{\phi} = \frac{\alpha}{4}$, i.e. $\delta = 1$. Then $y(t)$ is a solution of

$$dy(t) = \frac{\alpha}{2} dw(t) - \frac{\lambda}{2} y(t) dt,$$

i.e. $y(t)$ is an Ornstein–Uhlenbeck process, and the potential V is quadratic.
Here

$$\eta(t, q) = e^{\frac{\lambda t}{4} - \frac{\lambda q^2}{\alpha^2}}.$$

From

$$y(t) = e^{-\frac{\lambda t}{2}}(z_0 + \frac{\alpha}{2} \int_0^t e^{\frac{\lambda s}{2}} dw(s))$$

$$= e^{-\frac{\lambda t}{2}}(z_0 + \tilde{w}(\frac{\alpha^2(e^{\lambda t} - 1)}{4\lambda})),$$

($\tilde{w}$ denoting another Brownian motion), it appears that $y(t)$ follows a normal law ν_t with mean $e^{-\frac{\lambda t}{2}} z_0$ and variance $\frac{\alpha^2(1-e^{-\lambda t})}{4\lambda}$.

For each $T > 0$, $(y(t))_{0 \leq t \leq T}$ is a Bernstein process with $\mu_0 = \delta_{z_0}$ and $\mu_1 = \nu_T$.
The law of $y(t)$ therefore has density

$$p_t(q) = \frac{2\sqrt{\lambda}}{\alpha\sqrt{2\pi(1-e^{-\lambda t})}} e^{\left(-\frac{2\lambda(q - e^{-\frac{\lambda t}{2}} z_0)^2}{\alpha^2(1 - e^{-\lambda t})}\right)}.$$

whence

$$\eta_*(t,q) = \frac{\rho_t(q)}{\eta(t,q)}$$

$$= \frac{1}{\alpha}\sqrt{\frac{\lambda}{\pi \sinh(\frac{\lambda t}{2})}}e^{\left(\frac{-\lambda q^2 - \lambda q^2 e^{-\lambda t} + 4\lambda q z_0 e^{-\frac{\lambda t}{2}} - 2\lambda z_0^2 e^{-\lambda t}}{\alpha^2(1-e^{-\lambda t})}\right)}$$

and one may check that, as was to be expected, η_* satisfies $(\mathcal{E}'_V)$.

(2) $\tilde{\phi} = \frac{3\alpha}{4}$, i.e. $\delta = 3$.
Then

$$\eta(t,q) = q \, e^{\frac{\lambda}{\alpha^2}(\frac{3\alpha^2 t}{4} - q^2)}.$$

Let us define

$$s(t) = e^{\frac{\lambda t}{2}} \frac{1}{y(t)};$$

then an easy computation using Itô's formula gives

$$ds(t) = -\frac{\alpha}{2} e^{\frac{\lambda t}{2}} s(t)^2 dw(t);$$

in particular, $s(t)$ is a forward martingale.

It may be seen that, in case $X_0 = 0$,

$$X_t = e^{-\lambda t} Y\left(\frac{\alpha^2(e^{\lambda t} - 1)}{4\lambda}\right)$$

where Y is a $BESQ^3$(squared Bessel process with parameter 3) with $Y(0) = 0$. But, for fixed $t > 0$, Y_t has the same law as tY_1, and $Y_1 = ||B_1||^2$ is the square of the norm of a 3–dimensional Brownian motion (see [6]); the law of Y_1 is therefore

$$\frac{1}{\sqrt{2\pi}} e^{-\frac{u}{2}} \sqrt{u} \mathbf{1}_{u \geq 0} du.$$

Therefore the density $\rho_t(q)$ of the law of $y(t)$ is given by:

$$\rho_t(q) = \frac{1}{\sqrt{2\pi}} \frac{16\lambda^{\frac{3}{2}}}{\alpha^3(1-e^{-\lambda t})^{\frac{3}{2}}} q^2 e^{-\frac{2\lambda q^2}{\alpha^2(1-e^{-\lambda t})}}$$

and

$$\eta_*(t,q) = \frac{\rho_t(q)}{\eta(t,q)} = \frac{16\lambda^{\frac{3}{2}}}{\alpha^3 \sqrt{2\pi}}(1-e^{-\lambda t})^{-\frac{3}{2}}qe^{\frac{3\lambda t}{4} - \frac{\lambda q^2}{\alpha^2 \tanh(\frac{\lambda t}{2})}}.$$

For $X_0 \neq 0$, a Bessel function appears in the expression for η_*.
M. Houda and the first author ([10]) have extended these computations.

7 Generalized Brownian Bridges

Let us fix $T > 0$ and $\beta \in \mathbf{R}$. Mansuy's generalized Brownian Bridge ([18]) $X^{(\beta,\gamma,T,x)}$ is the solution $X(t)$ of the stochastic differential equation

$$dX(t) = \theta dw(t) - \beta \frac{X(t)}{T-t}dt$$

with initial value $X(0) = x$.
It is a Bernstein process for

$$\eta(t,q) := (T-t)^{-\frac{\beta}{2}} e^{-\frac{\beta q^2}{2\theta^2(T-t)}}$$

and

$$V(t,q) := \frac{\beta(\beta-1)q^2}{2(T-t)^2},$$

an example of *semi–classical potential*.

8 Conclusion

The class of Bernstein reciprocal processes allows a study of its symmetries founded on its quantum-like analogy. In particular, as suggested by Feynman, it should allow the construction of a dynamical theory where those symmetries should play a role as fundamental as in classical mechanics. Although a lot of progress has been made since [23] much more should be done. Using the symmetries, how can we define a stochastic version of classical integrability? Such a notion should inject in stochastic analysis key ideas of dynamical systems theory. For instance a stochastic deformation of Symplectic Geometry whose symplectomorphisms would precisely describe all transformations preserving dynamcal structures.

Since Schrödinger's original problem is now regarded as a foundational aspect of Optimal transportation theory it is likely that, in this area also, those processes will play an important role as natural regularizations of the traditional analytical and deterministic approaches.

Acknowledgements Both authors are extremely grateful to Professors Marco Fuhrman, Elisa Mastrogiacomo, Paola Morando and Stefania Ugolini for their excellent organization of the conference and their wonderful hospitality. Professor Jean-Claude Zambrini was kind enough to read a preliminary version of our manuscript; his comments and suggestions led to numerous improvements in the structure and style of our text. We thank him heartily for his efforts.

References

1. Albeverio, S., De Vecchi, F., Morando, P., Ugolini, S. et al.: Weak symmetries of stochastic differential equations driven by semimartingales with jumps. Electron. J. Prob. **25** (2020)
2. Beurling, A.: An automorphism of product measures. Ann. Math. **72**(1), 189–200 (1960)
3. Bluman, G., Cole, J.: General similarity solution of the heat equation. J. Math. Mech. **18**, 1025–1042 (1969)
4. De Vecchi, F., Morando, P., Ugolini, S.: Reduction and reconstruction of SDEs via Girsanov and quasi Doob symmetries. J. Phys. A: Math. Theor. **54**(18), 185203 (2021)
5. De Vecchi, F., Morando, P., Ugolini, S.: Symmetries of stochastic differential equations using Girsanov transformations. J. Phys. A: Math. Theor. **53**(13), 135204 (2020)
6. Göing-Jaeschke, A., Yor, M.: A survey and some generalizations of Bessel processes. Bernoulli **9**(2), 313–349 (2003)
7. Kent Harrison, B., Estabrook, F.B.: Geometric approach to invariance groups and solution of partial differential systems. J. Math. Phys. **12**, 653–666 (1971)
8. Hénon, S.: Un modèle de taux avec volatilité stochastique, Ph.D. thesis (2005)
9. Hochberg, K.J.: A signed measure on path space related to wiener measure. Ann. Probab. **6**(3), 433–458 (1978)
10. Houda, M., Lescot, P.: Some Bernstein processes similar to Cox-Ingersoll-Ross ones. Stoch. Dyn. **19**(6), 1950047 (2019)
11. Léonard, C.: A survey of the Schrödinger problem and some of its connections with optimal transport (2013). arXiv:1308.0215
12. Lescot, P.: Symmetries of the Black-Scholes equation. Methods Appl. Anal. **19**(2), 147–160 (2012)
13. Lescot, P., Quintard, H.: Symmetries of the backward heat equation with potential and interest rate models. C. R. Acad. Sci. Paris, Ser. I **352**, 525–528 (2014)
14. Lescot, P., Quintard, H., Zambrini, J.-C.: Solving stochastic differential equations with Cartan's exterior differential system (2015)
15. Lescot, P., Valade, L.: Symmetries of partial differential equations and stochastic processes in mathematical physics and in finance. J. Phys.: Conf. Ser. **2019**, 1194 012070 (2019)
16. Lescot, P., Zambrini, J.-C.: Isovectors for the Hamilton–Jacobi–Bellman Equation, Formal Stochastic Differentials and First Integrals in Euclidean Quantum Mechanics. Seminar on Stochastic Analysis, Random Fields and Applications IV, Progr. Probab. 58, Birkhäuser, Basel, , pp. 187–202 (2004)
17. Lescot, P., Zambrini, J.-C.: Probabilistic deformation of contact geometry, diffusion processes and their quadratures, Seminar on Stochastic Analysis, Random Fields and applications V, Progr. Probab. 59, Birkhäuser, Basel, pp. 203–226 (2008)
18. Mansuy, R.: On a one–parameter generalization of the brownian bridge and associated quadratic functionals. J. Theor. Prob. **17**(4), 1021–1029 (2004)

19. Olver, P.J.: Applications of Lie Groups to Differential Equations, Graduate Texts in Mathematics. Springer (1993)
20. Quintard, H.: Symétries d'équations aux dérivées partielles, calcul stochastique, applications à la physique mathématique et à la finance, Rouen (2015)
21. Valade, L.: Ph.D. Thesis, Rouen (2021)
22. Zambrini, J.C.: The research program of stochastic deformation (with a view toward geometric mechanics), Stochastic analysis: a series of lectures, pp. 359–393 (2015)
23. Zambrini, J.C.: Variational processes and stochastic versions of mechanics. J. Math. Phys. **27**(9), 2307–2330 (1986)

On the Positivity of Local Mild Solutions to Stochastic Evolution Equations

Carlo Marinelli and Luca Scarpa

Abstract We provide sufficient conditions on the coefficients of a stochastic evolution equation on a Hilbert space of functions driven by a cylindrical Wiener process ensuring that its mild solution is positive if the initial datum is positive. As an application, we discuss the positivity of forward rates in the Heath-Jarrow-Morton model via Musiela's stochastic PDE.

Keywords Positivity · Mild solutions · Stochastic evolution equations · HJM model.

1 Introduction

Let us consider a stochastic evolution equation of the type

$$du + Au\, dt = F(u)\, dt + B(u)\, dW, \qquad u(0) = u_0, \tag{1.1}$$

where A is a linear maximal monotone operator on a Hilbert space of functions H, the coefficients F and B satisfy suitable integrability assumptions, and W is a cylindrical Wiener process. Precise assumptions on the data of the Cauchy problem (1.1) are given in Sect. 2 below. Our goal is to establish a maximum principle for (local) mild solutions to (1.1), i.e. to provide sufficient conditions on the operator A and on the coefficients F and B such that positivity of the initial datum u_0 implies positivity of the solution u (see Theorem 2.2 below).

C. Marinelli (✉)
Dipartimento di Matematica, University College London, Gower Street,
London WC1E 6BT, UK
e-mail: c.marinelli@ucl.ac.uk

L. Scarpa
Faculty of Mathematics, Politecnico di Milano, Via E. Bonardi 9, 20133 Milano, Italy
e-mail: luca.scarpa@univie.ac.at

© Springer Nature Switzerland AG 2021
S. Ugolini et al. (eds.), *Geometry and Invariance in Stochastic Dynamics*,
Springer Proceedings in Mathematics & Statistics 378,
https://doi.org/10.1007/978-3-030-87432-2_12

A simpler problem was studied in [10], where the coefficients F and B are assumed to be Lipschitz continuous. Here we simply assume that F and B satisfy rather minimal integrability conditions and that a local mild solution exists. On the other hand, in [10] the linear operator A need only generate a positivity preserving semigroup, while here we require that A generates a sub-Markovian semigroup.

We refer to [10] for a discussion about the relation between other positivity results for solutions to stochastic partial differential equations and ours. It is however probably worth pointing out that most existing results seem to deal with equations in the variational setting (see, e.g., [1, 7, 8, 13]).

As an application, we provide an alternative, more direct proof of the positivity of forward rates in the Heath-Jarrow-Morton [5] framework than the one in [10]. This is obtained, as is now classical, viewing forward curves as solutions to the so-called Musiela stochastic PDE (see, e.g., [3, 11]).

2 Assumptions and Main Result

Let $(\Omega, \mathscr{F}, \mathbb{P})$ be a probability space endowed with a complete right-continuous filtration $(\mathscr{F}_t)_{t \in [0,T]}$, with $T > 0$ a fixed final time, on which all random elements will be defined. Identities and inequalities between random variables are meant to hold $\mathbb{P}$-almost surely, and two stochastic processes are declared equal, unless otherwise stated, if they are indistinguishable. The σ-algebra of progressively measurable subsets of $\Omega \times [0, T]$ will be denoted by $\mathscr{R}$. We shall denote a cylindrical Wiener process on a separable Hilbert space U by W. Standard notation and terminology of stochastic calculus for semimartingales will be used throughout (see, e.g., [12]). In particular, given an adapted process X and a stopping time τ, X^τ will denote the process X stopped at τ. Similarly, if X is also càdlàg, $X^{\tau-}$ stands for the process X pre-stopped at τ.

For any separable Hilbert spaces E_1 and E_2, we use the symbols $\mathscr{L}(E_1, E_2)$ and $\mathscr{L}^2(E_1, E_2)$ for the spaces of linear continuous and Hilbert-Schmidt operators from E_1 to E_2, respectively. The space of continuous bilinear maps from $E_1 \times E_1$ to E_2 will be denoted by $\mathscr{L}_2(E_1; E_2)$. The n-th order Fréchet and Gâteaux derivatives of a function $\Phi : E_1 \to E_2$ at a point $x \in E_1$ are denoted by $D^n \Phi(x)$ and $D^n_G \Phi(x)$, respectively, omitting the superscript if $n = 1$, as usual.

We shall work under the following standing assumptions.

(A1) There exists an open set $\mathcal{O}$ in $\mathbb{R}^d$, $d \geq 1$, and a Borel measure μ such that $H = L^2(\mathcal{O}, \mu)$.

The norm and the scalar product on H will be denoted by $\|\cdot\|$ and $\langle \cdot, \cdot \rangle$, respectively. For any $f, g \in H$, we shall write $f \geq g$ to mean that $f \geq g$ μ-almost everywhere.

(A2) A is a linear maximal monotone operator on H such that its resolvent is sub-Markovian and is a contraction with respect to the $L^1(\mathcal{O}, \mu)$-norm.

Recall that the resolvent of A, i.e. the family of linear continuous operators on H defined by

$$J_\lambda := (I + \lambda A)^{-1}, \quad \lambda > 0,$$

is said to be sub-Markovian if, for every $\lambda > 0$ and every $\phi \in H$ such that $0 \leq \phi \leq 1$, one has $0 \leq J_\lambda \phi \leq 1$.

(A3) $F : \Omega \times [0, T] \times H \to H$ and $B : \Omega \times [0, T] \times H \to \mathscr{L}^2(U, H)$ are $\mathscr{R} \otimes \mathscr{B}(H)$-measurable, and there exists a constant $C > 0$ such that

$$-\langle F(\omega, t, h), h^- \rangle + \frac{1}{2} \left\| 1_{\{h \leq 0\}} B(\omega, t, h) \right\|_{\mathscr{L}^2(U,H)}^2 \leq C \|h^-\|_H^2$$

for all $(\omega, t, h) \in \Omega \times [0, T] \times H$.

(A4) $u_0 \in L^0(\Omega, \mathscr{F}_0; H)$

Definition 2.1 A local mild solution to the Cauchy problem (1.1) is a pair (u, τ), where τ is a stopping time with $\tau \leq T$, and $u : [\![0, \tau[\![\to H$ is a measurable adapted process with continuous trajectories such that, for any stopping time $\sigma < \tau$, one has

(i) $S(t - \cdot)F(u) 1_{[\![0,\sigma]\!]} \in L^0(\Omega; L^1(0, t; H))$ for all $t \in [0, T]$;
(ii) $S(t - \cdot)B(u) 1_{[\![0,\sigma]\!]} \in L^0(\Omega; L^2(0, t; \mathscr{L}^2(U, H)))$ for all $t \in [0, T]$,

and

$$u = S(\cdot)u_0 + \int_0^{\cdot} S(\cdot - s)F(s, u(s))\,ds + \int_0^{\cdot} S(\cdot - s)B(s, u(s))\,dW(s).$$

The last identity is to be understood in the sense of indistinguishability of processes defined on the stochastic interval $[\![0, \tau[\![$. Here the stochastic convolution is defined on $[\![0, \sigma]\!]$, for every stopping time $\sigma < \tau$, as

$$\left(\int_0^t S(t - s) B(s, u(s)) 1_{[\![0,\sigma]\!]}(s)\, dW(s) \right)_{t \in [0,\sigma]}.$$

The main result is the following.

Theorem 2.2 Let (u, τ) be a local mild solution to the Cauchy problem (1.1) such that, for every stopping time $\sigma < \tau$, one has

i) $F(u) 1_{[\![0,\sigma]\!]} \in L^0(\Omega; L^1(0, T; H))$;
(ii) $B(u) 1_{[\![0,\sigma]\!]} \in L^0(\Omega; L^2(0, T; \mathscr{L}^2(U, H)))$.

If $u_0 \geq 0$, then $u^{\tau-}(t) \geq 0$ for all $t \in [0, T]$.

3 Auxiliary Results

The arguments used in the proof of Theorem 2.2 (see Sect. 4 below) rely on the following results, that we recall here for the reader's convenience. The first is a continuous dependence result for mild solutions to stochastic evolution equations in the form (1.1) with respect to the coefficients and the initial datum. This is a consequence of a more general statement proved in [9, Corollary 3.4]. Let

$$(u_{0n})_n \subset L^0(\Omega, \mathscr{F}_0; H),$$
$$(f_n)_n, f \subset L^0(\Omega; L^1(0, T; H)),$$
$$(G_n)_n, G \subset L^0(\Omega; L^2(0, T; \mathscr{L}^2(U, H)))$$

be such that the H-valued processes f_n, f, $G_n v$, and $G v$ are strongly measurable and adapted for all $v \in U$ and $n \in \mathbb{N}$. Then the Cauchy problems

$$du_n + A u_n \, dt = f_n \, dt + G_n \, dW, \qquad u_n(0) = u_{0n},$$

and

$$du + A u \, dt = f \, dt + G \, dW, \qquad u(0) = u_0,$$

admit unique mild solutions u_n and u, respectively.

Proposition 3.1 *Assume that*

$$\begin{aligned}
u_{0n} &\longrightarrow u_0 && \text{in } L^0(\Omega; H), \\
f_n &\longrightarrow f && \text{in } L^0(\Omega; L^1(0, T; H)), \\
G_n &\longrightarrow G && \text{in } L^0(\Omega; L^2(0, T; \mathscr{L}^2(U, H))).
\end{aligned}$$

Then $u_n \to u$ in $L^0(\Omega; C([0, T]; H))$.

The second result we shall need is a generalized Itô formula, the proof of which can be found in [10].

Proposition 3.2 *Let $G: H \to \mathbb{R}$ be continuously Fréchet differentiable and DG be Gâteaux differentiable, with $D_{\mathcal{G}}^2 G: H \to \mathscr{L}_2(H; \mathbb{R})$ such that $(\varphi, \zeta_1, \zeta_2) \mapsto D_{\mathcal{G}}^2 G(\varphi)[\zeta_1, \zeta_2]$ is continuous, and assume that G, DG, and $D_{\mathcal{G}}^2 G$ are polynomially bounded. Moreover, let the processes $f \in L^0(\Omega; L^1(0, T; H))$ and $\Phi \in L^0(\Omega; L^2(0, T; \mathscr{L}^2(U, H)))$ be measurable and adapted, and $v_0 \in L^0(\Omega, \mathscr{F}_0; H)$. Setting*

$$v := v_0 + \int_0^\cdot f(s) \, ds + \int_0^\cdot \Phi(s) \, dW(s),$$

one has

$$G(v) = G(v_0) + \int_0^\cdot \left(DG(v) f + \frac{1}{2} \operatorname{Tr}\left(\Phi^* D_{\mathcal{G}}^2 G(v) \Phi \right) \right)(s) \, ds$$
$$+ \int_0^\cdot DG(v(s)) \Phi(s) \, dW(s).$$

Finally, we recall an inequality for maximal monotone linear operators with sub-Markovian resolvent due to Brézis and Strauss (see [2, Lemma 2]).[1]

Lemma 3.3 *Let $\beta \colon \mathbb{R} \to 2^{\mathbb{R}}$ be a maximal monotone graph with $0 \in \beta(0)$. Let $\varphi \in L^p(\mathcal{O})$ with $A\varphi \in L^p(\mathcal{O})$, and $z \in L^q(\mathcal{O})$ with $z \in \beta(\varphi)$ a.e. in $\mathcal{O}$, where $p, q \in [1, +\infty]$ and $1/p + 1/q = 1$. Then*

$$\int_{\mathcal{O}} (A\varphi) z \geq 0.$$

We include a sketch of proof for the reader's convenience, assuming for simplicity that $\beta \colon \mathbb{R} \to \mathbb{R}$ is continuous and bounded. Let $j \colon \mathbb{R} \to \mathbb{R}_+$ be a (differentiable, convex) primitive of β and

$$A_\lambda := \frac{1}{\lambda}\big(I - (I + \lambda A)^{-1}\big) = \frac{1}{\lambda}(I - J_\lambda), \qquad \lambda > 0,$$

be the Yosida approximation of A. It is well known that A_λ is a linear maximal monotone bounded operator on H and that, for every $v \in D(A)$, $A_\lambda v \to Av$ as $\lambda \to 0$. Let $v \in D(A)$. The convexity of j implies, for every $\lambda > 0$, that

$$\langle A_\lambda v, \beta(v)\rangle_{L^2} = \frac{1}{\lambda}\langle v - J_\lambda v, j'(v)\rangle_{L^2}$$

$$\geq \frac{1}{\lambda}\left(\int_{\mathcal{O}} j(v) - \int_{\mathcal{O}} j(J_\lambda v)\right) = \frac{1}{\lambda}\big(\|j(v)\|_{L^1} - \|j(J_\lambda v)\|_{L^1}\big).$$

Since J_λ is sub-Markovian and j is convex, the generalized Jensen inequality for positive operators (see [4]) and the contractivity of J_λ in L^1 imply that

$$\|j(J_\lambda v)\|_{L^1} \leq \|J_\lambda j(v)\|_{L^1} \leq \|j(v)\|_{L^1},$$

i.e. that

$$\langle A_\lambda v, \beta(v)\rangle_{L^2} \geq 0$$

for every $\lambda \to 0$. Passing to the limit as $\lambda \to 0$ yields $\langle Av, \beta(v)\rangle_{L^2} \geq 0$.

4 Proof of Theorem 2.2

The proof is divided into two parts. First we show that a local mild solution u to (1.1) can be approximated by strong solutions to regularized equations. As a second step, we show that such approximating processes are positive, thanks to a suitable version of Itô's formula.

---

[1] For a related inequality cf. also [14, Lemma 5.1].

4.1 Approximation of the solution

Let (u, τ) be a local mild solution to (1.1). Let σ be a stopping time with $\sigma < \tau$, so that $u : [\![0, \sigma]\!] \to H$ is well defined, and set

$$\bar{u} := u^\sigma \in L^0(\Omega; C([0, T]; H)),$$
$$\bar{F} := F(\cdot, u)\mathbf{1}_{[\![0,\sigma]\!]} \in L^0(\Omega; L^1(0, T; H)),$$
$$\bar{B} := B(\cdot, u)\mathbf{1}_{[\![0,\sigma]\!]} \in L^0(\Omega; L^2(0, T; \mathscr{L}^2(U, H))).$$

In particular, One has

$$\bar{u}(t) := S(t)u_0 + \int_0^t S(t-s)\bar{F}(s)\,ds + \int_0^t S(t-s)\bar{B}(s)\,dW(s) \qquad (4.1)$$

for all $t \in [0, T]$ $\mathbb{P}$-a.s., or, equivalently, $\bar{u}$ is the unique global mild solution to the Cauchy problem

$$d\bar{u} + A\bar{u}\,dt = \bar{F}\,dt + \bar{B}\,dW, \qquad \bar{u}(0) = u_0.$$

Recalling that $J_\lambda \in \mathscr{L}(H, \mathsf{D}(A))$ for all $\lambda > 0$, one has

$$\bar{F}_\lambda := J_\lambda F(\cdot, u)\mathbf{1}_{[\![0,\sigma]\!]} = J_\lambda \bar{F} \in L^0(\Omega; L^1(0, T; \mathsf{D}(A))),$$
$$\bar{B}_\lambda := J_\lambda B(\cdot, u)\mathbf{1}_{[\![0,\sigma]\!]} = J_\lambda \bar{B} \in L^0(\Omega; L^2(0, T; \mathscr{L}^2(U, \mathsf{D}(A)))),$$
$$u_{0\lambda} := J_\lambda u_0 \in L^0(\Omega, \mathscr{F}_0; \mathsf{D}(A)),$$

where the second assertion is an immediate consequence of the ideal property of Hilbert-Schmidt operators. The process $u_\lambda : \Omega \times [0, T] \to H$ defined as

$$u_\lambda(t) := S(t)u_{0\lambda} + \int_0^t S(t-s)\bar{F}_\lambda(s)\,ds + \int_0^t S(t-s)\bar{B}_\lambda(s)\,dW(s), \qquad t \in [0, T], \qquad (4.2)$$

therefore belongs to $L^0(\Omega; C([0, T]; \mathsf{D}(A)))$ and is the unique global strong solution to the Cauchy problem

$$du_\lambda + Au_\lambda\,dt = \bar{F}_\lambda\,dt + \bar{B}_\lambda\,dW, \qquad u_\lambda(0) = u_{0\lambda},$$

i.e.
$$u_\lambda + \int_0^\cdot Au_\lambda(s)\,ds = u_{0\lambda} + \int_0^\cdot \bar{F}_\lambda(s)\,ds + \int_0^\cdot \bar{B}_\lambda(s)\,dW(s) \quad (4.3)$$

in the sense of indistinguishable H-valued processes. Furthermore, since J_λ is contractive and converges to the identity in the strong operator topology of $\mathscr{L}(H, H)$ as $\lambda \to 0$, i.e. $J_\lambda h \to h$ for every $h \in H$, one has

$$u_{0\lambda} \longrightarrow u_0 \text{ in } L^0(\Omega; H),$$
$$\bar{F}_\lambda \longrightarrow \bar{F} \text{ in } L^0(\Omega; L^2(0, T; H)),$$
$$\bar{B}_\lambda \longrightarrow \bar{B} \text{ in } L^0(\Omega; L^2(0, T; \mathscr{L}^2(U, H))),$$

where the second convergence follows immediately by the dominated convergence theorem, and the third one by a continuity property of Hilbert-Schmidt operators (see, e.g., [6, Theorem 9.1.14]). Finally, thanks to Proposition 3.1, we deduce that

$$u_\lambda \longrightarrow \bar{u} \text{ in } L^0(\Omega; C([0, T]; H)). \quad (4.4)$$

4.2 Positivity

Let us introduce the functional

$$G: H \longrightarrow \mathbb{R}_+$$
$$\varphi \longmapsto \frac{1}{2}\int_{\mathcal{O}} |\varphi^-|^2,$$

as well as the family of regularized functionals $(G_n)_{n \in \mathbb{N}}$ defined by

$$G_n: H \longrightarrow \mathbb{R}_+$$
$$\varphi \longmapsto \int_{\mathcal{O}} g_n(\varphi),$$

where $g_n : \mathbb{R} \to \mathbb{R}_+$ is a convex, twice continuously differentiable approximation of $r \mapsto (r^-)^2/2$ such that (g_n'') is uniformly bounded and converges to $mathbbm 1_{\mathbb{R}_-}$, and $g_n'(r) \to -r^-$ as $n \to \infty$ for every $r \in \mathbb{R}$. The existence of such an approximating sequence is well known (see, e.g., [15, Sect. 3] for an analogous construction). One can verify (see, e.g., [10]) that, for every $n \in \mathbb{N}$, G_n is everywhere continuously Fréchet differentiable with derivative

$$DG_n : H \longrightarrow \mathscr{L}(H, \mathbb{R}) \simeq H$$
$$\varphi \longmapsto g_n'(\varphi),$$

and that $DG_n \colon H \to H$ is Gâteaux differentiable with Gâteaux derivative given by

$$D_{\mathcal{G}}^2 G_n \colon H \longrightarrow \mathscr{L}(H, H) \simeq \mathscr{L}_2(H; \mathbb{R})$$
$$\varphi \longmapsto \left[(\zeta_1, \zeta_2) \mapsto \int_{\mathcal{O}} g_n''(\varphi) \zeta_1 \zeta_2 \right].$$

Furthermore, the map $(\varphi, \zeta_1, \zeta_2) \mapsto D_{\mathcal{G}}^2 G_n(\varphi)(\zeta_1, \zeta_2)$ is continuous. Proposition 3.2 applied to the process u_λ defined by (4.3) then yields

$$\begin{aligned}
G_n(u_\lambda) + \int_0^{\cdot} \langle Au_\lambda, DG_n(u_\lambda) \rangle(s) \, ds \\
= G_n(u_{0\lambda}) + \int_0^{\cdot} DG_n(u_\lambda(s)) \bar{B}_\lambda(s) \, dW(s) \\
+ \int_0^{\cdot} \left(DG_n(u_\lambda) \bar{F}_\lambda + \frac{1}{2} \operatorname{Tr}\bigl(\bar{B}_\lambda^* D_{\mathcal{G}}^2 G_n(u_\lambda) \bar{B}_\lambda\bigr) \right)(s) \, ds
\end{aligned} \quad (4.5)$$

Recalling that $g_n' \colon \mathbb{R} \to \mathbb{R}$ is increasing, Lemma 3.3 implies that

$$\langle Au_\lambda, DG_n(u_\lambda) \rangle = \langle Au_\lambda, g'(u_\lambda) \rangle \geq 0, \quad (4.6)$$

hence also, denoting a complete orthonormal system of U by (e_j),

$$\int_{\mathcal{O}} g_n(u_\lambda(t)) \leq \int_{\mathcal{O}} g_n(u_{0\lambda}) + \int_0^t g_n'(u_\lambda(s)) \bar{B}_\lambda(s) \, dW(s)$$
$$+ \int_0^t g_n'(u_\lambda(s)) \bar{F}_\lambda(s) \, ds + \frac{1}{2} \int_0^t \sum_{j=0}^{\infty} \int_{\mathcal{O}} g_n''(u_\lambda(s)) \bigl| \bar{B}_\lambda(s) e_j \bigr|^2 \, ds$$

for every $t \in [0, T]$ and $n \in \mathbb{N}$. We are now going to pass to the limit as $n \to \infty$ in this inequality. Recalling that (g_n'') is uniformly bounded and that the paths of u_λ belong to $C([0, T]; H)$ $\mathbb{P}$-a.s., the dominated convergence theorem yields

$$\int_{\mathcal{O}} g_n(u_\lambda(t)) \longrightarrow \frac{1}{2} \| u_\lambda^-(t) \|^2 \quad \forall t \in [0, T],$$
$$\int_{\mathcal{O}} g_n(u_{0\lambda}) \longrightarrow \frac{1}{2} \| u_{0\lambda}^- \|^2.$$

Note that u_0 is positive and J_λ is positivity preserving, hence $u_{0\lambda} = J_\lambda u_0$ is also positive and, in particular, $u_{0\lambda}^-$ is equal to zero a.e. in $\mathcal{O}$. Let us introduce the (real) continuous local martingales $(M^{\lambda,n})_{n \in \mathbb{N}}$, M^λ, defined as

$$M_t^{\lambda,n} := \int_0^t g_n'(u_\lambda(s))\bar{B}_\lambda(s)\,dW(s),$$

$$M_t^\lambda := -\int_0^t u_\lambda^-(s)\bar{B}_\lambda(s)\,dW(s).$$

One has, by the ideal property of Hilbert-Schmidt operators,

$$[M^{\lambda,n} - M^\lambda, M^{\lambda,n} - M^\lambda]_t = \int_0^t \|(g_n'(u_\lambda(s)) + u_\lambda^-(s))\bar{B}_\lambda(s)\|^2_{\mathscr{L}^2(U,\mathbb{R})}\,ds$$

$$\leq \int_0^t \|g_n'(u_\lambda(s)) + u_\lambda^-(s)\|^2 \|\bar{B}_\lambda(s)\|^2_{\mathscr{L}^2(U,H)}\,ds$$

for all $t \in [0, T]$. Recalling that $u_\lambda \in L^0(\Omega; C([0,T]; H))$ and $g_n'(r) \to -r^-$ for every $r \in \mathbb{R}$, it follows by the dominated convergence theorem that $[M^{\lambda,n} - M^\lambda, M^{\lambda,n} - M^\lambda] \to 0$, hence that $M^{\lambda,n} \to M^\lambda$, as $n \to \infty$, i.e. that

$$\int_0^\cdot g_n'(u_\lambda(s))\bar{B}_\lambda(s)\,dW(s) \longrightarrow -\int_0^\cdot u_\lambda^-(s)\bar{B}_\lambda(s)\,dW(s)$$

in the ucp topology. Similarly, the pathwise continuity of u_λ and the dominated convergence theorem yield

$$\int_0^t g_n'(u_\lambda(s))\bar{F}_\lambda(s)\,ds \longrightarrow -\int_0^t u_\lambda^-(s)\bar{F}_\lambda(s)\,ds$$

for all $t \in [0, T]$ as $n \to \infty$. Finally, the pointwise convergence $g_n'' \to \mathbb{1}_{\mathbb{R}_-}$ and the dominated convergence theorem imply that

$$\int_0^t \sum_{j=0}^\infty \int_\mathcal{O} g_n''(u_\lambda(s))|\bar{B}_\lambda(s)e_j|^2\,ds \longrightarrow \int_0^t \sum_{j=0}^\infty \int_\mathcal{O} \mathbb{1}_{\{u_\lambda(s)\leq 0\}}|\bar{B}_\lambda(s)e_j|^2\,ds$$

for all $t \in [0, T]$ as $n \to \infty$. We are thus left with

$$\|u_\lambda^-(t)\|^2 \leq \int_0^t \left(-2\langle u_\lambda^-(s), \bar{F}_\lambda(s)\rangle + \sum_{j=0}^\infty \int_\mathcal{O} \mathbb{1}_{\{u_\lambda(s)\leq 0\}}|\bar{B}_\lambda(s)e_j|^2\right)ds$$

$$-\int_0^t u_\lambda^-(s)\bar{B}_\lambda(s)\,dW(s).$$

Let us now take the limit as $\lambda \to 0$: if follows from the convergence property (4.4) and the continuous mapping theorem that

$$\|u_\lambda^-(t)\|^2 \longrightarrow \|\bar{u}^-(t)\|^2.$$

Recalling that $\bar{F}_\lambda = J_\lambda F(\bar{u})$, which converges pointwise to $\bar{F}$, one has

$$\int_0^t -2\langle u_\lambda^-(s), \bar{F}_\lambda(s)\rangle \, ds \longrightarrow \int_0^t -2\langle \bar{u}^-(s), \bar{F}(s)\rangle \, ds$$

Appealing again to (4.4), it is not difficult to check that

$$\mathbb{1}_{\{\bar{u}(s)\leq 0\}} \geq \limsup_{\lambda\to 0} \mathbb{1}_{\{u_\lambda(s)\leq 0\}}$$

Hence it follows from Fatou's lemma that

$$\limsup_{\lambda\to 0} \int_0^t \sum_{j=0}^\infty \int_\mathcal{O} \mathbb{1}_{\{u_\lambda(s)<0\}} |\bar{B}_\lambda(s)e_j|^2 \, ds \leq \int_0^t \sum_{j=0}^\infty \int_\mathcal{O} \mathbb{1}_{\{\bar{u}(s)<0\}} |\bar{B}(s)e_j|^2 \, ds.$$

Let us define the real continuous local martingale M as

$$M_t := -\int_0^t \bar{u}^-(s)\bar{B}(s) \, dW(s).$$

One has

$$[M^\lambda - M, M^\lambda - M]_t = \int_0^t \|u_\lambda^-(s)\bar{B}_\lambda(s) - \bar{u}^-(s)\bar{B}(s)\|^2_{\mathscr{L}^2(U,\mathbb{R})} \, ds,$$

where, by the ideal property of Hilbert-Schmidt operators and the contractivity of J_λ,

$$\|u_\lambda^- \bar{B}_\lambda - \bar{u}^- \bar{B}\|_{\mathscr{L}^2(U,\mathbb{R})} \leq \|(u_\lambda^- - \bar{u}^-)\bar{B}_\lambda\|_{\mathscr{L}^2(U,\mathbb{R})} + \|\bar{u}^-(\bar{B}_\lambda - \bar{B})\|_{\mathscr{L}^2(U,\mathbb{R})}$$
$$\leq \|u_\lambda^- - \bar{u}^-\| \|\bar{B}\|_{\mathscr{L}^2(U,H)} + \|\bar{u}^-\| \|\bar{B}_\lambda - \bar{B}\|_{\mathscr{L}^2(U,H)}.$$

Since u_λ converges to $\bar{u}$ in the sense of (4.4) and, as already seen, $\bar{B}_\lambda \to \bar{B}$ in $L^0(\Omega; L^2(0,T; \mathscr{L}^2(U,H)))$, the dominated convergence theorem yields, for every $t \in [0,T]$,

$$[M^\lambda - M, M^\lambda - M]_t \longrightarrow 0,$$

thus also

$$\int_0^\cdot u_\lambda^-(s)\bar{B}_\lambda(s) \, dW(s) \longrightarrow \int_0^\cdot \bar{u}^-(s)\bar{B}(s) \, dW(s).$$

in the ucp topology. Recalling assumption (A3), one obtains, for every $t \in [0,T]$,

$$\|\bar{u}^-(t)\|^2 \leq 2C \int_0^t \|\bar{u}^-(s)\|^2 \, ds - 2\int_0^t \bar{u}^-(s)B(s,\bar{u}(s)) \, dW(s),$$

thus also, integrating by parts,

$$e^{-2Ct}\|\bar{u}^-(t)\|^2 \leq -2\int_0^t e^{-2Cs}\bar{u}^-(s)B(s,\bar{u}(s))\,dW(s) =: \tilde{M}_t$$

The process $\tilde{M}$ is a positive local martingale, hence a supermartingale, with $\tilde{M}(0) = 0$, therefore M is identically equal to zero. This implies that $\|\bar{u}^-(t)\| = 0$ for all $t \in [0, T]$, hence, in particular, that $u(t)$ is positive for all $t \in [0, T]$. By definition of $\bar{u}$, we deduce that

$$u^\sigma \geq 0 \quad \forall t \in [0, T] \times \mathcal{O}$$

for every $\sigma < \tau$. Since σ is arbitrary, this readily implies that

$$u^{\tau-}(t) \geq 0 \quad \forall t \in [0, T],$$

thus completing the proof of Theorem 2.2.

Remark 4.1 In [10] the substantially weaker assumption was made that $-A$ generates a positive semigroup. This was possible because F and B were assumed to be Lipschitz continuous. In fact, in this case the process u_λ, strong solution of the equation obtained by replacing A with its Yosida approximation A_λ in (1.1), i.e.

$$du_\lambda + A_\lambda u_\lambda\,dt = F(u_\lambda)\,dt + B(u_\lambda)\,dW, \quad u_\lambda(0) = u_0,$$

converges to the unique mild solution u to (1.1), and the positivity of u_λ, for every $\lambda > 0$, was shown. In the more general situation considered here, where F and B are not supposed to be Lipschitz continuous, it is not even clear whether the above regularized equation admits a solution at all. For this reason we introduced a different approximation scheme in Sect. 4.1, that implies the need for an estimate such as (4.6), which in turn is satisfied if $-A$ generates a sub-Markovian semigroup, rather than just a positive one.

5 Positivity of Forward Rates

Musiela's stochastic PDE can be written as

$$du + Au\,dt = \beta(t,u)\,dt + \sum_{k=1}^\infty \sigma_k(t,u)\,dw^k(t), \quad u(0) = u_0, \qquad (5.1)$$

where $-A$ is (formally, for the moment) the infinitesimal generator of the semigroup of translations, $(w^k)_{k\in\mathbb{N}}$ is a sequence of independent standard Wiener processes, σ_k is a random, time-dependent superposition operator for each $k \in \mathbb{N}$, as well as β, and u takes values in a space of continuous functions, so that $u(t,x) := [u(t)](x)$,

$x \geq 0$, models the value of the forward rate prevailing at time t for delivery at time $t + x$. In order to exclude arbitrage (or, more precisely, in order for the corresponding discounted bond price process to be a local martingale), β needs to satisfy the so-called Heath-Jarrow-Morton no-arbitrage condition

$$\beta(t, v) = \sum_{k=1}^{\infty} \sigma_k(t, v) \int_0^{\cdot} [\sigma_k(t, v)](y) \, dy.$$

In order for (5.1) to admit a solution with continuous paths, a by now standard choice of state space is the Hilbert space H_α, $\alpha > 0$, which consists of absolutely continuous functions $\phi : \mathbb{R}_+ \to \mathbb{R}$ such that

$$\|\phi\|_{H_\alpha}^2 := \phi(\infty)^2 + \int_0^{\infty} |\phi'(x)|^2 e^{\alpha x} \, dx < \infty.$$

Under measurability, local boundedness, and local Lipschitz continuity conditions on (σ_k), one can rewrite (5.1) as

$$du + Au \, dt = \beta(t, u) \, dt + B(t, u) \, dW(t), \qquad u(0) = u_0, \qquad (5.2)$$

where $-A$ is the generator of the semigroup of translations on H_α, W is a cylindrical Wiener process on $U = \ell^2$, and $B : \Omega \times \mathbb{R}_+ \times H \to \mathscr{L}^2(U, H)$ is such that

$$\sum_{k=1}^{\infty} \int_0^{\cdot} \sigma_k(s, v(s)) \, dw^k(s) = \int_0^{\cdot} B(s, v(s)) \, dW(s).$$

Under such assumptions on (σ_k), (5.1) admits a unique local mild solution with values in H_α. If (σ_k) satisfy stronger (global) boundedness and Lipschitz continuity assumptions, then the local mild solution is in fact global. For details we refer to [3], as well as to [10].

Positivity of forward rates, i.e. of the mild solution to (5.1), is established in [10] by proving positivity of mild solutions in weighted L^2 spaces to regularized versions of (5.1). Such an approximation argument is employed because the conditions on (σ_k) ensuring (local) Lipschitz continuity of the coefficients in the associated stochastic evolution (5.2) equation in H_α do not imply (local) Lipschitz continuity of the coefficients if state space is changed to a weighted L^2 space.

Thanks to Theorem 2.2, we can give a much shorter, more direct proof of the (criterion for the) positivity of forward rates. Let $L^2_{-\alpha}$ denote the weighted space $L^2(\mathbb{R}_+, e^{-\alpha x} \, dx)$, and note that H_α is continuously embedded in $L^2_{-\alpha} =: H$. Let us check that assumptions (A1), (A2), and (A3) are satisfied. Assumption (A1) holds true with the choice $\mathcal{O} = \mathbb{R}_+$, endowed with the absolutely continuous measure $\mu(dx) := e^{-\alpha x} \, dx$. As far as assumption (A2) is concerned, a simple computation shows that $A + \alpha I$ is monotone on $L^2_{-\alpha}$, and, by standard ODE theory, one also verifies that the range of $A + \alpha I + I$ coincides with the whole space $L^2_{-\alpha}$, therefore $A + \alpha I$ is maximal monotone. Even though A itself is not maximal monotone, this is clearly not restrictive, as the "correction" term αI can be incorporated in β without loss of generality. To verify that the resolvent $J_\lambda \in \mathscr{L}(H)$ of $A + \alpha I$ is sub-Markovian, let $y \in H$, so that $J_\lambda y \in D(A)$ is the unique solution y_λ to the problem

$$y_\lambda - \lambda y_\lambda' + \lambda \alpha y_\lambda = y.$$

If $0 \leq y \leq 1$ a.e. in $\mathbb{R}_+$, then we have, multiplying both sides by $(y_\lambda - 1)^+$, in the sense of the scalar product of H, that

$$(1 + \lambda \alpha)\langle y_\lambda, (y_\lambda - 1)^+ \rangle_{-\alpha} - \lambda \langle y_\lambda', (y_\lambda - 1)^+ \rangle_{-\alpha}$$
$$= \langle y, (y_\lambda - 1)^+ \rangle_{-\alpha} \leq \langle 1, (y_\lambda - 1)^+ \rangle_{-\alpha}.$$

Here and in the following we denote the scalar product and norm of $L^2_{-\alpha}$ simply by $\langle \cdot, \cdot \rangle_{-\alpha}$ and $\| \cdot \|_{-\alpha}$, respectively. Since

$$\langle y_\lambda, (y_\lambda - 1)^+ \rangle_{-\alpha} - \langle 1, (y_\lambda - 1)^+ \rangle_{-\alpha} = \|(y_\lambda - 1)^+\|^2_{-\alpha}, \tag{5.3}$$

we obtain

$$\|(y_\lambda - 1)^+\|^2_{-\alpha} - \frac{\lambda}{2} \int_0^\infty \frac{d}{dx}((y_\lambda - 1)^+)^2(x) e^{-\alpha x}\, dx + \lambda \alpha \langle y_\lambda, (y_\lambda - 1)^+ \rangle_{-\alpha} \leq 0,$$

where, integrating by parts,

$$-\frac{\lambda}{2} \int_0^\infty \frac{d}{dx}((y_\lambda - 1)^+)^2(x) e^{-\alpha x}\, dx$$
$$= -\frac{\lambda \alpha}{2} \int_0^\infty ((y_\lambda(x) - 1)^+)^2 e^{-\alpha x}\, dx + \frac{\lambda}{2}((y_\lambda(0) - 1)^+)^2$$
$$= -\frac{\lambda \alpha}{2} \langle y_\lambda, (y_\lambda - 1)^+ \rangle_{-\alpha} + \frac{\lambda \alpha}{2} \langle 1, (y_\lambda - 1)^+ \rangle_{-\alpha} + \frac{\lambda}{2}((y_\lambda(0) - 1)^+)^2$$
$$\geq -\frac{\lambda \alpha}{2} \langle y_\lambda, (y_\lambda - 1)^+ \rangle_{-\alpha}.$$

Rearranging terms yields

$$\|(y_\lambda - 1)^+\|^2_{-\alpha} + \frac{\lambda \alpha}{2} \langle y_\lambda, (y_\lambda - 1)^+ \rangle_{-\alpha} \leq 0,$$

where the second term on the left-hand side is positive by (5.3). Therefore $\|(y_\lambda - 1)^+\|_{-\alpha} = 0$, which implies that $y_\lambda \leq 1$ a.e. in $\mathbb{R}_+$. A completely similar argument, i.e. scalarly multiplying the resolvent equation by y_λ^-, also shows that $y_\lambda \geq 0$ a.e. in $\mathbb{R}_+$, thus completing the proof that J_λ is sub-Markovian. We still need to show that J_λ is contractive in $L^1_{-\alpha}$. Let $y, z \in H$ and $y_\lambda := J_\lambda y$, $z_\lambda := J_\lambda z$, so that

$$(y_\lambda - z_\lambda) - \lambda(y_\lambda - z_\lambda)' + \lambda\alpha(y_\lambda - z_\lambda) = y - z. \tag{5.4}$$

Define the sequences of functions (γ_k), $(\hat{\gamma}_k) \subset \mathbb{R}^\mathbb{R}$ as

$$\gamma_k : r \mapsto \tanh(kr), \qquad \hat{\gamma}_k : r \mapsto \int_0^r \gamma_k(s)\,ds,$$

and recall that, as $k \to \infty$, γ_k converges pointwise to the sign function, and $\hat{\gamma}_k$ converges pointwise to the absolute value function. Scalarly multiplying (5.4) with $\gamma_k(y_\lambda - z_\lambda)$ yields

$$(1 + \lambda\alpha)\langle y_\lambda - z_\lambda, \gamma_k(y_\lambda - z_\lambda)\rangle_{-\alpha} - \lambda\langle(y_\lambda - z_\lambda)', \gamma_k(y_\lambda - z_\lambda)\rangle_{-\alpha}$$
$$= \langle y - z, \gamma_k(y_\lambda - z_\lambda)\rangle_{-\alpha} \leq \|y - z\|_{L^1_{-\alpha}},$$

where, integrating by parts,

$$\langle(y_\lambda - z_\lambda)', \gamma_k(y_\lambda - z_\lambda)\rangle_{-\alpha}$$
$$= \int_0^\infty \bigl(\gamma_k(y_\lambda - z_\lambda)(x)(y_\lambda - z_\lambda)'(x)\bigr)e^{-\alpha x}\,dx$$
$$= \int_0^\infty \frac{d}{dx}\hat{\gamma}_k(y_\lambda - z_\lambda)(x)e^{-\alpha x}\,dx$$
$$= -\hat{\gamma}_k(y_\lambda(0) - z_\lambda(0)) + \alpha\int_0^\infty \hat{\gamma}_k(y_\lambda - z_\lambda)(x)e^{-\alpha x}\,dx$$
$$\leq \alpha\int_0^\infty \hat{\gamma}_k(y_\lambda - z_\lambda)(x)e^{-\alpha x}\,dx.$$

This implies

$$\langle y_\lambda - z_\lambda, \gamma_k(y_\lambda - z_\lambda)\rangle_{-\alpha}$$
$$+ \lambda\alpha\langle y_\lambda - z_\lambda, \gamma_k(y_\lambda - z_\lambda)\rangle_{-\alpha} - \lambda\alpha\int_0^\infty \hat{\gamma}_k(y_\lambda - z_\lambda)(x)e^{-\alpha x}\,dx$$
$$\leq \|y - z\|_{L^1_{-\alpha}}.$$

Taking the limit as $k \to \infty$, the sum of the second and third term on the left-hand side converges to zero by the dominated convergence theorem, while the first term on the left-hand side converges to $\|y_\lambda - z_\lambda\|_{L^1_{-\alpha}}$, thus proving that

$$\|y_\lambda - z_\lambda\|_{L^1_{-\alpha}} \leq \|y - z\|_{L^1_{-\alpha}},$$

i.e. that the resolvent of $A + \alpha I$ is contractive in $L^1_{-\alpha}$. We have thus shown that assumption (A2) holds for $A + \alpha I$. Moreover, assumption (A3) is satisfied if, for example,

$$|\sigma_k(\omega, t, x, r)|\mathbb{1}_{\{r \le 0\}} \lesssim r^-.$$

for all $k \in \mathbb{N}$ and $(\omega, t, x) \in \Omega \times \mathbb{R}_+^2$ (see [10], where also slightly more general sufficient conditions are provided). Since all integrability assumptions of Theorem 2.2 are satisfied, as it follows by inspection of the proof of well-posedness in H_α (see [3, 10, 11]), we conclude that, under the above assumptions on (σ_k), forward rates are positive at all times.

Acknowledgements Large part of this paper was written while the first-named author was visiting the Interdisziplinäres Zentrum für Komplexe Systeme, Universität Bonn, Germany, whose hospitality is gratefully acknowledged. The second-named author was partially supported by the Vienna Science and Technology Fund (WWTF) through Project MA14-009 and by the Austrian Science Fund (FWF) through Project M 2876.

References

1. Barbu, V., Da Prato, G., Röckner, M.: Existence and uniqueness of nonnegative solutions to the stochastic porous media equation. Ind. Univ. Math. J. **57**(1), 187–211 (2008). MR 2400255
2. Brézis, H., Strauss, W.A.: Semi-linear second-order elliptic equations in L^1. J. Math. Soc. Jpn. **25**, 565–590 (1973). MR 0336050 (49 # 826)
3. Filipović, D.: Consistency problems for Heath-Jarrow-Morton interest rate models. Lecture Notes in Mathematics, vol. 1760. Springer, Berlin (2001)
4. Haase, M.: Convexity inequalities for positive operators. Positivity **11**(1), 57–68 (2007). MR 2297322 (2008d:39034)
5. Heath, D.C., Jarrow, R.A., Morton, A.: Bond pricing and the term structure of interest rates: a new methodology for contingent claims valuation. Econometrica **60**(1), 77–105 (1992)
6. Hytönen, T., van Neerven, J., Veraar, M., Weis, L.: Analysis in Banach spaces. Vol. II, Probabilistic methods and operator theory. Springer, Cham (2017). MR 3752640
7. Krylov, N.V.: Maximum principle for SPDEs and its applications, Stochastic differential equations: theory and applications, Interdiscip. Math. Sci., vol. 2, pp. 311–338. World Scientific Publishing, Hackensack, NJ (2007). MR 2393582
8. Krylov, N.V.: A relatively short proof of Itô's formula for SPDEs and its applications. Stoch. Partial Differ. Equ. Anal. Comput. **1**(1), 152–174 (2013). MR 3327504
9. Kunze, M., van Neerven, J.: Continuous dependence on the coefficients and global existence for stochastic reaction diffusion equations. J. Differ. Equ. **253**(3), 1036–1068 (2012). MR 2922662
10. Marinelli, C.: Positivity of mild solution to stochastic evolution equations with an application to forward rates. arXiv:1912.12472
11. Marinelli, C.: Local well-posedness of Musiela's SPDE with Lévy noise. Math. Finance **20**(3), 341–363 (2010). MR 2667893
12. Métivier, M.: Semimartingales. Walter de Gruyter & Co., Berlin (1982). MR 688144 (84i:60002)
13. Pardoux, E.: Equations aux derivées partielles stochastiques nonlinéaires monotones, Ph.D. thesis, Université Paris XI (1975)
14. Röckner, M., Wang, F-Y.: Non-monotone stochastic generalized porous media equations. J. Differ. Equ. **245**(12), 3898–3935 (2008). MR 2462709
15. Scarpa, L., Stefanelli, U.: An order approach to SPDEs with antimonotone terms. Stoch. Partial Differ. Equ. Anal. Comput. **8**(4), 819–832 (2020). https://doi.org/10.1007/s40072-019-00161-7

Invariance of Poisson Point Processes by Moment Identities with Statistical Applications

Nicolas Privault

Abstract This paper reviews nonlinear extensions of the Slivnyak-Mecke formula as moment identities for functionals of Poisson point processes, and some of their applications. This includes studying the invariance of Poisson point processes under random transformations, as well as applications to distribution estimation for random sets in stochastic geometry, random graph connectivity, and density estimation for neuron membrane potentials in Poisson shot noise models.

Keywords Poisson point process · Moments · Stochastic geometry · Random-connection model · Filtered shot noise processes · Gram-charlier expansions · Neuron membrane potentials

Mathematics Subject Classification (2010): 60G57 · 60G55 · 60D05 · 60G40 · 60G48 · 60H07

1 Introduction

Computing the moments and cumulants of random variables has important applications in probability and statistics, e.g. for the estimation of distributions. In this paper we review the computation of moments of random functionals of Poisson point processes via combinatorial identities that extend the Slivnyak-Mecke formula to higher order moments, see [16, 19], and present some applications.

For this, we derive moment identities for stochastic integrals using sums over partitions. Those identities are used to derive criteria for invariance of Poisson random measures under random transformations, and for distribution estimation of the cardinality of random sets based on a Poisson point process.

N. Privault (✉)
School of Physical and Mathematical Sciences, Nanyang Technological University, Nanyang Link 637371, Singapore
e-mail: nprivault@ntu.edu.sg

© Springer Nature Switzerland AG 2021
S. Ugolini et al. (eds.), *Geometry and Invariance in Stochastic Dynamics*, Springer Proceedings in Mathematics & Statistics 378,
https://doi.org/10.1007/978-3-030-87432-2_13

The moments of random functionals can be used to estimate random graph connectivity in the random connection model using probability generating functions. This approach is based on the computation of the moments of k-hop path counts as sums over non-flat partitions, using extensions of moment identities to multiparameter processes, see [2].

The calculation of moments can also be applied to estimate the skewness and kurtosis of probability distributions, and to approximate probability densities via Edgeworth and Gram-Charlier expansions. Examples are provided using stochastic differential equations in Poisson shot noise models, with an application to the estimation of probability densities of neuron membrane potentials.

This paper is organized as follows. In Sect. 2 we review moment identities for Poisson stochastic integrals with random integrands. In Sect. 5, such identities are specialized to indicator functions of random sets, for application in stochastic geometry. Section 6 deals with applications to the statistics of k-hop counts in the random-connection model, using multiparameter stochastic integrals for the analysis of random graph connectivity. Section 8 considers the moments of Poisson shot noise processes, with an application to the modeling of membrane potential distributions.

2 Moments of Poisson Point Processes

We consider a Poisson point process with intensity measure $\sigma(dx)$ on the space

$$\Omega^\mathbb{X} := \{\xi = \{x_i\}_{i \in I} \subset \mathbb{X} \ : \ \#(A \cap \xi) < \infty \text{ for all compact } A \in \mathcal{B}(\mathbb{X})\}$$

of locally finite configurations on a subset $\mathbb{X} \subset \mathbf{R}^d$, where $\xi(A) = \#\{k \ : \ x_k \in A\}$ denotes the count of configuration points that belong to a measurable subset $A \subset \mathbb{X}$.

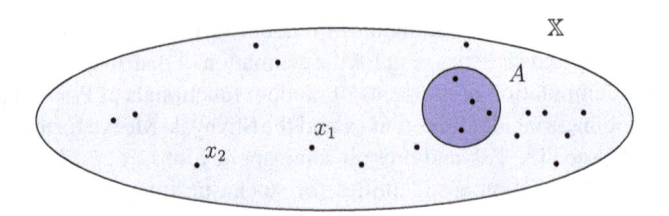

For all compact disjoint subsets $A_1, \ldots, A_n$ of $\mathbb{X}$, $n \geq 1$, the mapping

$$\xi \mapsto (\xi(A_1), \ldots, \xi(A_n))$$

is a vector of independent Poisson distributed random variables on $\mathbf{N}$ with respective intensities $\sigma(A_1), \ldots, \sigma(A_n)$. As a consequence, the Poisson stochastic integral with respect to the Poisson random measure with intensity $\sigma(dx)$ on $\mathbb{X}$ has the moment generating function

$$\mathbb{E}\left[\exp\left(\sum_{x\in\xi}h(x)\right)\right] = \exp\left(\int_{\mathbb{X}}(e^{h(x)}-1)\sigma(dx)\right). \tag{1}$$

3 Slivnyak-Mecke Identity

The Slivnyak-Mecke [11, 23] identity allows one to compute the moments of first order stochastic integrals of random integrands as

$$\mathbb{E}\left[\sum_{x\in\xi}u(x,\xi)\right] = \mathbb{E}\left[\int_{\mathbb{X}}\varepsilon_x^+ u(x,\xi)\sigma(dx)\right], \tag{2}$$

where ε_x^+ is the addition operator defined on random variables F on $\Omega^{\mathbb{X}}$ as

$$\varepsilon_x^+ F(\xi) = F(\xi \cup \{x\}), \qquad x \in \mathbb{X}.$$

4 Nonlinear Slivnyak-Mecke Identities

Next, we show how the Slivnyak-Mecke identity can be used to derive a covariance formula with random integrands. We have

$$\mathbb{E}\left[\sum_{x_1\in\xi}u_1(x_1,\xi)\sum_{x_2\in\xi}u_2(x_2,\xi)\right] = \mathbb{E}\left[\sum_{x_1\in\xi}\left(\sum_{x_2\in\xi}u_2(x_2,\xi)\right)u_i(x_1,\xi)\right]$$

$$= \mathbb{E}\left[\int_{\mathbb{X}}\epsilon_{x_1}^+\left(\sum_{x_2\in\xi}u_2(x_2,\xi)u_1(x_1,\xi)\right)\sigma(dx_1)\right],$$

with

$$\epsilon_{x_1}^+\sum_{x_2\in\xi}u_2(x_2,\xi) = \sum_{x_2\in\xi}\epsilon_{x_1}^+ u_2(x_2,\xi) + \epsilon_{x_1}^+ u_2(x_1,\xi).$$

Hence, another application of (2) yields

$$\mathbb{E}\left[\sum_{x_1 \in \xi} u_1(x_1, \xi) \sum_{x_2 \in \xi} u_2(x_2, \xi)\right]$$

$$= \mathbb{E}\left[\int_{\mathbb{X}} \sum_{x_2 \in \xi} \varepsilon_{x_1}^+(u_1(x_1, \xi) u_2(x_2, \xi)) \sigma(dx_1)\right] + \mathbb{E}\left[\int_{\mathbb{X}} \varepsilon_{x_1}^+(u_1(x_1, \xi) u_2(x_1, \xi)) \sigma(dx_1)\right]$$

$$= \mathbb{E}\left[\int_{\mathbb{X}^2} \varepsilon_{x_1}^+ \varepsilon_{x_2}^+(u_1(x_1, \xi) u_2(x_2, \xi)) \sigma(dx_1) \sigma(dx_2)\right] + \mathbb{E}\left[\int_{\mathbb{X}} \varepsilon_{x_1}^+(u_1(x_1, \xi) u_2(x_1, \xi)) \sigma(dx_1)\right].$$

Proposition 1 below, see Theorem 1 in [19], can be regarded as a nonlinear extension of the Slivnyak-Mecke formula (2) with random integrands $u : \mathbb{X} \times \Omega^{\mathbb{X}} \longrightarrow \mathbb{R}$. The sum (3) runs over the partitions $\pi_1, \ldots, \pi_k$ of $\{1, \ldots, n\}$, where $|\pi_i|$ denotes the cardinality of the block π_i, $i = 1, \ldots, k$. Given $\mathfrak{z}_n = (z_1, \ldots, z_n) \in \mathbb{X}^n$ we will use the shorthand notation $\varepsilon_{\mathfrak{z}_n}^+$ for the operator

$$(\varepsilon_{\mathfrak{z}_n}^+ F)(\xi) = F(\xi \cup \{z_1, \ldots, z_n\}), \qquad \xi \in \Omega^{\mathbb{X}},$$

for F a random variable on $\Omega^{\mathbb{X}}$.

Proposition 1 Let $u : \mathbb{X} \times \Omega^{\mathbb{X}} \longrightarrow \mathbb{R}$ be a (measurable) process. For all $n \geq 1$ we have

$$\mathbb{E}\left[\left(\sum_{x \in \xi} u(x, \xi)\right)^n\right] = \sum_{\rho \in \Pi[n]} \mathbb{E}\left[\int_{\mathbb{X}^{|\rho|}} \varepsilon_{\mathfrak{z}_{|\rho|}}^+ \prod_{l=1}^{|\rho|} u^{|\rho_l|}(z_l) \sigma^{\otimes |\rho|}(d\mathfrak{z}_{|\rho|})\right], \quad (3)$$

where the sum runs over all partitions ρ of $\{1, \ldots, n\}$ with cardinality $|\rho|$.

See [9] for an extension of (3) to point processes admitting Papangelou intensities, and [2] for an extension to multiparameter processes. This result can be more generally stated as the next joint moment identity for Poisson stochastic integrals with random integrands, cf. Proposition 7 in [19].

Proposition 2 Let $u_1, \ldots, u_p : \mathbb{X} \times \Omega^{\mathbb{X}} \longrightarrow \mathbb{R}$ be random processes, $p \geq 1$. For all $n_1, \ldots, n_p \geq 0$ and $n := n_1 + \cdots + n_p$, We have

$$\mathbb{E}\left[\left(\sum_{x_1 \in \xi} u_1(x_1, \xi)\right)^{n_1} \cdots \left(\sum_{x_p \in \xi} u_p(x_p, \xi)\right)^{n_p}\right]$$

$$= \sum_{k=1}^{n} \sum_{\pi_1 \cup \cdots \cup \pi_k = \{1, \ldots, n\}} \mathbb{E}\left[\int_{\mathbb{X}^k} \varepsilon_{x_1}^+ \cdots \varepsilon_{x_k}^+ \left(\prod_{j=1}^{k} \prod_{i=1}^{p} u_i^{l_{i,j}^n}(x_j, \xi)\right) \sigma(dx_1) \cdots \sigma(dx_k)\right],$$

where the sum runs over all partitions $\pi_1, \ldots, \pi_k$ of $\{1, \ldots, n\}$ and the power $l_{i,j}^n$ is the cardinality

$$l_{i,j}^n := |\pi_j \cap (n_1 + \cdots + n_{i-1}, n_1 + \cdots + n_i]|, \quad i = 1, \ldots, k, \quad j = 1, \ldots, p.$$

Proposition 2 implies in particular the next joint moment identity. Let $f_1, \ldots, f_p : \mathbb{X} \longrightarrow \mathbb{R}$ be deterministic functions, $p \geq 1$. Then, for any bounded random variable F and $n_1, \ldots, n_p \geq 0$ and $n := n_1 + \cdots + n_p$, we have

$$\mathbb{E}\left[F\left(\sum_{x_1 \in \xi} f_1(x_1)\right)^{n_1} \cdots \left(\sum_{x_p \in \xi} f_p(x_p)\right)^{n_p}\right]$$

$$= \sum_{k=1}^{n} \sum_{\pi_1 \cup \cdots \cup \pi_k = \{1,\ldots,n\}} \int_{\mathbb{X}^k} \mathbb{E}\left[\epsilon_{x_1}^+ \cdots \epsilon_{x_k}^+ F\right] \prod_{j=1}^{k} \prod_{i=1}^{p} f_i^{l_{i,j}^n}(x_j) \sigma(\mathrm{d}x_1) \cdots \sigma(\mathrm{d}x_k).$$

5 Random Sets in Stochastic Geometry

We consider possibly random sets $A(\xi)$ such that

$$\{\xi \in \Omega^{\mathbb{X}} : A(\xi) \subset K\} \in \mathcal{F} \quad \text{for all } K \in \mathcal{K}(\mathbb{X}),$$

and let $N(A(\xi))$ denote the cardinality of $\xi \cap A(\xi)$. The next proposition is a factorial moment identity for $N(A)$, see Proposition 2.1 in [3].

Proposition 3 *Let $A(\xi)$ be a random measurable subset of $\mathbb{X}$. For all $n \geq 1$ and sufficiently integrable random variable F, we have*

$$\mathbb{E}\left[F \, N(A)_{(n)}\right] = \mathbb{E}\left[\int_{\mathbb{X}^n} \varepsilon_{\underline{x}_n}^+ (F \mathbf{1}_{A^n}(x_1, \ldots, x_n)) \sigma^{\otimes n}(dx_1, \ldots, dx_n)\right],$$

where $N(A)_{(n)} = N(A)(N(A) - 1)(N(A) - n + 1)$ denotes the descending factorial of $N(A)$, $n \geq 1$.

Given K in the collection $\mathcal{K}(\mathbb{X})$ of compact subsets of $\mathbb{X}$, let

$$\mathcal{F}_K := \sigma(\xi(U) : U \subset K, \sigma(U) < \infty)$$

denote the sigma-algebra generated by $\xi \mapsto \xi(U)$, with $U \subset K$ and $\sigma(U) < \infty$. We recall that a random compact set S is called a *stopping set* if

$$\{\xi \in \Omega^{\mathbb{X}} : S(\xi) \subset K\} \in \mathcal{F}_K \quad \text{for all } K \in \mathcal{K}(\mathbb{X}).$$

In other words, modifying the configuration ξ outside of $S(\xi)$ does not affect $S(\xi)$ itself, see [24] and Definition 2.27 page 335 of [12].

In the sequel, we consider stopping sets S satisfying the following monotonicity and stability conditions.

(i) The stopping set S is non-increasing in the sense that

$$S(\xi \cup \{x\}) \subset S(\xi), \quad \xi \in \Omega^X, \quad x \in X.$$

(ii) The stopping set S is stable in the sense that

$$x \in S(\xi) \implies x \in S(\xi \cup \{x\}), \quad \xi \in \Omega^X, \quad x \in X.$$

Examples of stopping sets satisfying the above conditions can be given as follows:

- The minimal closed ball $S = B_m$ centered at 0 and containing exactly $m \geq 1$ points, see Fig. 1a.
- The complement S of the open convex hull $\overline{S}$ of a Poisson point process inside a convex subset of finite σ-measure in $\mathbf{R}^d$, see Fig. 1b.
- The Voronoi flower S, which is the union of closed balls centered at the vertices of the Voronoi polygon, containing the point 0 and exactly two other process points, see Fig. 1c.
- The complement S of the union of open cones generated by a Boolean-Poisson model on a set of finite σ-measure in $\mathbf{R}^d$, see Fig. 2.
- Other examples of stopping sets include the Voronoi sausage or the Delaunay lunes, see e.g. [6, 7].

From (5) and Proposition 3 we obtain the next factorial moment identity.

Proposition 4 *Let $\overline{S}$ be the complement of a stable, non-increasing stopping set S. For all $n \geq 1$, we have*

$$\mathbb{E}\left[F N(\overline{S})_{(n)}\right] = \mathbb{E}\left[\int_{\overline{S}^n} \varepsilon_{x_1}^+ \cdots \varepsilon_{x_n}^+ F \, \sigma(dx_1) \cdots \sigma(dx_n)\right],$$

for F a bounded random variable.

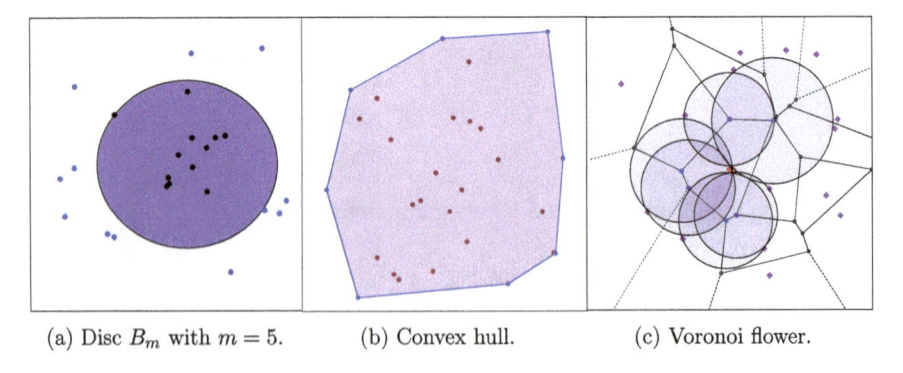

(a) Disc B_m with $m = 5$.   (b) Convex hull.   (c) Voronoi flower.

Fig. 1 Examples of stopping sets

Given S a stopping set, we consider the stopped sigma-algebra generated by S defined as

$$\mathcal{F}_S := \sigma(B \in \mathcal{F} \ : \ B \cap \{\xi \in \Omega^X \ : \ S(\xi) \subset K\} \in \mathcal{F}_K, \ K \in \mathcal{K}(X)),$$

see Definition 1 in [24]. As a consequence of Proposition 4, we obtain the following invariance result, see Propositions 4.1–4.2 and Corollary 5.2 in [18].

Corollary 1 *Consider* $S(\xi)$ *a stable and non-increasing stopping set and* $F(\xi)$ *a non-negative* $\mathcal{F}_S$-*measurable random variable with* $\mathbb{E}\left[e^{z\sigma(S)}(1+z)^{\xi(S)}F(\xi)\right] < \infty$ *for some* $z > 0$. *We have the Girsanov identity*

$$\mathbb{E}[F(\xi)] = \mathbb{E}\left[e^{-z\sigma(S)}(1+z)^{\xi(S)}F(\xi)\right]. \tag{4}$$

Relation (4) yields the following conditional Laplace transform for $S(\xi)$ a stable and non-increasing stopping set:

$$\mathbb{E}[e^{-z\sigma(S)} \mid \xi(S) = n] = \frac{1}{(1+z)^n}\frac{\mathbb{P}_z(\xi(S) = n)}{\mathbb{P}(\xi(S) = n)}, \quad z > 0, \quad n \in \mathbb{N},$$

where $\mathbb{P}_z$ denotes the Poisson point process distribution with intensity $z\sigma(dx)$, which is consistent with the gamma-type results of Theorem 2 of [13] and Theorem 2 of [24], and this recovers the gamma distribution of $\sigma(S)$ conditionally to $\xi(S) = n$, when $\mathbb{P}_z(\xi(S) = n)$ does not depend on $z > 0$.

Corollary 2 *Let S be a non-increasing and stable stopping set. Then the complement* $\overline{S}$ *of S satisfies*

$$\mathbb{P}(N(\overline{S}) = n \mid \mathcal{F}_S) = \frac{e^{-(\sigma(\overline{S}))}}{n!}(\sigma(\overline{S}))^n, \quad n \geq 0.$$

Proof We note that the complement $\overline{S}$ of a stable and non-increasing stopping set S fulfills the condition

$$\varepsilon^+_{\underline{y}_n}(1_{\overline{S}}(x_1)\cdots 1_{\overline{S}}(x_n)) = 1_{\overline{S}}(x_1)\cdots 1_{\overline{S}}(x_n), \quad x_1,\ldots, x_n \in X, \quad n \geq 1, \tag{5}$$

and apply the factorial moment identity of proposition 4.

Corollary 2 shows in particular that, given the stopping set S, the count $N(\overline{S})$ is a Poisson random variable with intensity $\sigma(\overline{S})$, see Theorem 3.1 of [1], and [17], when S is the closed complement of the Poisson convex hull $\overline{S}$. From Corollary 2 we can construct an alternative estimator

$$\mathbb{P}(N(\overline{S}) = n \mid \mathcal{F}_S) = \frac{(\sigma(\overline{S}))^n}{n!}e^{-\sigma(\overline{S})}. \tag{6}$$

of the distribution $\mathbb{P}(N(\overline{S}) = n)$ of the number of Poisson vertices inside the complement $\overline{S}$ of a stopping set S, in addition to the standard sampling estimator $\mathbf{1}_{\{N(\overline{S})=n\}}$, see [22] for numerical experiments where the performances of the estimators $\mathbf{1}_{\{N(\overline{S})=n\}}$ and (6) are compared via their respective variances given by $\mathbb{P}(N(\overline{S})=n)(1-\mathbb{P}(N(\overline{S})=n))$, and $\mathbb{E}\big[(\sigma(\overline{S}))^{2n}\mathrm{e}^{-2\sigma(\overline{S})}\big]/n!^2 - (\mathbb{P}(N(\overline{S})=n))^2$.

6 Multiparameter Integrals in Random Graphs

In this section we consider joint moment identities for multiparameter processes $(u_{z_1,\ldots,z_r})_{(z_1,\ldots,z_r)\in\mathbb{X}^r}$.

- Let $\Pi[n\times r]$ denote the set of partitions of

$$\Delta_{n\times r} := \{1,\ldots,n\}\times\{1,\ldots,r\} = \big\{(k,l) \,:\, k=1,\ldots,n,\; l=1,\ldots,r\big\}.$$

- Given $\rho = \{\rho_1,\ldots,\rho_m\}$ a partition of $\Delta_{n\times r}$, let $\zeta^\rho \,:\, \Delta_{n\times r} \longrightarrow \{1,\ldots,m\}$ let

$$\zeta^\rho(k,l) = p \text{ if and only if } (k,l)\in\rho_p,$$

denote the index of the block ρ_p containing (k,l).

In the next proposition, see Theorem 3.1 in [2], we use the notation

$$\epsilon^+_{\mathfrak{z}_k}u(z_1,\ldots,z_k,\xi) := u(z_1,\ldots,z_k,\xi\cup\{z_1,\ldots,z_k\}),\quad \mathfrak{z}_n = (z_1,\ldots,z_n)\in\mathbb{X}^n, \tag{7}$$

for $(u(z_1,\ldots,z_k,\xi))_{z_1,\ldots,z_k\in\mathbb{X}}$ a multiparameter process.

Proposition 5 *We have*

$$\mathbb{E}\left[\left(\sum_{z_1,\ldots,z_r\in\xi}u(z_1,\ldots,z_r,\xi)\right)^n\right] = \sum_{\rho\in\Pi[n\times r]}\mathbb{E}\left[\int_{\mathbb{X}^{|\rho|}}\epsilon^+_{\mathfrak{z}_{|\rho|}}\prod_{k=1}^n u(z^\rho_{\pi_k})\sigma^{\otimes|\rho|}(d\mathfrak{z}_{|\rho|})\right],$$

where $z^\rho_{\pi_k} := (z_{\zeta^\rho(k,1)},\ldots,z_{\zeta^\rho(k,r)})$ *and* $\pi_k := \{(k,1),\ldots,(k,r)\}$, $k=1,\ldots,n$.

When $n=1$, this yields the multivariate version of the Georgii, [14] identity

$$\mathbb{E}\left[\sum_{z_1,\ldots,z_r\in\xi}u(z_1,\ldots,z_r,\xi)\right] = \sum_{\rho\in\Pi[1\times r]}\mathbb{E}\left[\int_{\mathbb{X}^r}\varepsilon^+_{\mathfrak{z}_{|\rho|}}u(z_{\zeta^\rho(1,1)},\ldots,z_{\zeta^\rho(1,r)})\sigma^{\otimes|\rho|}(d\mathfrak{z}_{|\rho|})\right].$$

We write $\pi\preceq\sigma$ when a partition $\pi\in\Pi[n\times r]$ is finer than another partition $\sigma\in\Pi[n\times r]$, i.e. when every block of π is contained in a block of σ. We also write $\rho\wedge\pi = \hat{0}$ when $\mu = \hat{0} := \{\{1,1\},\ldots,\{n,r\}\}$ is the only partition $\mu\in\Pi[n\times r]$ such that $\mu\preceq\pi$ and $\mu\preceq\rho$, i.e. $|\pi_k\cap\rho_l|\leq 1$ for $k=1,\ldots,n$, $l=1,\ldots,|\rho|$. The moment identity in the next proposition is written as a sum over partitions $\rho\in\Pi[n\times r]$ such that the partition diagram $\Gamma(\pi,\rho)$ is *non-flat*, see Chap. 4 of [15], where $\pi := (\pi_1,\ldots,\pi_n)\in\Pi[n\times r]$ is given by $\pi_k := \{(k,1),\ldots,(k,r)\}$, $k=1,\ldots,n$.

Invariance of Poisson Point Processes by Moment Identities ...

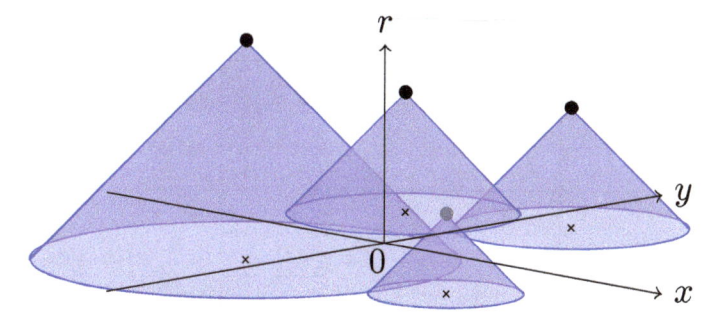

Fig. 2 Cones generated by a Boolean-Poisson model

Fig. 3 Example of a non-flat partition of $\Pi[3 \times 2]$

Proposition 6 *Assume that* $u(z_1, \ldots, z_r, \xi) = 0$ *whenever* $z_i = z_j$, $1 \leq i \neq j \leq r$, $\xi \in \Omega^{\mathbb{X}}$. *We have*

$$\mathbb{E}\left[\left(\sum_{z_1,\ldots,z_r \in \xi} u(z_1, \ldots, z_r, \xi)\right)^n\right] = \sum_{\substack{\rho \in \Pi[n \times r] \\ \rho \wedge \pi = \hat{0}}} \mathbb{E}\left[\int_{\mathbb{X}^{|\rho|}} \epsilon^+_{\hat{3}|\rho|} \prod_{k=1}^{n} u(z^{\rho}_{\pi_k}) \sigma^{\otimes|\rho|}(d\hat{3}|\rho|)\right],$$

where the sum is over non-flat *partition diagrams* $\Gamma(\pi, \rho)$, *with* $z^{\rho}_{\pi_k} := (z_{\zeta^{\rho}(k,1)}, \ldots, z_{\zeta^{\rho}(k,r)})$ *and* $\pi_k := \{(k, 1), \ldots, (k, r)\}$, $k = 1, \ldots, n$.

Figure 3 shows an example of a non-flat partition of $\Pi[n \times r]$ with $n = 3$ and $r = 2$, which is tagged using the four symbols $\triangle, \square, \diamondsuit, \bigcirc$, with $\pi_3 = \{(3, 1), (3, 2)\}$, $\pi_2 = \{(2, 1), (2, 2)\}$, $\pi_1 = \{(1, 1), (1, 2)\}$, and $\triangle = \{(1, 2), (2, 1), (3, 2)\}$, $\bigcirc = \{(1, 1), (3, 1)\}$, $\diamondsuit = \{(2, 2)\}$.

Figure 4 illustrates the non-flat partition technique in the case $n = 3$ and $r = 2$, by displaying 6 out of the 87 multigraphs occurring in the computation of the case of the third moment of the 3-hop count based on possible combinations of common nodes in the product (8), together with each corresponding non-flat partition of $[3 \times 2] = \{(1, 1), (1, 2), (2, 1), (2, 2), (3, 1), (3, 2)\}$, and every path in the multigraph is followed from the blue node x to the red node y.

Fig. 4 Matching of non-flat partitions of [3 × 2] to multigraphs with identification of common nodes

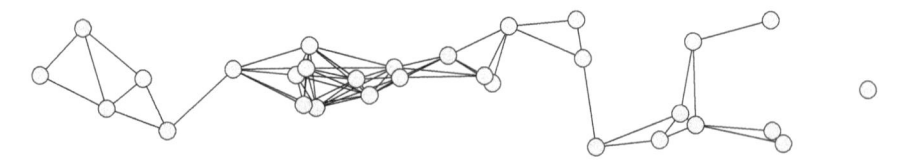

Fig. 5 Random-connection graph

7 Random-Connection Model

In the random-connection model, two vertices $x \neq y$ of the Poisson point process ξ of nodes on $\mathbb{X} \subset \mathbb{R}^d$ are independently connected with the probability $H(x, y)$ given ξ in the probability space $\Omega^{\mathbb{X}}$, where $H : X \times X \longrightarrow [0, 1]$ is a connection function, see Fig. 5. In particular the 1-hop count $\mathbb{1}_{\{x \leftrightarrow y\}}$, where $x \leftrightarrow y$ means that $x \in X$ is connected to $y \in X$, is a Bernoulli random variable with parameter $H(x, y)$ and we have the relation

$$\mathbb{E}\left[\epsilon^+_{\hat{3}_r} \prod_{i=0}^{r} \mathbb{1}_{\{z_i \leftrightarrow z_{i+1}\}}(\xi) \bigg| \xi \right] = \prod_{i=0}^{r} H(z_i, z_{i+1})$$

for any subset $\{z_0, \ldots, z_{r+1}\}$ of distinct elements of $\mathbb{X}$, where $\epsilon^+_{\hat{3}_r}$ is the addition operator of point process nodes at the locations $\hat{3}_r = \{z_1, \ldots, z_r\}$, see (7).

Given $x, y \in \mathbb{X}$ two vertices in $\mathbb{X}$, the count $N^{x,y}_r$ of $(r + 1)$-hop paths from x to y as particular cases, i.e. the number of $(r + 1)$-hop sequences $z_1, \ldots, z_r \in \xi$ of vertices connecting x to y in the random graph is the multiparameter stochastic integral

$$N^{x,y}_{r+1} = \sum_{z_1, \ldots, z_r \in \xi} u(z_1, \ldots, z_r)$$

over the vertices of the point process ξ, of the multiparameter r-process

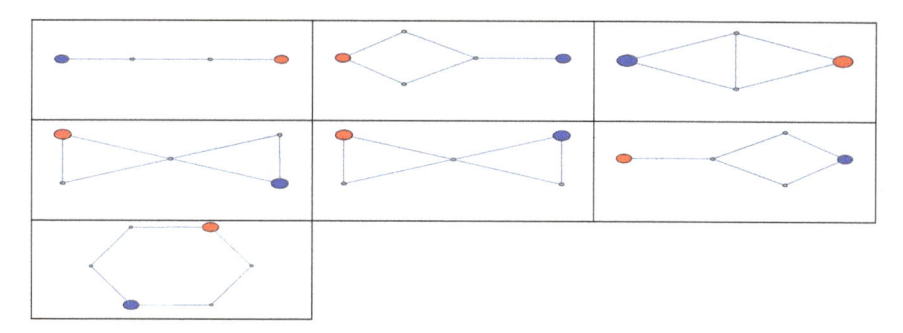

Fig. 6 The seven possible ways to join two nodes via two 3-hop paths and their common nodes

$$u(z_1, \ldots, z_r, \xi) := 1\!\!1_{\{z_i \neq z_j, \ 1 \leq i < j \leq r\}} 1\!\!1_{\{z_1, \ldots, z_r \in \xi\}} \prod_{i=0}^{r} 1\!\!1_{\{z_i \leftrightarrow z_{i+1}\}}(\xi)$$

which vanishes on the diagonals in $\mathbb{X}^r$, with $z_0 := x$ and $z_{r+1} := y$. Computing the moments of N_r requires to raise N_r to a given power, creating product terms of the form

$$\prod_{l=1}^{n} u(z_1^{(l)}, \ldots, z_r^{(l)}; \xi) \tag{8}$$

where $(z_1^{(l)}, \ldots, z_r^{(l)})$ denotes the sequence of points appearing in the l-th product term. For example, computing the second moment of a 3-hop count requires to identify and count the 7 possible multigraphs that can connect x to y via two 3-hop paths with possible common nodes as in Fig. 6, see also Fig. 2 in [10], in which every path in each multigraph is followed from the blue node x to the red node y. The difficulty in dealing with common nodes is that they break the independence property in the product (8), and as such they have to be dealt with separately.

The next proposition, which is a direct consequence of Proposition 5, provides a general expression for the moments of the count $N_{r+1}^{x,y}$ of $(r+1)$-hop paths, see [20].

Proposition 7 *The moment of order n of the $(r+1)$-hop count between $x, y \in \mathbb{X}$ is given by*

$$\mathbb{E}\left[(N_{r+1}^{x,y})^n\right] = \sum_{\substack{\rho \in \Pi[n \times r] \\ \rho \wedge \pi = \hat{0}}} \mathbb{E}\left[\int_{\mathbb{X}^{|\rho|}} \prod_{l=1}^{n} \prod_{i=0}^{r} H^{1/n_{l,i}^\rho}(z_{\zeta^\rho(l,i)}, z_{\zeta^\rho(l,i+1)}) \sigma^{\otimes|\rho|}(d\mathfrak{z}_{|\rho|})\right],$$

where $z_0 = x$, $z_{r+1} = y$, $\zeta^\rho(l, 0) = 0$, $\zeta^\rho(l, r+1) = r+1$, and

$$n_{l,i}^\rho := \#\{(p, j) \in \{1, \ldots, n\} \times \{0, \ldots, r\} \ : \ \{\zeta^\rho(l, i), \zeta^\rho(l, i+1)\} = \{\zeta^\rho(p, j), \zeta^\rho(p, j+1)\}\}.$$

In particular, the first order moment of the $(r+1)$-hop count between $x \in \mathbb{X}$ and $y \in Y$ is given as

$$H^{(r+1)}(z_0, z_{r+1}) := \mathbb{E}\left[\sum_{z_1,\ldots,z_r \in \xi} u(z_1, \ldots, z_r, \xi)\right]$$

$$= \int_{\mathbb{R}^d} \cdots \int_{\mathbb{R}^d} \prod_{i=0}^{r} H(z_i, z_{i+1}) \sigma(dz_1) \cdots \sigma(dz_r), \quad z_0, z_{r+1} \in \mathbb{R}^d.$$

The 2-hop count between $x \in \mathbb{X}$ and $y \in Y$ is given by the first order integral

$$\sum_{z \in \xi} u(z, \xi) = \sum_{z \in \xi} \mathbb{1}_{\{x \leftrightarrow z\}} \mathbb{1}_{\{z \leftrightarrow y\}}(\xi) = \sum_{z \in \xi} \mathbb{1}_{\{x \leftrightarrow z\}} \mathbb{1}_{\{z \leftrightarrow y\}},$$

and its moment of order n is

$$\mathbb{E}\left[\left(\sum_{z \in \xi} u(z, \xi)\right)^n\right] = \sum_{\rho \in \Pi[n \times 1]} \int_{\mathbb{X}^{|\rho|}} \prod_{l=1}^{|\rho|} (H(x, z_l) H(z_l, y)) \sigma^{\otimes |\rho|}(dz_1, \ldots, dz_{|\rho|})$$

$$= \sum_{k=1}^{n} S(n, k) \left(\int_{\mathbb{R}^d} H(x, z) H(z, y) \sigma(dz)\right)^k$$

$$= \sum_{k=1}^{n} S(n, k) \left(H^{(2)}(x, y)\right)^k,$$

which shows that the 2-hop count between $x \in \mathbb{X}$ and $y \in Y$ is a Poisson random variable with mean $H^{(2)}(x, y)$.

Variance of 3-hop counts

When $n = 2$ and $r = 3$ Proposition 7 allows us to compute the variance of the 3-hop count between $x \in \mathbb{X}$ and $y \in Y$, as follows:

$$\text{Var}\left[N_3^{x,y}\right] = H^{(3)}(x, y) + 2\int_{\mathbb{X}} H(x, z_1) H^{(2)}(z_1, y) H^{(2)}(z_1, y) \sigma(dz_1) \qquad (9)$$

$$+ 2\int_{\mathbb{X}} H(x, z_1) H^{(2)}(x, z_1) H^{(2)}(z_1, y) H(z_1, y) \sigma(dz_1)$$

$$+ \int_{\mathbb{X}^2} H(x, z_1) H(z_1, z_2) H(z_2, y) H(x, z_2) H(z_1, y) \sigma^{\otimes 2}(dz_1, dz_2),$$

In the case of a Poisson point process with flat intensity $\sigma(dx) = \lambda dx$ on $\mathbb{X}$, $\lambda > 0$ with a Rayleigh fading function $H(x, y)$ of the form

$$H_\beta(x, y) := e^{-\beta \|x-y\|^2}, \quad x, y \in \mathbb{R}^d, \quad \beta > 0,$$

we have

$$H_\beta^{(2)}(x, y) = \lambda \int_{\mathbb{R}^d} H_\beta(x, z) H_\beta(z, y) dz = \lambda \left(\frac{\pi}{2\beta}\right)^{d/2} e^{-\|x-y\|^2/2},$$

and (9) recovers the variance

$$\text{Var}\left[N_3^{x,y}\right] = 2\lambda^3 \left(\frac{\pi^3}{8\beta^3}\right)^{d/2} e^{-\beta\|x-y\|^2/2} + \lambda^2 \left(\frac{\pi^2}{3\beta^2}\right)^{d/2} e^{-\beta\|x-y\|^2/3}$$

$$+ 2\lambda^3 \left(\frac{\pi^3}{12\beta^3}\right)^{d/2} e^{-3\beta\|x-y\|^2/4} + \lambda^2 \left(\frac{\pi^2}{8\beta^2}\right)^{d/2} e^{-\beta\|x-y\|^2}$$

of 3-hop counts between $x \in \mathbb{X}$ and $y \in \mathbb{Y}$, see Theorem II.2 of [10]. The knowledge of moments can provide accurate numerical estimates of the probability $P(N_k^{x,y} > 0)$ of at least one k-hop path by expressing it as a series of factorial moments, see [10].

8 Moments of Poisson Shot Noise Processes

We consider a Poisson point process $\xi(dx)$ with intensity measure $\sigma(dt, d\theta)$ on $\mathbb{X} = \mathbb{R} \times S$, where $S = [0, N]$, and the N shot noise processes given by

$$Q_k(t, \xi) = \sum_{(s_j, \theta_j) \in \xi} g_k(t - s_j, \theta_j), \quad k = 1, \ldots, N,$$

where the shot noise kernels $g_k(u, \theta)$ are such that $g_k(u, \theta) = 0$ for all $u < 0$ and $\theta \in S$. In this framework, the moment generating function of $Q_k(t, \xi)$ is given from (1) as

$$\mathbb{E}\left[\exp(Q_k(t, \xi))\right] = \exp\left(\int_{(-\infty, t] \times S} (e^{g_k(t-u, \theta)} - 1) \sigma(du, d\theta)\right).$$

Consider the Poisson shot noise stochastic differential equation

$$\tau \frac{dY_N}{dt}(t, \xi) = -Y_N(t, \xi) + \sum_{k=1}^{N} (w_k - Y_N(t, \xi)) Q_k(t, \xi), \quad (10)$$

where $\tau > 0$ and $w_1, \ldots, w_n \in \mathbb{R}$, whose solution is the filtered shot noise process

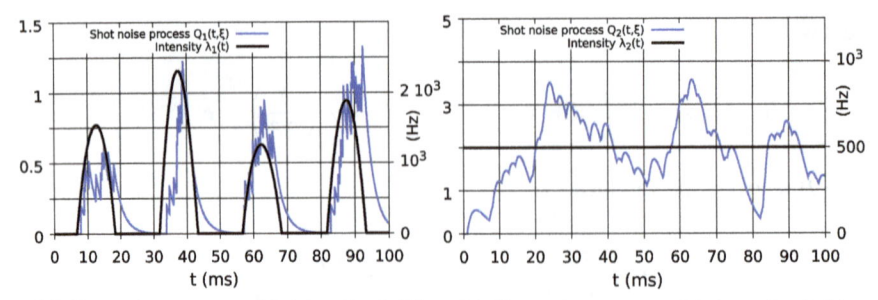

(a) Shot noise process with intensity $\lambda_1(t)$. (b) Shot noise process with intensity $\lambda_2(t)$.

Fig. 7 Shot noise processes $Q_1(t, \xi)$ and $Q_2(t, \xi)$

$$Y_N(t, \xi) = \frac{1}{\tau} \sum_{k=1}^{N} w_k \int_{-\infty}^{t} Q_k(z, \xi) e^{-\int_z^t Q_0(u,\xi) du} dz \tag{11}$$

$$= \frac{1}{\tau} \int_{-\infty}^{t} e^{-\int_z^t Q_0(s,\xi) ds} \sum_{(s_j, \theta_j) \in \xi} f^{(w)}(z - s_j, \theta_j) dz, \quad t \in \mathbf{R},$$

where

$$Q_0(u, \xi) := \frac{1}{\tau} + \frac{1}{\tau} \sum_{k=1}^{N} Q_k(u, \xi) = \frac{1}{\tau} + \frac{1}{\tau} \sum_{(s_j, \theta_j) \in \xi} f(u - s_j, \theta_j),$$

with

$$f(z, \theta) := \sum_{k=1}^{N} g_k(z, \theta) \quad \text{and} \quad f^{(w)}(z, \theta) := \sum_{k=1}^{N} w_k g_k(z, \theta), \quad z \in \mathbf{R}, \ \theta \in S,$$

see e.g. Sect. 2.1 of [4] and [5]. The following numerical examples use the parameters of the double source model of [4] for the modeling of neuron membrane potentials, where $N = 2$, $\lambda_2(t) = 500$Hz and $\lambda_1(t)$ is a periodic function of time, $t \in [0, 100]$ (Fig. 7).

Figure 8 presents the graphs of the intensities $\lambda_1(t)$, $\lambda_2(t)$ and a numerical simulation of $V_2(t, \xi)$ in the double source model.

9 Computation of Joint Moments

The next proposition gives a general formula for the computation of the joint moments of $Y_N(t_1, \xi), \ldots, Y_N(t_n, \xi)$ in the multiple source model as a direct consequence of (11).

(a) Intensities $\lambda_1(t)$ and $\lambda_2(t)$.

(b) Mean and standard deviation of $V_2(t,\xi)$.

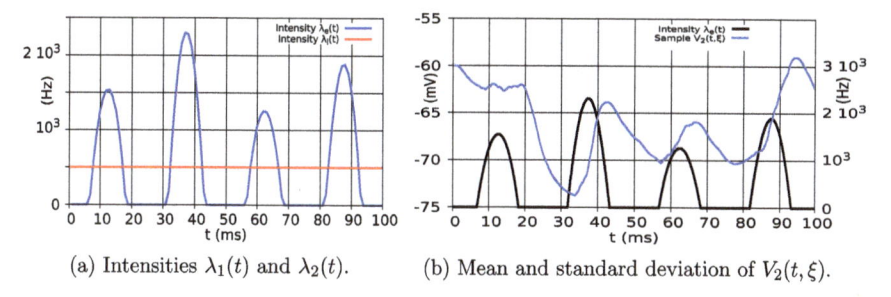

Fig. 8 Sample of $V_2(t,\xi)$ with mean, standard deviation and intensities $\lambda_1(t), \lambda_2(t)$

Proposition 8 *We have the joint moment identity*

$$\mathbb{E}[Y_N(t_1,\xi)\cdots Y_N(t_n,\xi)] = \frac{1}{\tau^n} \int_{-\infty}^{t_1}\cdots\int_{-\infty}^{t_n} m_{n,N}(z_1,\ldots,z_n;t_1,\ldots,t_n)dz_1\cdots dz_n,$$

where

$$m_{n,N}(z_1,\ldots,z_n;t_1,\ldots,t_n) := \mathbb{E}\left[\prod_{k=1}^n \left(e^{-\int_{z_l}^{t_l} Q_0(u,\xi)du} \sum_{(u_j,\theta_j)\in\xi} f^{(w)}(z_k - u_j, \theta_j)\right)\right].$$

The functions $m_{n,N}(z_1,\ldots,z_n;t_1,\ldots,t_n)$ can be evaluated from Proposition 2 as a sum over the set $\Pi[n]$ of partitions $\pi = \{\pi_1,\ldots,\pi_k\}$ of $\{1,\ldots,n\}$ with cardinality $k = |\pi| = 1,\ldots,n$, as

$m_{n,N}(z_1,\ldots,z_n;t_1,\ldots,t_n)$

$$= \mathbb{E}\left[e^{-\sum_{l=1}^n \int_{z_l}^{t_l} Q_0(u,\xi)du}\right] \sum_{\pi\in\Pi[n]} \prod_{j=1}^{|\pi|} \int_{(-\infty,\hat{z}_{\pi_j}]\times S} \prod_{l=1}^n e^{-\frac{1}{\tau}\int_{z_l}^{t_l} f(u-y,\eta)du} \prod_{i\in\pi_j} f^{(w)}(z_i - y, \eta)\sigma(dy, d\eta),$$

$(z_1,\ldots,z_n) \in (-\infty, t_1] \times \cdots \times (-\infty, t_n]$, with $\hat{z}_{\pi_j} = \min_{i\in\pi_j} z_i$, where, by (1), we have

$$\mathbb{E}\left[e^{-\sum_{l=1}^n \int_{z_l}^{t_l} Q_0(u,\xi)du}\right]$$

$$= e^{-\frac{1}{\tau}\sum_{l=1}^n (t_l - z_l)} \exp\left(\int_{(-\infty,\max(t_1,\ldots,t_n)]\times S} \left(e^{-\frac{1}{\tau}\sum_{l=1}^n \int_{z_l}^{t_l} f(u-s,\theta)du} - 1\right)\sigma(ds, d\theta)\right).$$

Figures 9, 10 and 11 present the evolutions of the mean κ_1, variance κ_2, third and fourth cumulants κ_3, κ_4, and skewness and excess kurtosis

$$\frac{\kappa_3}{(\kappa_2)^{3/2}} = \frac{\mathbb{E}[(V_2 - \mathbb{E}[V_2])^3]}{(\mathbb{E}[(V_2 - \mathbb{E}[V_2])^2])^{3/2}} \quad \text{and} \quad \frac{\kappa_4}{(\kappa_2)^2} = \frac{\mathbb{E}[(V_2 - \mathbb{E}[V_2])^4]}{(\mathbb{E}[(V_2 - \mathbb{E}[V_2])^2])^2} - 3 \tag{12}$$

of the potential $V_2(t, \xi)$, computed from Proposition 8 as functions of the arrival intensity parameter λ at $t = 0.2$.

10 Gram-Charlier Expansions

The Gram-Charlier expansion of the continuous probability density function $\phi_X(x)$ of a random variable X, see Sect. 17.6 of [8], is given by

$$\phi_X(x) = \frac{1}{\sqrt{\kappa_2}} \varphi\left(\frac{x - \kappa_1}{\sqrt{\kappa_2}}\right) + \frac{1}{\sqrt{\kappa_2}} \sum_{n=3}^{\infty} c_n H_n\left(\frac{x - \kappa_1}{\sqrt{\kappa_2}}\right) \varphi\left(\frac{x - \kappa_1}{\sqrt{\kappa_2}}\right), \qquad (13)$$

where $\varphi(x)$ is the standard normal density function, $H_n(x)$ denotes the Hermite polynomial of degree n, and the sequence $(c_n)_{n \geq 3}$ is given from the cumulants $(\kappa_n)_{n \geq 1}$ of X. In particular, the coefficients c_3 and c_4 can be expressed from the skewness $\kappa_3/\kappa_2^{3/2}$ and the excess kurtosis κ_4/κ_2^2 as $c_3 = \kappa_3/(3!\kappa_2^{3/2})$ and $c_4 = \kappa_4/(4!\kappa_2^2)$, which are computed from (12). Figures 12 and 13 present the Gram-Charlier density expansions (13) at different times for the estimation of the probability density function of

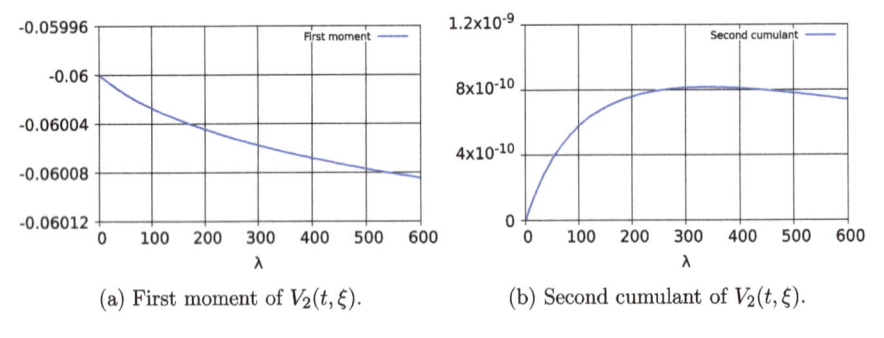

(a) First moment of $V_2(t, \xi)$.

(b) Second cumulant of $V_2(t, \xi)$.

Fig. 9 First and second cumulants of $V_2(t, \xi)$ at $t = 0.2$

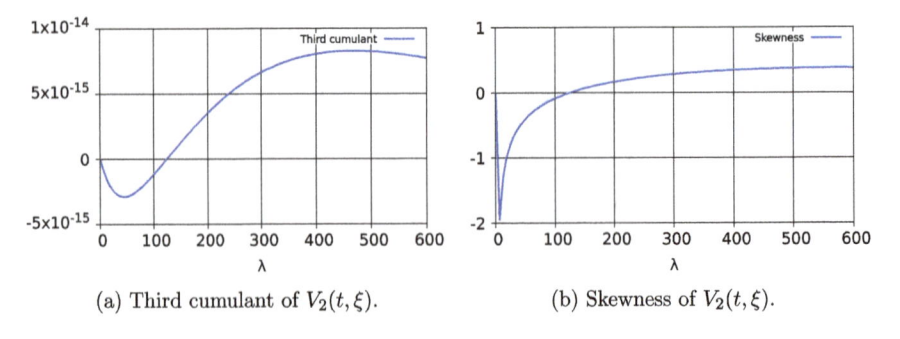

(a) Third cumulant of $V_2(t, \xi)$.

(b) Skewness of $V_2(t, \xi)$.

Fig. 10 Third cumulant and skewness of $V_2(t, \xi)$ at $t = 0.2$

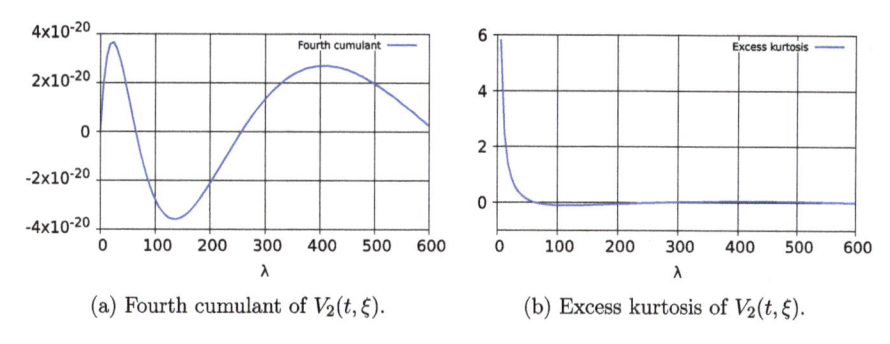

Fig. 11 Fourth cumulant and excess kurtosis of $V_2(t,\xi)$ at $t = 0.2$

(a) Fourth cumulant of $V_2(t,\xi)$.

(b) Excess kurtosis of $V_2(t,\xi)$.

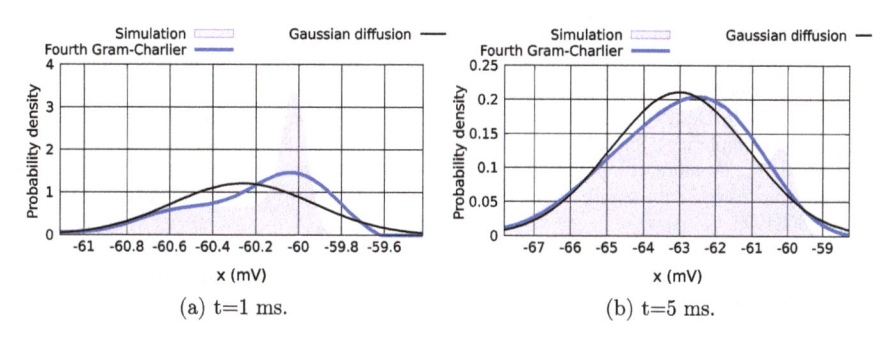

(a) t=1 ms.

(b) t=5 ms.

Fig. 12 Gram-Charlier density expansions vs simulated densities

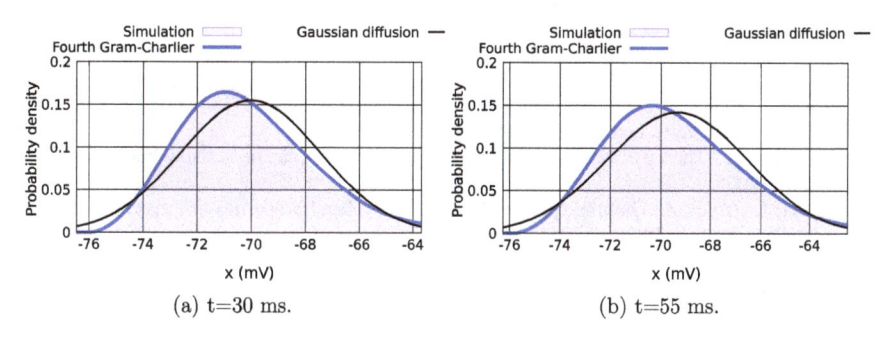

(a) t=30 ms.

(b) t=55 ms.

Fig. 13 Gram-Charlier density expansions vs simulated densities

the membrane potential $V_2(t,\xi)$ in the double source model (10) of Fig. 8, see [21] for details.

In comparison with the Gaussian diffusion approximation with matching mean and variance, the fourth-order Gram-Charlier approximations provide a better fit of the actual probability densities obtained by Monte Carlo simulation of (11) (purple areas), which show time-varying skewness.

Acknowledgements The Mathematica code used to produce Figs. 4 and 6 was provided by A.P. Kartun-Giles.

References

1. Baldin, N., Reiß, M.: Unbiased estimation of the volume of a convex body. Stoch. Process. Appl. **126**, 3716–3732 (2016)
2. Bogdan, K., Rosiński, J., Serafin, G., Wojciechowski, L.: Lévy systems and moment formulas for mixed Poisson integrals. In: Stochastic Analysis and Related Topics, Program Probability, vol. 72 pp. 139–164. Birkhäuser/Springer, Cham (2017)
3. Breton, J.-C., Privault, N.: Factorial moments of point processes. Stoch. Process. Their Appl. **124**(10), 3412–3428 (2014)
4. Brigham, M., Destexhe, A.: The impact of synaptic conductance inhomogeneities on membrane potential statistics. Preprint (2015)
5. Brigham, M., Destexhe, A.: Nonstationary filtered shot-noise processes and applications to neuronal membranes. Phys. Rev. E **91**, 062102 (2015)
6. Cowan, R.: A more comprehensive complementary theorem for the analysis of Poisson point processes. Adv. Appl. Probab. **38**(3), 581–601 (2006). https://doi.org/10.1239/aap/1158684993
7. Cowan, R., Quine, M., Zuyev, S.: Decomposition of gamma-distributed domains constructed from Poisson point processes. Adv. Appl. Probab. **35**(1), 56–69 (2003). https://doi.org/10.1239/aap/1046366099
8. Cramér, H.: Mathematical Methods of Statistics. Princeton University Press, Princeton, NJ (1946)
9. Decreusefond, L., Flint, I.: Moment formulae for general point processes. J. Funct. Anal. **267**, 452–476 (2014)
10. Kartun-Giles, A.P., Kim, S.: Counting k-hop paths in the random connection model. IEEE Trans. Wirel. Commun. **17**(5), 3201–3210 (2018)
11. Mecke, J.: Stationäre zufällige Masse auf lokalkompakten Abelschen Gruppen. Z. Wahrscheinlichkeitstheorie Verw. Geb. **9**, 36–58 (1967)
12. Molchanov, I.: Theory of random sets. Probability and its Applications (New York). Springer, London (2005)
13. Møller, J., Zuyev, S.: Gamma-type results and other related properties of Poisson processes. Adv. Appl. Probab. **28**(3), 662–673 (1996)
14. Nguyen, X.X., Zessin, H.: Integral and differential characterization of the Gibbs process. Math. Nachr. **88**, 105–115 (1979)
15. Peccati, G., Taqqu, M.: Wiener Chaos: Moments, Cumulants and Diagrams: A survey with Computer Implementation. Springer, Bocconi & Springer Series (2011)
16. Privault, N.: Moments of Poisson stochastic integrals with random integrands. Probab. Math. Stat. **32**(2), 227–239 (2012)
17. Privault, N.: Invariance of Poisson measures under random transformations. Ann. Inst. H. Poincaré Probab. Statist. **48**(4), 947–972 (2012)
18. Privault, N.: Laplace transform identities for the volume of stopping sets based on Poisson point processes. Adv. Appl. Probab. **47**, 919–933 (2015)
19. Privault, N.: Combinatorics of Poisson stochastic integrals with random integrands. In: Peccati, G., Reitzner, M. (eds.) Stochastic Analysis for Poisson Point Processes: Malliavin Calculus. Wiener-Itô Chaos Expansions and Stochastic Geometry, Bocconi & Springer Series, vol. 7, pp. 37–80. Springer, Berlin (2016)
20. Privault, N.: Moments of k-hop counts in the random-connection model. J. Appl. Probab. **56**(4), 1106–1121 (2019)
21. Privault, N.: Nonstationary shot-noise modeling of neuron membrane potentials by closed-form moments and Gram-Charlier expansions. Biol. Cybern. **114**, 499–518 (2020)

22. Privault, N.: Cardinality estimation for random stopping sets based on Poisson point processes. ESAIM Probab. Stat. **25**, 87–108 (2021)
23. Slivnyak, I.M.: Some properties of stationary flows of homogeneous random events. Theory Probab. Appl. **7**(3), 336–341 (1962)
24. Zuyev, S.: Stopping sets: gamma-type results and hitting properties. Adv. Appl. Probab. **31**(2), 355–366 (1999). https://doi.org/10.1239/aap/1029955139

CPSIA information can be obtained
at www.ICGtesting.com
Printed in the USA
BVHW060341160223
658591BV00006B/323

9 783030 874346